For a complete listing of titles in the
Artech House GNSS Technologies and Applications,
turn to the back of this book.

Location-Based Services in Cellular Networks from GSM to 5G NR

Location-Based Services in Cellular Networks from GSM to 5G NR

Adrián Cardalda García

Stefan Maier

Abhay Phillips

ARTECH HOUSE

BOSTON | LONDON

artechhouse.com

Library of Congress Cataloging-in-Publication Data
A catalog record for this book is available from the U.S. Library of Congress.

British Library Cataloguing in Publication Data
A catalogue record for this book is available from the British Library.

Cover design by John Gomes

ISBN 13: 978-1-63081-634-6

ARTECH HOUSE
685 Canton Street
Norwood, MA 02062

10 9 8 7 6 5 4 3 2 1

To Nico and Anna, for the many weekends I have stolen from them to complete this book.
A. C. G.

To Marie and Stefanie, for your encouragement and motivation.
S. M.

To Sabine, for her patience and support.
A. P.

Contents

Preface

When you order a ride-sharing service using your smartphone, upon getting the request the driver already has your location and can drive directly to pick you up. You come back home after an afternoon of shopping and your phone asks you your opinion about one of the shops you have visited. These are examples of how location-based services (LBS) have become ubiquitous in our lives and are an essential aspect of wireless communications. However, for the regular user, positioning determination may still sound like magic, like in a movie, where the police department manages to locate the terrorist using some mysterious signal triangulation algorithm. This book aims to make LBS accessible and understandable to the public, distinguishing myths from facts and explaining the technology behind the positioning algorithms.

LBS do not only address taxi and ride-sharing services; they are also a cornerstone of safety-of-life applications, like the E911 regulations in the United States. In the event of an emergency call, the mobile device and the cellular network calculate the caller's position and send it to the emergency dispatcher, playing a vital role in the timely arrival of the first responders. Furthermore, a large number of commercial applications, like the Internet of Things (IoT) (e.g., bike sharing), also rely on LBS. With the advent of 5G, the number of applications demanding precise positioning technologies has greatly increased. Its use is not only in E911, but also autonomous driving, unmanned aerial vehicles (UAVs, drones), eHealth, Industrial IoT (IIoT), traffic monitoring and control, and a large list of other services. Some of them, such as autonomous driving, are safety-critical and require high reliability and low latency. This book provides an overview of the LBS requirements for both regulatory and commercial applications.

This book starts with a basic introduction to positioning fundamentals, followed by the applications of positioning to cellular networks both for emergency services and commercial use cases. The different localization technologies are then described, starting with A-GNSS and high-accuracy GNSS extensions (e.g., RTK or PPP) and continuing with terrestrial radio positioning (ECID, DL-TDOA, multi-RTT, or angle-based techniques). The book goes beyond to non-RF positioning methods and other technologies useful for autonomous driving, giving an overview of inertial navigation and sensor fusion. In the last part of the book, the different positioning protocols used in GSM, CDMA, UMTS, LTE, and 5G are analyzed, with special emphasis on the LTE Positioning Protocol (LPP), which is used for both LTE and 5G positioning.

The main objective of the book is to highlight the importance of accurate positioning and provide an understanding of the different technologies used to achieve this. The focus of this book is on location-based services, their applications in cellular networks, and the communication protocols associated with them. This includes examples of call flows and detailed definitions of the different concepts involved in LTE or 5G positioning sessions. Each part of the book builds on the previous one: Part I explores the location requirements and use cases, Part II details the different technologies that can be used to address those requirements, and Part III focuses on the implementation of the selected technologies to cover the cellular network positioning requirements.

This book is aimed at people involved in the fields of mobile communications, the automotive industry, and the Internet of Things (IoT), who require an understanding of localization techniques. It is also aimed at university students of engineering, mathematics, physics, or the equivalent. The book gives an overview of the positioning basics and builds onto complex, high-accuracy positioning technologies needed for future applications. Previous knowledge in positioning or navigation is not required, although beneficial. Familiarity with basic engineering mathematics and geometry will be helpful for understanding some of the equations and derivations in this book. After reading this book, the reader should have a clear picture of the process initiated between a mobile phone and the cellular network related to a localization session and an understanding of the different technologies and algorithms involved. This knowledge will prove crucial in the highly connected 5G ecosystem, where LBS will play a fundamental role in its roll-out.

Part I

Positioning Overview, Applications, and Use Cases

Chapter 1

Introduction to Positioning in Cellular Networks

1.1 INTRODUCTION

Cellular network location-based services (LBS) are rapidly gaining momentum in the 5G New Radio (5G NR) ecosystem. Initially deployed for emergency services in the United States, mobile device positioning has now become crucial for a wide variety of commercial applications. Starting with Long Term Evolution (LTE) and now more prominently with 5G NR, the improvement in mobile positioning accuracy has become an enabler of futuristic technologies such as augmented reality (AR) or autonomous driving.

This book will take a close look at cellular network positioning requirements, the technologies that help fulfill those requirements, network architecture, and protocol call flows. Before diving into location-based services in cellular networks, this introductory chapter will briefly present the history of both cellular networks and navigation to help set up the context for the chapters to come.

1.2 HISTORY OF CELLULAR NETWORKS

The early 1980s saw the deployment of the first cellular systems in the United States, called Advanced Mobile Phone Service (AMPS) [1] and also referred to as the First Generation (1G) mobile phone system. Wireless communications prior to this consisted of a few transmitters covering a large area and supporting a limited

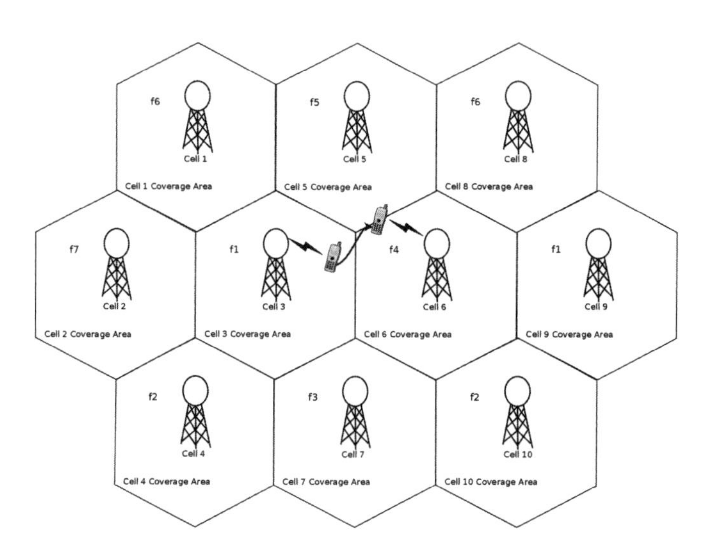

Figure 1.1 Illustration of a cellular network.

number of connections. These systems were limited by the number of users they could support using a few communication channels.

The AMPS system overcame the limitations of other early analog systems by dividing large areas into smaller cells, each of which consists of a transmitting station using frequencies different than that of its neighboring cells, as shown in Figure 1.1. When a user moves from one cell coverage area to another, a handover procedure is triggered, and the user is served afterward by the new cell. The area covered by each cell corresponds to the area where the transmitted signal is visible to the mobile device. The signals are transmitted with relatively low power in order to serve the cell area, but without interfering with the neighbor cells. This enables frequency reuse; that is, relatively close cells can transmit on the same frequency channels, as seen in Figure 1.1, where Cell 3 and Cell 9 both use the same frequency $f1$. This improves the spectral efficiency and enables unlimited network scalability. Depending on the capacity needs, cell sizes can be adjusted (i.e. urban areas with smaller cell sizes and lower transmit powers and rural areas with bigger sizes and higher transmit powers). Cellular networks offer mobility between cells, allowing users to roam between cells using coordination techniques between the cells; this allows operators to offer nationwide mobile service with roaming.

The AMPS system uses a frequency spectrum divided into subbands of 30 kHz called channels. The available channels are divided into forward channels, assigned for transmitting downlink signals to the mobile device and reverse channels, to receive the uplink signals transmitted by the mobile. AMPS uses a technique called frequency division multiple access (FDMA), in which the channels are assigned individually to each user to carry the voice signal. The voice signal in this system was analog and the whole channel was dedicated to one voice call.

1.2.1 2G

The early 1990s saw the rise of Second Generation (2G) mobile phone systems, whose growth was propelled by the increasing demand for wireless services needing to achieve higher spectral efficiency and better coverage. The first of the 2G networks in the United States was an improvement on the AMPS system, called Digital Advanced Mobile Phone Service (D-AMPS). D-AMPS was standardized as IS-54 [2,3]. The D stands for digital, and the main upgrade was the conversion of the analog voice signals into digital signals. The digitization of the voice signal allowed an increase in the capacity of the network. Furthermore, the time division multiple access (TDMA) technique allowed to divide every frequency into three time slots, each of which is dedicated to a different user.

However, D-AMPS was neither the first nor the only 2G cellular network. 2G network is a term that encompasses multiple technologies developed in parallel in different countries. These technologies improved over the 1G systems significantly and offered features like encrypting of user data and basic data services like the short message service (SMS). Apart from D-AMPS, the most prominent second generation mobile networks are Global System for Mobile (GSM) in Europe, Interim Standard 95 (IS-95) in the United States and Personal Digital Cellular (PDC) in Japan. GSM is also a TDMA-based technology created by the European Telecommunication Standard Institute (ETSI) [4]. It achieved great success and worldwide deployment, and is still active today. The second most popular 2G technology, IS-95, was developed by Qualcomm Technologies in USA and later incorporated by the Telecom Industry Association (TIA). IS-95 is also often referred to by its duplexing scheme, code division multiple access (CDMA). It multiplexes several channels on the same frequency resources by separating them using orthogonal codes for each channel.

1.2.2 3G

The middle of the 1990s saw large-scale adoption of mobile phones in many countries and the price of mobiles phone calls reduced drastically. This led to much larger capacity demands, especially in urban areas. At the same time, the internet was deployed worldwide and the need for mobile data increased. In order to adapt to the mobile data requirements, the 2G technologies were updated for supporting data services. This upgrade is popularly categorized as the 2.5G mobile system. The most popular of these technologies are the General Packet Radio Service (GPRS) and IS-95B. GPRS is an update to GSM technology, which transmits data on the same time slots as used by GSM voice channels. IS-95B is the corresponding update to CDMA technology.

The amount of data traffic that can be supported with 2.5G was restricted by radio technology and design limitations of the second generation networks. Furthermore, as the number of people using mobile phones rapidly increased, other needs such as mobility and interoperability between the different network standards also arose. Soon, it became clear that a completely new generation of technologies was needed in order to support the rising data traffic and the mobility and compatibility requirements. The success of cellular networks was crucially tied to the harmonization between different manufacturers and countries. The International Telecommunication Union (ITU), an agency of the United Nations (UN), recognized this need and established the International Mobile Telecommunication 2000 (IMT-2000) project with the goal of defining the Third Generation (3G) network. IMT-2000 aimed to establish a global standard for common understanding of frequency spectrum usage, access network specification, and cellular standards, providing an update path from existing 2G networks. Furthermore, 3G networks were developed to support not only voice channels but also data services such as browsing, email services, and data download.

The main goal of the 3G networks was to provide, in addition to circuit switched (CS) voice call services, packet switched (PS) services with low, medium, and high data rate connections. The circuit switched network components were reused from the 2G network elements and they were enhanced with components to support packet switched connections, connections to the internet, and support for multimedia services. Two main standardization bodies were created to realize the IMT-2000 project, the Third Generation Partnership Project (3GPP) and Third Generation Partnership Project 2 (3GPP2).

The 3GPP was established in 1998 by the ETSI and included other standard development organizations from around the world, for example, Association of Radio Industries and Businesses (ARIB) (Japan) and Alliance for Telecommunications Industry Solutions (ATIS) (United States). The initial goal of 3GPP was to standardize the 3G technologies while offering interoperability with the existing components of the legacy GSM networks. As a result of this effort, the Universal Mobile Telecommunication System (UMTS) standard was defined [6]. UMTS provided a radio access technology based on CDMA technology, with two major implementations that became popular around the world:

- Wideband Code Division Multiple Access (WCDMA) [7], which uses frequency division duplex (FDD) technology. WCDMA is also known as UTRA-FDD and is widely adopted in Europe, Asia, and parts of North America.
- time division synchronous code division multiple access (TD-SCDMA) [8], which uses time division duplex (TDD) techniques. It is also known as UTRA-TDD and is primarily deployed in China.

In parallel, North America saw the creation of the 3GPP2 that focused on the 3G standards as an evolution to the legacy IS-95 network. The new technology, published in IS-2000, was known as Code Division Multiple Access 2000 (CDMA2000) or C2K. As the CDMA2000 network was a natural evolution of the IS-95 based networks, it was primarily deployed in North America.

Both 3GPP and 3GPP2 also standardized updates to the 3G networks with the aim of providing higher data rates. These updates are often referred to as 3.5G technologies. The 3GPP technology for 3.5G was the high-speed packet access (HSPA) [9] and for 3GPP2 was the Evolution-Data Optimized (EVDO).

1.2.3 4G

The next cellular network revolution originated in the middle of the 2000 decade with the arrival of the smartphones. This was enabled by the increase in mobile phone processing power, improvement in display resolution, and system-on-chip (SoC) technology. Ever-increasing transistor and memory density lead to compact mobiles with very large computing power. Voice-centered mobiles phones mutated to devices combining a computer and a mobile phone. A plethora of data-hungry mobile applications started to appear, demanding higher quality of service (QoS). These applications required much higher wireless data capacity per user and per cell

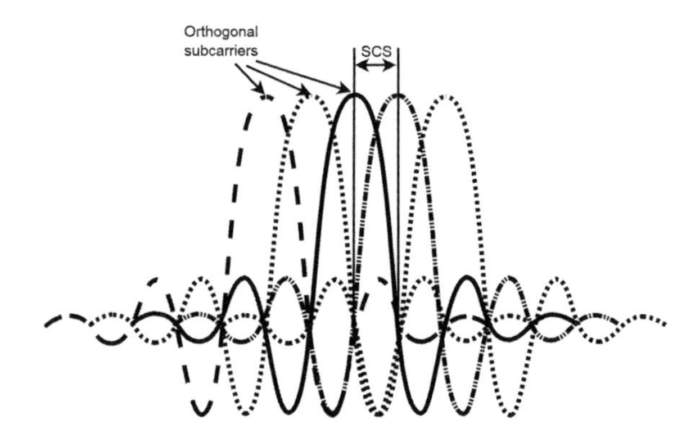

Figure 1.2 LTE OFDM signal.

than what 3G and 3.5G networks could offer. Furthermore, 3G and 3.5G networks updates were inefficient due to the need to establish dedicated connections for voice channels.

The convergence of voice and data on one network was one of the major drivers for 4G systems. This possibility was fueled by the adoption of technologies such as IPv6 and voice over IP (VoIP). As with 3G, both 3GPP and 3GPP2 started to work in parallel toward the standardization of a 4G technology. The 3GPP proposed a technology called LTE [10]. LTE is based on orthogonal frequency-division multiple access (OFDMA) in the downlink and single carrier frequency-division multiple access (SC-FDMA) in the uplink. OFDMA consists of multiple low data rate orthogonal subcarriers, as shown in Figure 1.2. For LTE, the subcarrier spacing (SCS) is 15 kHz. The subcarriers can be assigned to multiple users for data transmissions.

The 3GPP2 started to work on a CDMA2000 technology evolution, known as the Ultra Mobile Broadband (UMB). However, UMB was eventually dropped and did not see real commercial deployment. Another organization joined the 4G race, the Institute of Electrical and Electronics Engineers (IEEE), with its 802.16 standards [11]. This technology is known as Worldwide Interoperability for Microwave Access (WiMAX), initially conceived as a wireless alternative to the wired connections for the last mile broadband access. Nonetheless, WiMAX did not find much commercial success as a cellular technology and LTE became the worldwide cellular industry standard for 4G.

This consolidation of all 4G cellular technologies under LTE resulted in real global mobility. Due to LTE's success, many operators started to reuse the spectrum by refarming 2G and 3G frequency bands and dedicating them to LTE. With the VoIP feature, LTE became the first data-only cellular network, offering peak data rates of 300 Mbps. The later evolution of LTE, such as the addition of carrier aggregation, can reach data rates of up to 1 Gbps. These upgrades are collected under the term LTE Advanced (LTE-A).

The LTE network also incorporated significant updates addressing markets different than the traditional mobile phone business. For instance, device-to-device (D2D) allows devices to directly communicate with one another using a local wireless channel rather than over regular network resources. D2D technology laid the foundations for the vehicle-to-everything (V2X) technology introduced in 3GPP Release 14. V2X includes connectivity modules in vehicles that allows them to communicate with other vehicles and road infrastructure, catering to road safety, assisted driving, and autonomous driving. Another significant example is the creation of standards for the Internet of Things (IoT), including machine type communications (MTC) and its successors (eMTC, feMTC, etc.) and the narrowband IoT (NB-IoT). These standards, introduced as of 3GPP Release 13, aimed at connecting a whole new category of devices other than smartphones. This expansion toward new markets, together with the usual demand for increased data rates, planted the seed of the 5G evolution.

1.2.4 5G

LTE and LTE-A introduced mobile broadband access to smartphones. However, toward the last part of the 2010s, the LTE networks started to reach their limits. The reasons for this are manifold, including the tremendous growth in smartphone penetration and the increase in bandwidth-hungry applications, as well as the large number of connected devices created for IoT business. Furthermore, V2X and other safety-critical applications increased the need for highly reliable connections with very low latency. Thus, the ITU gathered all these new requirements under the IMT-2020 project [12].

The 3GPP initiated the definition of the Fifth Generation technology, under the name 5G NR, often referred to by either of the acronyms 5G or NR. The 5G NR requirements have been structured under three main categories:

- Enhanced mobile broadband (eMBB), whose aim is to provide wireless connectivity with very high bandwidth.

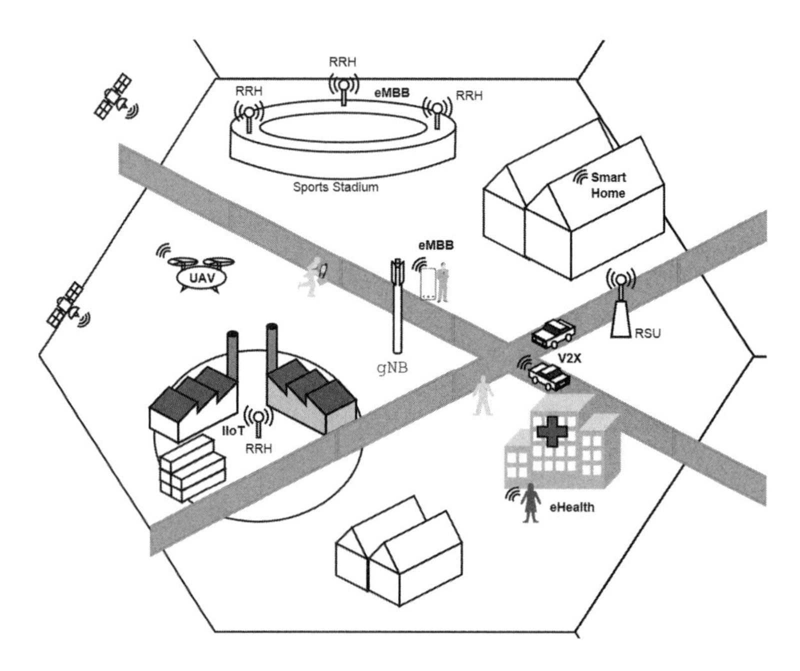

Figure 1.3 Example of a typical 5G network.

- Massive machine type communications (mMTC), providing connectivity to a large number of IoT devices such as smart meters, watches, or wearables. mMTC requires very large cell and network capacity.
- Ultra-reliable low latency communications (URLLC), targeted to provide low latency robust communication links for V2X, remote surgery, and other safety-critical applications.

The 5G network is the first cellular network that is not designed with the sole focus on mobile phones. It includes other devices and use cases, referred to as industry verticals. Another significant difference with respect to previous cellular networks is the use of higher frequencies in the millimeter-wave (mm-wave) spectrum, starting at 24 GHz. 5G also brought important features such as massive multiple input, multiple output (MIMO), beamforming, cloud computing, and network virtualization. All these features help increase the scalability and modularity of the network and to reach peak data rates of 20 Gbps with very high user density. Figure 1.3 shows how a typical 5G network could look, with new use cases such as eHealth, V2X, unmanned aerial vehicle (UAV), smart home, or Industrial Internet of Things (IIoT), in addition to the traditional mobile phone business. Figure 1.3 also introduces the concept of small cell deployments, with remote radio heads (RRHs) to serve high density areas such as a sports stadium. Comparing Figure 1.3 to Figure 1.1 can give a good overview of how the complexity of the network has increased.

1.2.5 Summary of the Cellular Technologies

Table 1.1 shows a comparison of the different network technology per generation, including radio access technology and the peak data rates and their main applications.

Table 1.1

Comparison of Cellular Technologies per Generation

Name	Gen.	RAT	Peak Data Rate	Main Applications
AMPS	1G	FDMA	N.A.	Analog voice
GSM	2G	TDMA	14.4 kbps	Voice, SMS
CDMA	2G	CDMA	14.4 kbps	Voice, SMS
GPRS	2.5G	TDMA	171.2 kbps	Email, browsing
WCDMA	3G	CDMA	2 Mbps	Voice, video, browsing
TD-SCDMA	3G	CDMA	2 Mbps	Voice, video, browsing
CDMA2000	3G	CDMA	3.1 Mbps	Voice, video, browsing
EV-DO	3.5G	CDMA	DL: 14.7 Mbps UL: 5.4 Mbps	High-speed data
HSPA	3.5G	CDMA	14.4 Mbps	High-speed data
WiMAX	4G	S-OFDMA	70 Mbps	Mobile broadband
LTE-A	4G	DL:OFDMA UL:SC-FDMA	DL: 3 Gbps UL: 1 Gbps	Mobile broadband, IoT, V2X
NR	5G	DL:CP-OFDM UL:DFT-s-OFDM	DL : 20 Gbps UL: 10 Gbps	eMBB, mMTC, URLLC

1.3 HISTORY OF NAVIGATION

Navigation on open oceans was one very important challenge for early sailors; they used multiple techniques including wind direction, the location of islands, and the location of celestial bodies for navigation purposes. The position of the sun and the time of day were typically used to estimate the latitude of one's location. The invention of the compass was a big leap in finding the direction of a ship. Later sailors adopted dead reckoning techniques, where the speed of the ship was calculated and used to estimate the distance traveled. This information was combined with the previously known location to calculate a new location at the end of each day. This location was then used as the last known location for the next day and so forth. Another important evolution in navigation was the invention of the sextant. The sextant assisted sailors in measuring the angle of a celestial object with respect to the horizon. It was used to accurately measure the angle of the sun, moon, stars, and star constellations. Comparing these measurements with known measurements at a specific time allowed to estimate the ship's position. Basic maps and the location of known islands helped to correct the location estimate regularly. These methods have obvious disadvantages during bad weather conditions and clouds, when the visibility of the stars was obscured.

The next major step in determining the current location of a ship was the development of the marine chronometer in the eighteenth century. The chronometer was able to precisely measure the time over sea voyages. By comparing the time measured at a known past location against the current time and also using the angle of the sun or the stars, sailors were able to estimate the current longitude.

With the invention of radio waves, radio beacons were applied for navigation purposes. A loop antenna was used to find the direction of a specific radio station. Subsequently, the antenna was tuned to another station in a different direction. Both these radio stations were installed at known locations and the location of the ship was calculated by triangulation. During the first and the second world wars, multiple radio-wave location techniques were developed and deployed effectively. One of the most notable technologies using radio beacons was the Long Range Navigation (LORAN) system, developed by the United States during the Second World War. It used low-frequency radio waves, propagating over very large distances, such as the whole Atlantic Ocean. The system consisted of multiple broadcast stations, one of which was the primary station. The stations were usually located a few hundred kilometers apart, close to the coastline, and transmitted radio pulses at high power. By measuring the time difference of the pulsed signals received from two stations, the ship could draw a hyperbola on the map representing its possible locations. With two pairs of stations, the ship could determine its position at the intersection of the two hyperbolae. A variant of the LORAN system called the LORAN-C system was used until the beginning of the twenty-first century for navigation on oceans. Its successor, the eLORAN, is being considered as a backup system for satellite navigation.

1.3.1 Satellite Navigation

In the late 1950s, the Soviet Union launched the Sputnik satellites into space. U.S. scientists realized that by utilizing the Doppler effect, they could track the orbit of the Sputnik's satellites. As the satellite got closer to the observers, the signal frequency and signal strength increased. Conversely, when the satellite moved away, its frequency and signal strength reduced. Later in the mid 1960s, the U.S. navy conducted experiments for locating submarines using Doppler frequency measurements on satellites orbiting the poles. The measured Doppler signals on the satellites lead to accurate estimates of the submarine's position. In the early 1970s, the U.S. Department of Defense decided to establish a more robust plan for providing navigation services using satellites. This plan resulted in the 1978 launch of the first satellite of what would be called the NAVSTAR Global Positioning System (GPS), today known simply as GPS. A total of 24 satellites were launched until 1993, the year in which the initial GPS constellation was completed. The satellites orbit at an altitude of approximately 20.200 km, part of medium Earth orbit (MEO) range. Operating at this orbit, and with the constellation of 24 satellites, GPS became a worldwide location determination service. The development of highly accurate atomic clocks and mechanisms to accurately predict the satellite orbits were some of the notable technological breakthroughs necessary for the robust functioning of GPS.

GPS satellites use CDMA technology with a code determined by a pseudo-random noise (PRN) number. The space vehicle number (SVN) identifies the satellite. GPS offers two different service types. The Standard Positioning Signal (SPS) is intended for civilian purposes and is transmitted on the L1 band (1.575 GHz). It consists of the coarse/acquisition (C/A) code, which is a short code modulated in the 1-MHz bandwidth. For military purposes, GPS offers an additional code, more precise and robust, which can only be decoded by authorized military receivers.

In the late 1970s, the Soviet Union developed a navigation system called Globalnaya Navigatsionnaya Sputnikovaya Sistema (GLONASS). The system became operational in the early 1990s. In the early 2000s, the Russian Federation developed the GLONASS system further to provide worldwide coverage with 24 active satellites. The GLONASS satellites orbit at slightly lower orbits than GPS, approximately 19.100 km above the Earth's surface. GLONASS provides better coverage in higher latitude regions compared to GPS, due to the higher orbital inclination of its satellites. Such inclinations provide better coverage in countries like Argentina and Australia. The GLONASS satellites each transmit using the same code but are separated from one another using FDMA. The initial system transmitted on two frequency ranges, GLONASS L1 (1598 MHz – 1605 MHz) and GLONASS L2 (1243 MHz – 1249 MHz). Future updates to GLONASS, known as GLONASS-K, plan to use CDMA (like GPS) instead of FDMA. In order to differentiate from the previous GLONASS signals, the L1 in CDMA is referred to as L1OC, while the FDMA one is now also called L1OF.

The European Union also developed its own global navigation satellite system (GNSS), called Galileo, which became operational in 2020. Galileo was planned to provide higher accuracy than GPS and GLONASS. It consists of 30 satellites orbiting around 23.200 km slightly is higher than GPS. The Galileo signal also uses CDMA and is defined to interwork closely with GPS signals. It operates the following civilian frequencies: E1(1575.42 MHz), E5a(1176.45 MHz), E5b(1207.14 MHz), and E6(1278.75 MHz).

Over time, additional satellite systems appeared across the globe that were developed by different countries. These satellite systems aim specifically to provide better regional coverage. The choice of higher satellite orbits, such as geosynchronous orbit (GSO) and geostationary orbit (GEO), is popular for regional systems because they are visible for a much longer time and hence require fewer satellites for navigation purposes.

The People's Republic of China started development of the BeiDou Navigation Satellite System (BDS) (also known as COMPASS) in 2004 and completed the initial constellation of 14 satellites by the end of 2012. BDS operates at two different altitudes. The system includes 5 GEO satellites and 5 inclined geosynchronous orbit (IGSO) satellites, both operating at 35.500 km, and 4 MEO satellites at 21.528 km. The satellites use CDMA technology and the civilian signal operates on frequencies B1-I (1561.098 MHz) and B2-I (1207.14 MHz). BDS has planned to provide worldwide coverage as of late 2020.

The Indian Regional Navigation Satellite System (IRNSS), also referred to as NAVIC, became operational in 2018. Currently, it consists of a constellation of 7 satellites with 3 GEO and 4 GSO satellites. The civilian operation uses the L5 (1176.45 MHz) and the S (2400 MHz) bands.

The Quasi-Zenith Satellite System (QZSS) from Japan covers mostly Japan and Southeast Asia. It will eventually be formed by 7 IGSO satellites. The QZSS signal is similar to GPS and operates on frequencies such as L1 C and L1CA (1575.42 MHz), L2C (1227.6 MHz), L5 (1176.45 MHz), and L6 (1278.75 MHz).

Another type of satellite system is the regional satellite-based augmentation system (SBAS), which is used in combination with other satellite systems such as

GPS to improve performance over a particular region. These systems typically consist of ground stations, which continuously monitor the satellite transmission, and GEO-stationary satellites. SBASs provide regional coverage and transmit assistance data to improve the accuracy of the location fixes and to supply information about the integrity of the satellites. Example of SBASs are the Wide Area Augmentation System (WAAS), used in the United States, and the European Geostationary Navigation Overlay Service (EGNOS), used in Europe. At the time of this writing, they operate on the L1 frequency, while L5 support is planned as well.

Table 1.2 gives an overview on the different satellite constellations available for civilian use, their characteristics, and frequency of operation. "Mod." stands for modulation.

Table 1.2

Civil Satellite Constellations as of 2020

System	**Country**	**Coverage**	**Orbits**	**Mod.**	**Signals**
GPS	USA	Global	MEO	CDMA	L1 C/A, L2C, L5
GLONASS	Russia	Global	MEO	FDMA	L1, L2
Galileo	EU	Global	MEO	CDMA	E1, E5a, E5b, E6
BeiDou	China	Regional	GEO, GSO, MEO	CDMA	B1-I, B2-I
NAVIC	India	Regional	GEO, GSO	CDMA	L5
QZSS	Japan	Regional	GSO	CDMA	L1, L2C, L5, L6
WAAS	USA	SBAS	GEO	CDMA	L1, L5
EGNOS	EU	SBAS	GEO	CDMA	L1, L5

1.3.2 Cellular Network Positioning

Early GPS receivers were bulky and required a lot of time to synchronize to the satellites for obtaining fixes. Thus, their usage in handheld equipment such as mobile phones was rather limited. Improvements in electronic circuitry and computing power allowed the reduction of the size of GPS receivers and the first mobiles phones containing the GPS modules started appearing in the late 1990s. The need to be able to locate an emergency wireless caller accurately also propelled the widespread adoption of this technology in mobile phones.

A stand-alone GPS receiver calculates its position by reading the navigation data from the satellites. This procedure can take up to several minutes, especially under bad signal conditions. The assisted GPS (A-GPS) technology was developed in order to reduce the time to acquire navigation data. A-GPS uses high-speed cellular network connections to provide the navigation data necessary for position calculation. The 3GPP later adopted and standardized the A-GPS technology, defining standard messaging protocols that enabled interworking between different devices and network implementations. The A-GPS technology drastically reduces the time required for a position fix, becoming the technology of choice for location determination in mobile phones, especially in the context of emergency calls. Additionally, computational intensive parts of the position calculation can be carried out in the mobile network rather than in the device, reducing the device complexity. With the deployment of further satellite systems, A-GPS was extended to include other GNSS constellations such as GLONASS, BeiDou, Galileo, QZSS, and IRNSS. Today, it is commonly referred to as assisted GNSS (A-GNSS).

However, A-GNSS technologies have limitations for indoor positioning or when the satellite signals are impaired, such as in dense urban scenarios. Thus, cellular network positioning technologies were defined in order to complement A-GNSS. The first cellular network positioning technologies designed for GSM networks were Cell ID and enhanced observed time difference (E-OTD). Similar methods and upgrades were defined for later cellular generations, such as the LTE positioning technologies observed time difference of arrival (OTDOA), enhanced cell ID (ECID) and uplink time difference of arrival (U-TDOA). These positioning technologies can be used on their own, or combined together with A-GNSS in a hybrid algorithm.

In the mid-2010s, the Federal Communication Commission (FCC) of the United States started to define indoor and vertical accuracy positioning requirements. The cellular positioning technologies and A-GNSS proved insufficient to achieve the FCC regulatory requirements. Thus, new positioning technologies based on Wi-Fi and Bluetooth access point measurements were incorporated into the standards. These technologies benefit from crowd-sourcing databases to map Wi-Fi and Bluetooth measurements to specific locations. The altitude determination problem was addressed by standardizing atmospheric pressure positioning measurements with the help of a barometric sensor.

The next big leap in cellular network positioning requirements came with the advent of autonomous driving and V2X. Inertial measurement unit (IMU) measurements were defined to bridge the gap between A-GNSS outages. IMU measurements can also be used, together with radar, lidar, and other sensors, in sensor fusion algorithms. Furthermore, high-accuracy GNSS methods, such as real-time kinematics (RTK) and precise point positioning (PPP), have been integrated

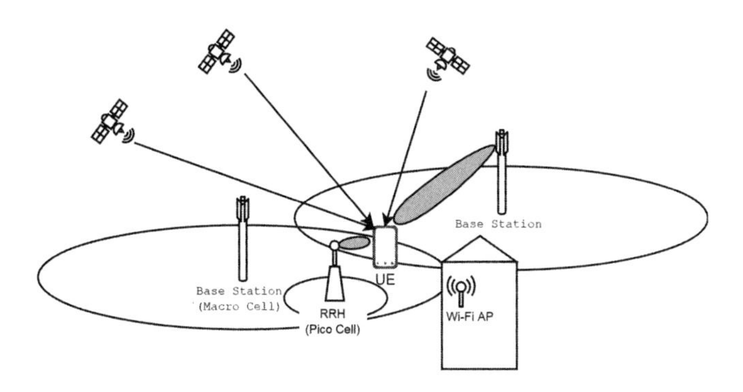

Figure 1.4 Example of a positioning scenario with A-GNSS, cellular technologies, and Wi-Fi.

into the LTE network. RTK and PPP enable centimeter-level positioning accuracy by using correction data obtained from reference ground stations.

With the advent of 5G, new location-based services commercial applications, such as augmented/virtual reality, safety-critical applications, UAV remote control, or factory automation arose, raising the bar of positioning requirements. All the radio access technology (RAT) independent positioning technologies (A-GNSS, Wi-Fi, Bluetooth, barometric sensors, RTK, etc.) are also supported in 5G. Furthermore, new cellular network technologies have been defined as well. These include downlink time difference of arrival (DL-TDOA), multi-round-trip time (multi-RTT), and angle-based techniques, among others. 5G positioning requirements do not need only high positioning accuracy, but also rely on other performance indicators such as availability, reliability, and energy consumption.

A typical positioning scenario is shown in Figure 1.4, including satellite, cellular-based, and RAT independent positioning. Explaining all these positioning technologies and their corresponding performances will be the main objective of this book.

1.4 ABOUT THIS BOOK

The goal of this book is to present location technologies in 2G, 3G, 4G, and 5G cellular networks. A basic idea of the cellular network architecture and working principle is given wherever appropriate. Nonetheless, prior knowledge of cellular technologies is an advantage. This book covers the basics of GNSS and other noncellular network technologies (e.g., Wi-Fi, radar, or sensor fusion) to the extent that is relevant for location-based services in cellular networks. The reader is advised to augment this book with specialized reading on GNSS, Wi-Fi, sensor fusion, and other noncellular network positioning technologies.

A reader not familiar with location-based services is advised to read this book sequentially, since the book starts with the basic concepts and then moves to more advanced topics. An advanced reader can select specific topics and directly move to the corresponding chapter.

The book is structured in three parts. Part I of the book introduces basic positioning concepts and the cellular network architecture, positioning use cases, and positioning requirements. This part covers Chapters 2 to 5.

Chapter 2 of this book introduces basic positioning terminology, such as the coordinate frames, the different type of positioning measurements, and the corresponding positioning algorithms. An understanding of the concepts presented in this chapter is essential to comprehend the rest of the book. Chapter 3 presents the positioning architecture for emergency services in cellular networks and the corresponding regulatory requirements. The regulatory requirements were the main driver for positioning accuracy in GSM, WCDMA, and LTE. The chapter also covers the E911 architecture in the United States and the eCall and Advanced Mobile Location (AML) architectures in the European Union. Chapter 4 describes the LTE positioning architecture, including the architecture for the commercial LBS and its applications. The V2X and IoT use cases and their impact on positioning are part of this chapter. The last chapter of this first part, Chapter 5, introduces the 5G network, explaining its differences from legacy networks and the E911 and commercial LBS architectures. Furthermore, it explores the new positioning use cases and applications envisaged for 5G NR.

Part II of the book covers Chapters 6 to 11 and describes the different positioning technologies that are used in cellular networks. Chapter 6 starts with a basic introduction to GNSS to later focus on the A-GNSS technology. High-accuracy GNSS enhancements, such as RTK and PPP, are explained in Chapter 7, including observation state representation (OSR) and state space representation (SSR) and the design for large-scale deployments of these techniques over cellular networks. Chapter 8 of the book covers the cellular network-based positioning technologies in legacy networks, from GSM to LTE. From very basic technologies like Cell ID, to more advanced technologies such as OTDOA, all are included in this chapter. The chapter also explores hybrid algorithms to combine A-GNSS and cellular network positioning technologies together. Chapter 9 focuses on the noncellular network-based positioning technologies, such as Wi-Fi and Bluetooth, and their addition to the LTE location-based services. Chapter 10 analyzes the 5G NR native positioning methods defined by the 3GPP in Release 16. These methods are partly an evolution of the LTE techniques, such as DL-TDOA, and partly completely new technologies, such as the angle-based positioning methods. Chapter 11 gives a comparison of the performance of the different positioning methods in real deployments with regard to their accuracy, reliability, ease of deployment, network capacity usage, energy efficiency and other relevant key performance indicators (KPIs).The overview of the different positioning technologies concludes in Chapter 12 with topics related to sensors and their integration in positioning systems. Basic concepts of IMU, radar, and lidar are discussed, especially in relation to cellular applications. This chapter also provides an overview of advanced sensor fusion techniques.

Part III of the book, from Chapters 13 to 16, focuses on the different communication protocols used in cellular network positioning. Chapter 13 gives a general overview of the location service (LCS) protocol for legacy networks in 2G and 3G, and introduces the fundamental transactions of a positioning session. Chapter 14 describes the LTE Positioning Protocol (LPP), and its evolution from the first LTE release (Release 9) to Release 15, detailing the positioning call flows and the most important information elements exchanged. Chapter 15 covers the LPP enhancements for 5G location-based services support, giving an overview of the new procedures and protocol messages to cover 5G positioning technologies. Finally, Chapter 16 explores the different positioning protocols used for communication between the radio access network (RAN) and the core network (CN), especially focusing on the LPPa and NRPPa protocols. The chapter also points out the differences between the 5G network with respect to LTE, introducing the topics of network virtualization and cloud computing.

References

[1] AT&T Archives, *Testing the First Mobile Phone Network*, https://techchannel.att.com/play-video.cfm/2011/6/13/AT%26T-Archives-AMPS%3A-coming-of-age, accessed on February 2020.

[2] Huurdeman, A. A., *The Worldwide History of Telecommunications*, Wiley Interscience, 2003.

[3] Wireless Communication, *IS-54 and IS-136 (Digital AMPS)*, http://www.wirelesscommunication.nl/reference/chaptr01/telephon/is54.htm, accessed on February 2020.

[4] ETSI, *Global System for Mobile communications*, https://www.etsi.org/technologies/mobile/2g, accessed on February 2020.

[5] Decker, P., and Walke, B., "A General Packet Radio Service proposed for GSM," *ETSI SMG Workshop "GSM in a Future Competitive Environment,"* Helsinki, Finland, Oct. 13, 1993, pp. 1-20.

[6] 3GPP, *UMTS*, https://www.3gpp.org/technologies/keywords-acronyms/103-umts, accessed on February 2020.

[7] 3GPP TS 25.001, *User Equipment (UE) Radio Transmission and Reception (FDD)*, V16.1.0, April 2019.

[8] 3GPP TS 25.002, *User Equipment (UE) Radio Transmission and Reception (TDD)*, V15.0.0, July 2018.

[9] 3GPP, *HSPA*, https://www.3gpp.org/technologies/keywords-acronyms/99-hspa, accessed on February 2020.

[10] 3GPP, *LTE*, https://www.3gpp.org/technologies/keywords-acronyms/98-lte, accessed on February 2020.

[11] IEEE 802.16, *Working Group on Broadband Wireless Access Standards*, http://ieee802.org/16/, accessed on February 2020.

[12] ITU-T SG13, *Focus Group IMT-2020 Deliverables*, December 2016 https://www.itu.int/en/ITU-T/focusgroups/imt-2020/Pages/default.aspx, accessed on February 2020.

Chapter 2

Positioning Fundamentals

2.1 INTRODUCTION

Positioning is the ability to determine the location of a particular object. However, how is that location defined? In order to describe the location of any object, a reference is needed. The reference can be a global reference, such as geodetic coordinates, or a local reference, such as the distance to a known object. The distance to a known object can be calculated by measuring the time needed by a signal to travel from the object to locate to the known object used as reference, or vice versa. It can also be calculated by power ranging (i.e., measuring the power lost by a signal to travel between the two objects). In fact, any physical value dependent on the location can in principle be used for positioning.

This chapter will introduce all this basic positioning terminology, starting with the coordinate frames used to describe a particular position, following with the different types of measurements and how these can be combined to form positioning algorithms, and concluding with a series of useful concepts and definitions related to positioning systems. Understanding of these basic concepts will be crucial to better comprehend the more complex positioning technologies defined in later chapters.

2.2 COORDINATE FRAMES

A coordinate frame is the basic instrument required to define the location of an object. Without a common coordinate frame, it is not possible to describe a position in a universal way. It can be seen as the language of positioning. This section will

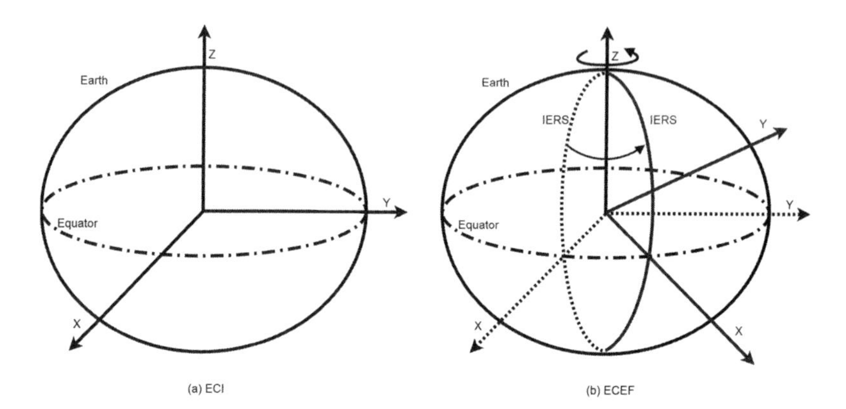

Figure 2.1 (a) Earth-centered inertial coordinate frame. (b) Earth-centered Earth-fixed coordinate frame.

introduce the various types of coordinate systems, the differences between them, and the impact of choosing one over the others for positioning applications.

The most generic and simple frame of reference is the Earth-centered inertial (ECI), whose origin is the center of the mass of the Earth. The x-y plane of the system is the equatorial plane and the z-axis is perpendicular to the equatorial plane and crosses the North Pole. As any inertial frame, the ECI is time homogeneous, space homogeneous, isotropic, and time independent. In order to fulfill the time independence property, the x-axis in ECI remains permanently fixed in the *celestial sphere*, which is an imaginary sphere of arbitrarily large radius, Earth-centered, and not following the Earth's rotation movement. Thus, the coordinates of any given point on the surface of the Earth vary with time for the ECI system. In consequence, the ECI is not useful for calculating a position on the surface of the Earth.

2.2.1 World Geodetic System

In order to overcome the disadvantages of the inertial frame and provide every point on the Earth's surface with a unique, time-invariant representation, a different Earth frame needs to be defined. This is known as Earth-centered Earth-fixed (ECEF) coordinate frame. The difference between ECI and ECEF is shown in Figure 2.1. On the left side of the image, it can be seen that the x-axis of ECI is fixed in the celestial sphere and does not rotate with the Earth's rotation. On the right side of the image, the x-axis of ECEF is fixed from the center of the Earth to the intersection between the equator and the International Earth Rotation Service (IERS) reference meridian. When the Earth rotates, the x-axis rotates as well. In the image, the dotted lines represent the position of the IERS, the x-axis, and the y-axis at a certain time instant. The solid lines represent the same IERS, x-axis, and y-axis in a different time instant in the future, after the Earth has rotated a certain angle.

The most commonly used ECEF system in navigation is the World Geodetic System (WGS), in particular its latest revision from 1984, WGS84. The WGS84 comprises a standard coordinate frame for the Earth, a standard spheroidal reference surface (the reference ellipsoid), and a gravitational equipotential surface that defines the nominal sea level. The coordinate origin of WGS84 is the mass center of the Earth and the x-y plane is the Earth's equatorial plane. However, in this case the x-axis is fixed to a point in the surface of the Earth: the intersection between the Equator line and the IERS Reference Meridian. It is worth noticing that the IERS Reference Meridian is 102.5 m east of the Greenwich Prime Meridian at the latitude of the Royal Observatory. The z-axis is the same as in ECI, perpendicular to the equator and crossing the North Pole.

The coordinates of a point in an ECEF system are usually represented by the geodetic coordinates latitude, longitude, and altitude:

- Latitude: angle between the x-y plane (the equatorial plane) and the line that crosses the point to locate perpendicularly to the surface of the reference ellipsoid. The latitude can take values in the interval [-90°, 90°] and is considered as positive if the point is north of the x-y plane.
- Longitude: angle between the reference meridian and the meridian that contains the point to locate. The longitude can take values [-180°, 180°] and is positive if the point is east of the prime meridian.
- Altitude: distance in meters over the nominal sea level.

An equivalence between geodetic coordinates and $\{XYZ\}_{\text{ecef}}$ can be defined as in Equation 2.1, where λ stands for latitude, μ for longitude, h for altitude, A is the Earth semimajor axis in meters, and e is the Earth eccentricity [1].

$$
\begin{aligned}
X_{ecef} &= (N + h) \cdot cos(\lambda) \cdot cos(\mu) \\
Y_{ecef} &= (N + h) \cdot cos(\lambda) \cdot sin(\mu \\
Z_{ecef} &= (N \cdot (1 - e^2) + h) \cdot sin(\lambda) \\
N &= \frac{A}{\sqrt{1 - (E^2 \cdot sin^2(\lambda))}}
\end{aligned}
\tag{2.1}
$$

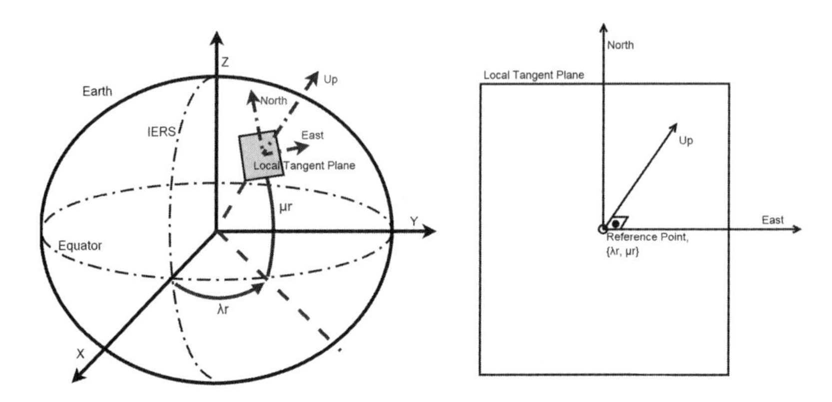

Figure 2.2 East North Up coordinate frame represented on a local tangent plane on the surface of the Earth.

2.2.2 East North Up

Some applications, for example navigation, usually take place on a region on the surface of the Earth that is relatively small compared to the whole surface. Thus, a more intuitive and practical coordinate system can be used instead of WGS84. The surface of the Earth is spherical, but a small section of the surface can be seen as planar [2]. That approximation can be exploited by defining a tangent plane to the surface of the Earth on a certain point. Navigation distances of up to a few kilometers are small enough to assume that the Earth can be approached by that plane without making a significant error. The Cartesian coordinate system east north up (ENU) is defined on that plane, called Local Tangent Plane. The ENU system is often used in targeting and tracking applications and it is also referred to as navigation frame. By convention, the x-axis is aligned in the east direction, the y-axis coincides with the north, and the up is represented by the z-axis.

The ENU system is illustrated in Figure 2.2. On the left part of the image, the ENU is superposed against the Earth. The local tangent plane is represented by a solid plane whose center is at a reference point of latitude and longitude defined by λ_r and μ_r. The east, north and up axes are defined relative to this reference point. On the right side of the image the local tangent plane is zoomed in.

Throughout this book, to avoid confusion between the ECEF x, y, and z and the x, y, and z from the navigation frame, the former will be represented as X_{ecef}, Y_{ecef} and Z_{ecef} or X, Y, and Z, and the latter will be referred as East, North and Up or E, N and U.

The conversion from ECEF to ENU requires a reference point (the point on which the tangent plane is defined) and a rotation matrix. The method is shown in Equation 2.2 and the inverse conversion, from ENU to ECEF, in Equation 2.3.

$$\begin{bmatrix} \mathrm{E} \\ \mathrm{N} \\ \mathrm{U} \end{bmatrix} = \begin{bmatrix} -\sin(\mu_r) & \cos(\lambda_r) & 0 \\ -\cos(\lambda_r)\sin(\mu_r) & -\sin(\mu_r)\sin(\lambda_r) & \cos(\mu_r) \\ \cos(\lambda_r)\cos(\mu_r) & \cos(\mu_r)\sin(\lambda_r) & \sin(\mu_r) \end{bmatrix} \begin{bmatrix} X - X_r \\ Y - Y_r \\ Z - Z_r \end{bmatrix} \tag{2.2}$$

$$\begin{bmatrix} X \\ Y \\ Z \end{bmatrix} = \begin{bmatrix} -\sin(\mu_r) & -\cos(\lambda_r)\sin(\mu_r) & \cos(\lambda_r)\cos(\mu_r) \\ \cos(\lambda_r) & -\sin(\mu_r)\sin(\lambda_r) & \cos(\mu_r)\sin(\lambda_r) \\ 0 & \cos(\mu_r) & \sin(\mu_r) \end{bmatrix} \begin{bmatrix} \mathrm{E} \\ \mathrm{N} \\ \mathrm{U} \end{bmatrix} + \begin{bmatrix} X_r \\ Y_r \\ Z_r \end{bmatrix} \tag{2.3}$$

For some navigation applications such as aviation, the ENU system can be replaced by a similar Cartesian coordinate system called north east down (NED), as it can be more convenient to represent certain navigation scenarios. The conversion between both systems is direct, given by Equation 2.4.

$$C_{\mathrm{ENU}}^{\mathrm{NED}} = C_{\mathrm{NED}}^{\mathrm{ENU}} = \begin{bmatrix} 0 & 1 & 0 \\ 1 & 0 & 0 \\ 0 & 0 & -1 \end{bmatrix} \tag{2.4}$$

2.2.3 Dispatchable Location

The WGS84 and ENU coordinate systems are suitable to uniquely define any point on the surface of the Earth in an absolute manner or relative to a reference location. Nonetheless, they are not necessarily the most convenient system for a person to understand. When a person reads 40.7128993°North, 74.015807°West, he or she will not immediately associate these coordinates with New York. However, a person will quickly identify a physical address such as 285 Fulton Street, 10007, New York City. This is what the FCC has defined as a dispatchable location: "the verified or corroborated street address plus additional information such as floor, suite, apartment or similar" [3].

The dispatchable location can be used for instance in E911 calls to direct the emergency services to the person in need of assistance faster than by delivering a geodetic location.

2.3 POSITIONING MEASUREMENTS

Throughout history, there are and have been many different positioning systems. However, most of them rely on the same basic set of measurements. Most of the positioning measurements are based on the calculation of certain signal parameters, such as the phase, the power, or the time of flight. This section describes those basic measurements that will be the core of the more complex positioning technologies explained later in this book.

2.3.1 Time of Arrival

A time of arrival (ToA) measurement calculates the time lapse between the transmission and the reception of a particular signal, as shown in Equation 2.5 [4]. Based on this time measurement, and knowing the propagation speed of the signal, v_p, the distance D between transmitter and receiver can be determined using Equation 2.6.

$$ToA = t_{rx} - t_{tx} \tag{2.5}$$

$$D = (t_{rx} - t_{tx}) \cdot v_p \tag{2.6}$$

Geometrically, the locus of points with constant distance to a known point is a circumference in 2-D or a sphere in 3-D. Assuming the position of either the transmitter or the receiver is known, the ToA measurement defines a circumference or a sphere of location. That means the receiver will be placed on a circumference or a sphere of radius equal to the distance measurement, whose origin is the transmitter or vice versa, as illustrated in Figure 2.3.

The center of the circumference/sphere is, in the example in Figure 2.3, the transmitter and the radius r is equal to the distance calculated in Equation 2.6. The equation of the sphere is defined in Equation 2.7 (for 2-D, it can be simplified removing the term with z), where a, b, and c are the coordinates of the center (in this example, the coordinates of the transmitter).

$$(x - a)^2 + (y - b)^2 + (z - c)^2 = r^2 \tag{2.7}$$

Replacing the radius r by the distance D calculated in Equation 2.6 results in Equation 2.8, where e_{ToA} has been added to represent the error associated to the ToA measurement.

$$(x - a)^2 + (y - b)^2 + (x - c)^2 = ((t_{rx} - t_{tx}) \cdot v_p)^2 + e_{toa}^2 \tag{2.8}$$

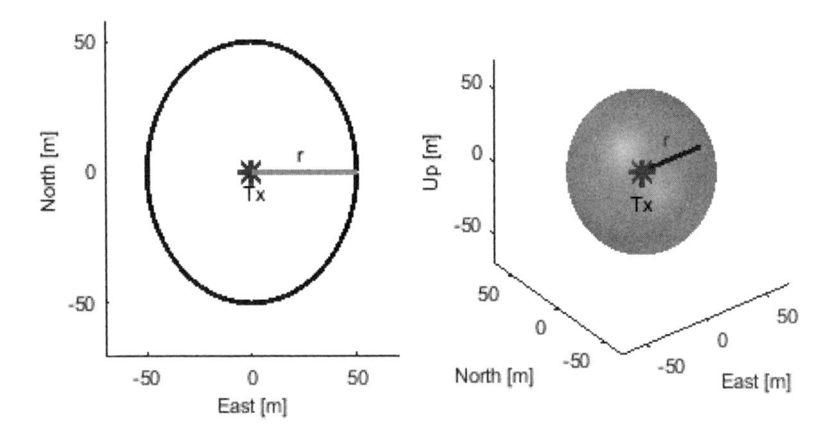

Figure 2.3 Circumfence and sphere of location defined by a ToA measurement.

A ToA system requires that both transmitter and receiver are synchronized in order to be able to calculate correctly the time measurement. In other words, the receiver needs to know exactly at what time instant the signal has been transmitted, according to its own receiver clock. Otherwise, the synchronization error τ_{rx-tx} needs to be included in Equation 2.5. In navigation, the synchronization error is typically represented by the transmitter and receiver clock biases, τ_{tx} and τ_{rx}, respectively, as shown in Equation 2.9.

$$ToA = t_{rx} - t_{tx} + \tau_{rx} - \tau_{tx} \tag{2.9}$$

2.3.2 Time Difference of Arrival

A time difference of arrival (TDoA) measurement calculates the time lapse between the reception of two signals coming from two different transmitters [4]. This is shown in Equation 2.10, where t_{rxi} is the time at which the signal coming from the transmitter i has been received. The time difference measurement is used to calculate the difference in the distance between the receiver and each of the transmitters, as it is represented in Equation 2.11, where d_1 denotes the distance between the receiver and the first transmitter and d_2 denotes the distance between the receiver and the second transmitter.

$$TDOA = t_{rx2} - t_{rx1} \tag{2.10}$$

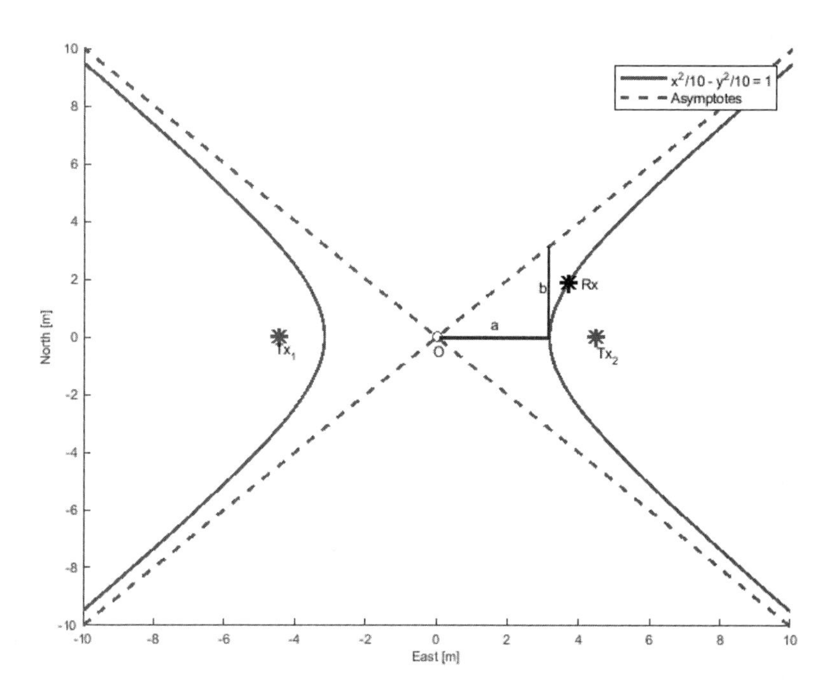

Figure 2.4 Hyperbola of location defined by a TDoA measurement.

$$(d_2 - d_1) = (t_{rx2} - t_{rx1}) \cdot v_p \tag{2.11}$$

Geometrically, the locus of points for which the differences of the distances to two fixed points is constant is called a hyperbola, if the location is done in a two-dimensional space. The two fixed points (the two transmitters) are called the foci of the hyperbola. The standard equation defining the hyperbola is shown in Equation 2.12.

$$\frac{x^2}{a^2} - \frac{y^2}{b^2} = 1 \tag{2.12}$$

The parameters a and b define the shape of the hyperbola, as it can be seen in Figure 2.4. For the example in Figure 2.4, $a = b = \sqrt{10}$. The transmitters Tx_1 and Tx_2 are at the focal points, while the receiver Rx can be located at any point on the right side of the curve. The dashed lines are the asymptotes of the hyperbola (i.e. the lines to which the hyperbola branches tend when $x, y \to \infty$). The distance

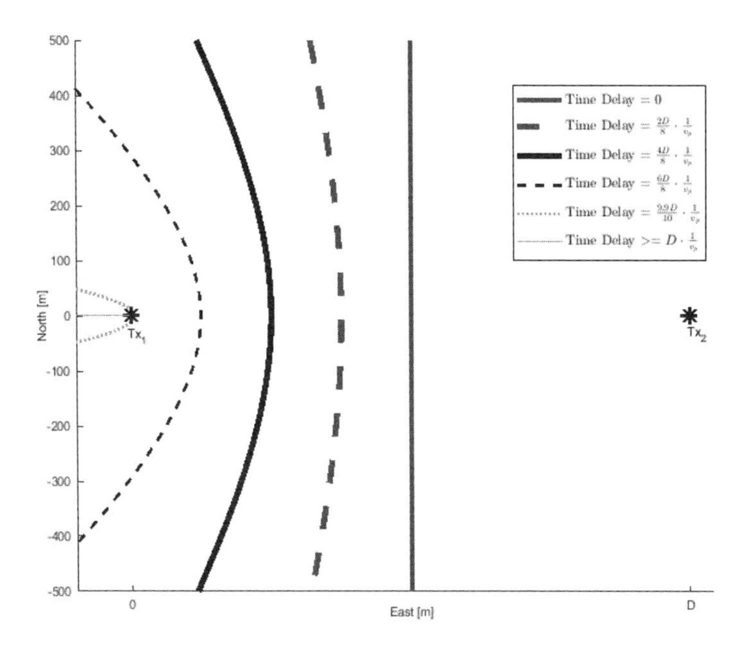

Figure 2.5 Localization hyperbola in function of the TD measurement.

between the center of the hyperbola (O) and each of the focal points is $\sqrt{a^2 + b^2}$. Finally, the *eccentricity* of a hyperbola is defined in Equation 2.13.

$$\epsilon = \frac{\sqrt{a^2 + b^2}}{a^2} \tag{2.13}$$

The hyperbola is defined by the time delay measurement, TD. Depending on the value of TD, the shape of the hyperbola varies as shown in Figure 2.5, where the D represents the distance between the two transmitters. As it can be seen, for TD = 0 and TD $\geqslant$ D, the result is not a hyperbola. These are the two degenerate cases for which the mathematical calculation of the hyperbola diverges to a straight line.

The generalization of the hyperbola for the 3-D case is a hyperbolic hyperboloid, often called just hyperboloid, defined by Equation 2.14. If the right side of the equation has a positive sign, the hyperboloid has only one sheet. A two-sheet hyperboloid is defined with a minus sign.

$$\frac{x^2}{a^2} + \frac{y^2}{b^2} - \frac{z^2}{c^2} = \pm 1 \tag{2.14}$$

In order to use TDoA for positioning, the hyperboloid can also be defined as in Equation 2.15, where $\overline{RxTx_i}$ is the distance between the receiver and the transmitter i and e_{tdoa} is the error associated with the TDoA measurement.

$$\overline{RxTx_2} - \overline{RxTx_1} = (t_{rx2} - t_{rx1}) \cdot v_p + e_{tdoa} \tag{2.15}$$

The distance between the receiver and the transmitter i can be generically written as in Equation 2.16.

$$\overline{RxTx_i} = \sqrt{(x - x_i)^2 + (y - y_i)^2 + (z - z_i)^2} \tag{2.16}$$

In a TDoA system, the receiver does not need to know the absolute time at which the signals have been transmitted. Thus, the receiver does not need to be synchronized. However, the time offset between the transmission of the two signals needs to be known. Hence, the transmitters still need to be synchronized to each other. A synchronization error between the transmitters will result in an error in the TDoA measurement, as given by Equation 2.17, where τ_{tx1} is the clock bias of transmitter i.

$$TDOA = t_{rx2} - t_{rx1} - \tau_{tx2} + \tau_{tx1} \tag{2.17}$$

2.3.3 Round-Trip Time

A round-trip time (RTT) measurement is a special type of ToA measurement. It calculates the time lapse between the instant when a signal has been transmitted and the instant when the response or acknowledgment to that transmitted signal is received back. Hence, in the ideal case, the RTT measurement is two times the ToA measurement.

As both transmission of the original signal and reception of the acknowledgment are done at the same point, with the same clock, there are no synchronization requirements between transmitter and receiver, as shown in Equation 2.18, where $t_{tx,s}$ is the transmission time of the signal s, $t_{rx,ack}$ is the reception time for the acknowledgment to signal s, and τ_{tx} is the clock bias of the transmitter that has sent signal s and received the acknowledgment.

$$RTT = (t_{rx,ack} + \tau tx) - (t_{tx,s} + \tau tx) = t_{rx} - t_{tx} \tag{2.18}$$

However, in reality there is an additional error induced by the processing time at the receiver (i.e., the time since the signal is received until the acknowledgment is sent). This processing time produces a certain delay that will add to the RTT measurement. The processing time can be calculated or estimated and subtracted from the total measurement. This is done for instance in LTE for ECID measurements.

2.3.4 Phase of Arrival

The phase of arrival (PoA) measurement calculates the phase of the received signal and calculates the distance between transmitter and receiver based on this phase measurement. If the transmitted signal is a sine wave with initial phase $\phi_0 = 0$, the received signal $s(t)$ can be represented by Equation 2.19, where A is the amplitude and f is the frequency of the signal.

$$s(t) = A \cdot \sin(2\pi \cdot f \cdot t + \phi_{rx}) \tag{2.19}$$

The received phase ϕ_{rx} depends on the distance traveled by the signal, as given by Equation 2.20, where k is a positive integer. Assuming that the propagation speed $v_p = c$, the speed of light and clearing the distance d, Equation 2.21 is obtained, where $\lambda = c/f$ is the wavelength of the signal.

$$\phi_{rx} + 2\pi \cdot k = \frac{2\pi \cdot f \cdot d}{v_p} \tag{2.20}$$

$$d = \lambda \cdot (\frac{\phi_{rx}}{2\pi} + k) \tag{2.21}$$

From Equation 2.21, it can be seen that the PoA measurement does not solve the distance unambiguously, as k can take any integer value bigger than 0. The reason for this is that the phase of the signal is periodic with a cycle of 2π. Given a measured phase ϕ_{rx}, it is not possible to distinguish between ϕ_{rx} and any multiple of $\phi_{rx} + 2\pi k$, as it is shown in Figure 2.6. Hence, the PoA measurement is useful to determine in which fraction of the wavelength a signal has been received, but it cannot tell how many wavelengths the signal has actually traveled.

Due to this limitation, pure PoA measurements can only be used to calculate short distances (within a wavelength). There are further techniques to enhance PoA, for example, the use of PRN sequences or the combination with other measurements to increase the measurement accuracy, as for instance in differential GNSS methods. These techniques will be explained in more detail in Chapters 6 and 7.

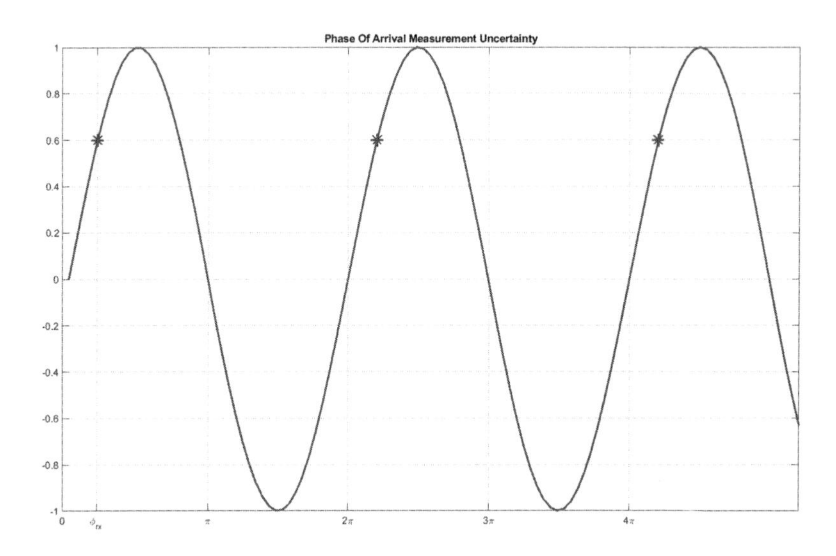

Figure 2.6 Illustration of the PoA measurement uncertainty.

2.3.5 Angle of Arrival

An angle of arrival (AoA) measurement determines the angle from which a certain signal is received. Knowing the incoming angle of the received signal, the receiver can estimate the direction in which the transmitter is, as shown in Figure 2.7. The transmitter can be at any point along the straight line defined by the angle α, represented in the figure by a dashed line.

There are predominantly two classical methods to determine the AoA of a signal. The first method determines the angle of arrival by measuring the phase difference of the signal received at two or more different antennas at the receiver. This method can also be referred to as phase difference of arrival (PDoA). It requires an antenna array of at least two elements at the receiver's side. Another precondition is that the receiver needs to be in the far field with respect to the transmitter. The far field is the region of the electromagnetic field where the incoming wave can be seen as a plane wave, as shown in Figure 2.8.

Imagine an antenna transmitting a spherical wave at a wavelength λ, represented by the dotted lines in the figure. The far field is the region in which the wave front can be considered as planar with respect to the size of the receiver antenna, D. The farther away the transmitter and receiver are from each other, the better

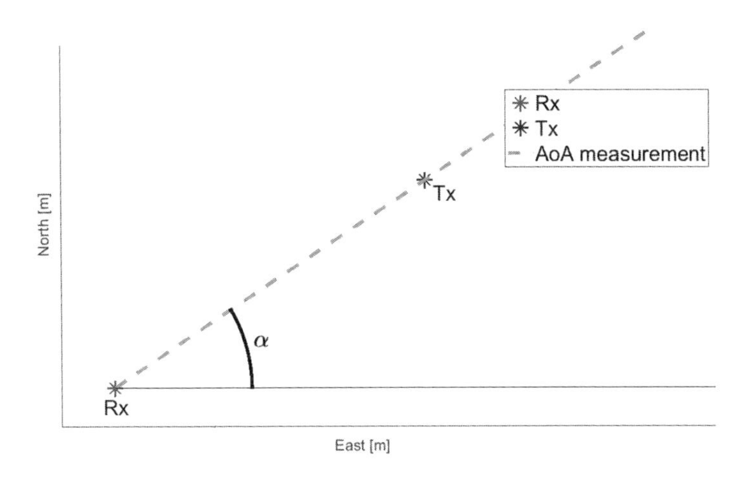

Figure 2.7 Example of localization using an angle of arrival measurement.

is the approximation of the spherical wave by a plane. The boundary between the near field and the far field depends on the electromagnetic size of the antennas. If the antennas are electromagnetically short (i.e., the antenna aperture is shorter than $\frac{\lambda}{2}$), the far field starts from a distance between the antennas of 2λ. If the antennas are electromagnetically long, the boundary is given by the Fraunhofer distance, d_f, defined by Equation 2.22.

$$d_f = \frac{2 \cdot D^2}{\lambda} \tag{2.22}$$

If the far field condition is met, the signals arriving at the different antenna elements of the receiver are parallel to each other and the phase difference between them can be used to calculate the angle of arrival. This is explained in Figure 2.9.

The receiver has two antennas separated by a distance d. The first antenna receives the incoming signal with a phase ϕ_0 and the second antenna receives the incoming signal with a phase ϕ_1. The difference between both phases is related to the additional travel distance x as given by Equation 2.23, where k is a positive integer.

$$\phi_1 - \phi_0 = 2\pi \cdot \frac{x + k}{k\lambda} \tag{2.23}$$

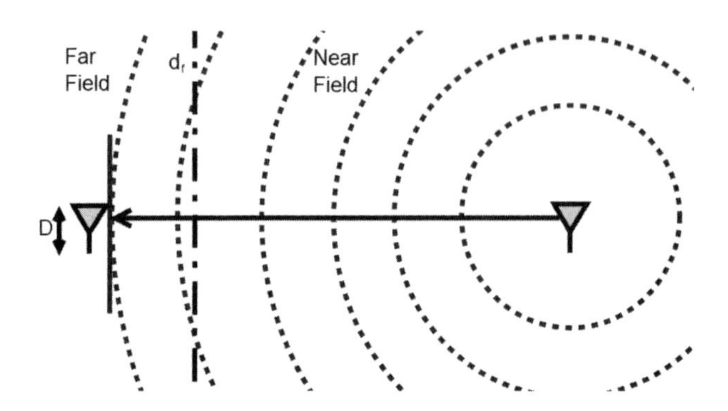

Figure 2.8 Definition of far field distance.

Applying the trigonometric relations, it can be seen that $x = d \cdot \cos\alpha$. Replacing in Equation 2.23 and clearing the angle of arrival, Equation 2.24 is obtained.

$$\alpha = \arccos(\frac{\lambda}{d} \cdot (\frac{\phi_1 - \phi_0}{2\pi} - k)) \tag{2.24}$$

The second method to measure the AoA of a signal is to use a directional antenna with a steerable beam to search the angle from where the incoming signal is arriving [5, 6]. The receiver beam is turned gradually to determine from which direction the received signal power is the highest (known as the beam peak). This is shown in Figure 2.10. The left part of the figure depicts the receiver antenna pattern with the main lobe pointing toward 180°. The beam is steered clockwise measuring the received power for the incoming signal, coming in this example from an AoA of 90°. The right part of the figure shows the received power versus the beam direction. As it can be seen, the highest received power is at 90°AoA, the same direction from which the incoming signal is received. Alternatively, the same approach can be used for finding the direction from which the minimum power is received (known as null). The main advantage of null finding when compared to beam peak finding is that the antenna pattern is usually steeper around the null, making it easier to detect.

More modern AoA estimation techniques include subspace-based methods [6,7]. These methods decompose the problem space into signal and noise subspaces and exploit the particular characteristics of each of these subspaces to calculate the AoA. The main advantage over classical techniques is that the subspace-based methods obtain better angular resolution. Some examples of subspace-based

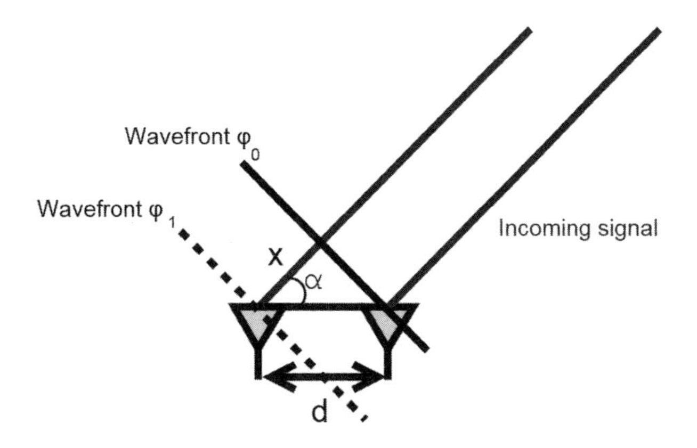

Figure 2.9 Calculation of the AoA by measuring the phase difference of the incoming signal.

methods are the multiple signal classification (MUSIC) algorithm [8] and the estimation of signal parameters via rotational invariant techniques (ESPRIT) [9].

2.3.6 Received Signal Strength

The received signal strength (RSS) method is based on the measurement of the received signal power. If the transmission power is known, the reception power can be used to estimate the path loss of the signal. The path loss depends on the distance between the transmitter and the receiver. Thus, this distance can be calculated using the Friis transmission equation in Equation 2.25, where λ is the wavelength of the signal and G_{tx} and G_{rx} are the gains of the transmitter and receiver antenna, respectively. The contemporary form of the Friis transmission equation is derived from the original form published by Friis, using the effective aperture of the antenna instead of the gains [10].

$$\frac{P_{rx}}{P_{tx}} = G_{tx} \cdot G_{rx}(\frac{\lambda}{4\pi \cdot d})^2 \tag{2.25}$$

The Friis transmission equation is more common in decibel form as represented by Equation 2.26, where the powers are in dBm and the gains of the antenna are in dB.

$$P_{rx} = P_{tx} + G_{tx} + G_{rx} + 20\log_{10}(\frac{\lambda}{4\pi \cdot d}) \tag{2.26}$$

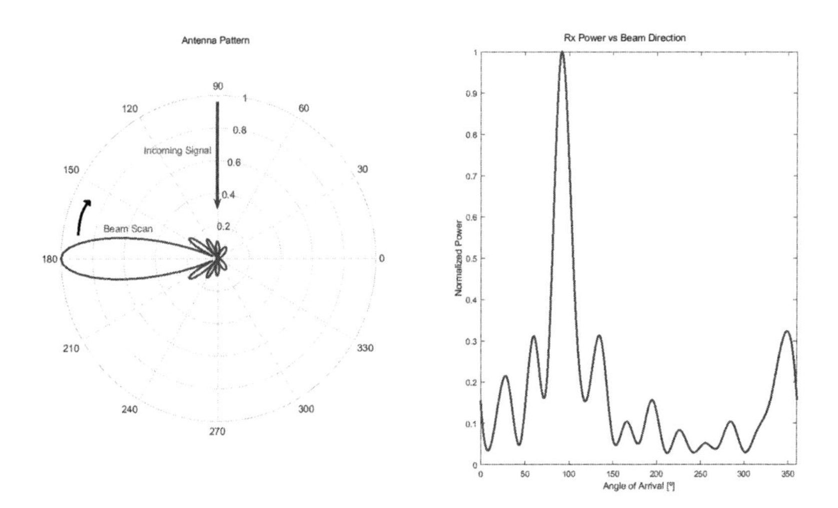

Figure 2.10 Calculation of the AoA by beam-steering.

The Friis transmission equation assumes that the signal propagates in the free space without any obstructions (reflections, diffraction, etc.); that the transmitter and the receiver are in the far field with respect to each other, as defined in Figure 2.8; that both the transmitter and receiver antenna have the same polarization and are aligned, and that the bandwidth of the signal is sufficiently narrow that a single wavelength value can be considered.

The calculated distance between the transmitter and the receiver, *d*, defines a sphere of location similar to the ToA measurement as seen in Figure 2.3. The RSS measurement can be described with Equation 2.27, based on the sphere equation seen in Equation 2.7.

$$(x-a)^2+(y-b)^2+(x-c)^2=(\frac{\lambda}{4\pi})^2\cdot\frac{P_{tx}\cdot D_{tx}\cdot D_{rx}}{P_{rx}}+e_{rss}^2. \tag{2.27}$$

In Equation 2.27, e_{rss} is the error associated to the RSS measurement. In reality, the signal will not propagate in ideal free space, but it will be distorted by reflections, diffraction, multipath, and other effects. As well, the signal might travel through solid materials like walls. Furthermore, in a typical cellular network environment, the transmitter and receiver antennas beams and polarization will be

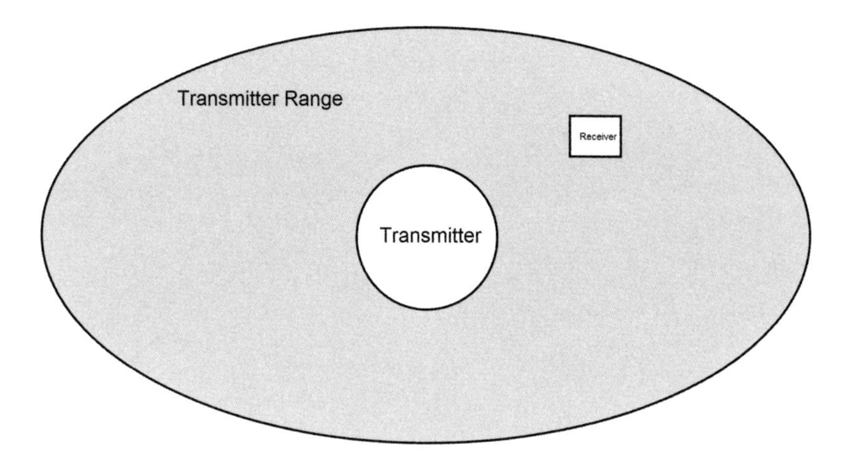

Figure 2.11 Example of location by proximity.

misaligned. All of these effects will add to the error of the RSS measurement, and the calculated distance will not be as accurate as for techniques based on ToA.

2.4 POSITIONING METHODS

The positioning measurements described in the previous section can be combined together in different methods to calculate the location of the desired object. This section explains some of those methods, which are used as a basis for cellular network positioning.

2.4.1 Proximity

Proximity is the simplest way to determine the location of an object. In fact, it is so simple that in most cases it is not considered a positioning technology at all. However, it is a method long used in cellular networks (cell ID) and also Wi-Fi and Bluetooth positioning systems. Thus, it is worth a brief explanation of its working principle.

Proximity is based on the knowledge that the object to be located is *near* a reference object whose position is known. For instance, if the object to be located is receiving a certain signal, it can be assumed that the receiver is within the range

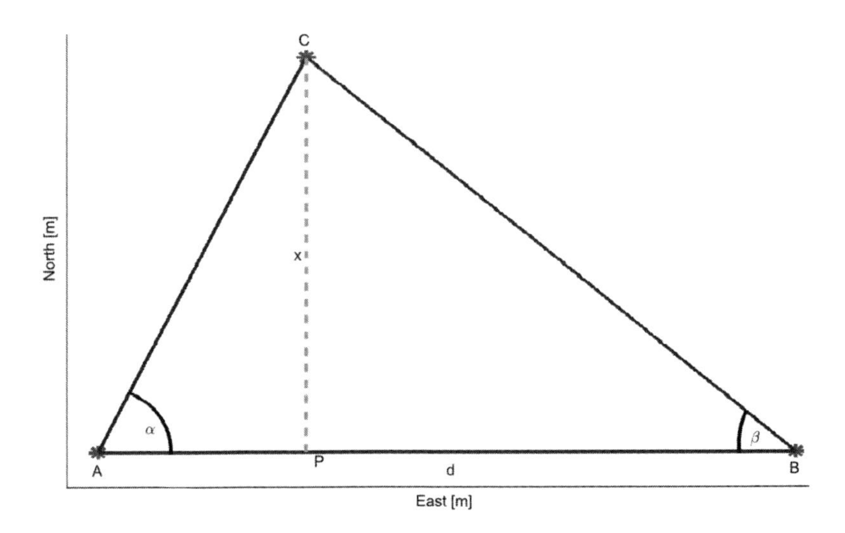

Figure 2.12 Triangulation.

of the transmitter, as seen in Figure 2.11. The receiver is located somewhere within the transmitter range, represented by the shadowed area in the figure.

Hence, the accuracy of proximity as a positioning method depends basically on the range of the signal used. In cellular networks, this range can be up to a few hundred meters or even kilometers. However, in Wi-Fi or Bluetooth networks this range will be typically of a few tens of meters, while in radio frequency identification (RFID), the maximum range will be of a few meters.

2.4.2 Triangulation

Triangulation is the technique to determine the position of an object by measuring angles to this object from known points. It is a very ancient method that was used by the Greek philosopher Thales to estimate the height of the pyramids in Egypt.

Triangulation can be explained by Figure 2.12 [4]. The distance from point P to a point C needs to be calculated, represented by x and a dashed line in the figure. The known points A and B will be used for that purpose by measuring the angles α and β. Applying the trigonometric identities, the known distance d between A and B can be related to x as given by Equation 2.28.

$$d = \frac{x}{\tan \alpha} + \frac{x}{\tan \beta} \tag{2.28}$$

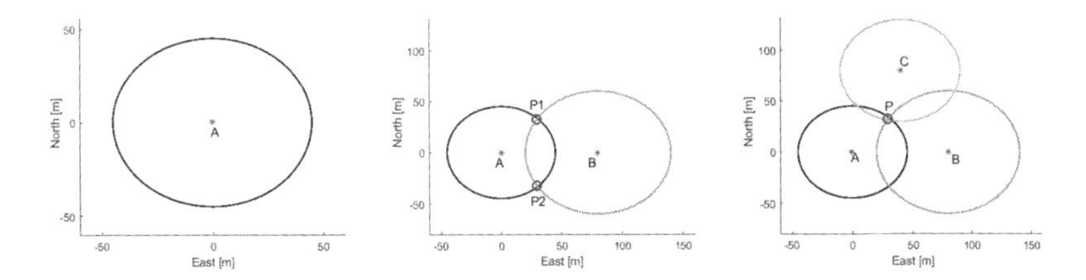

Figure 2.13 Trilateration.

As $\tan\phi = \frac{\sin\phi}{\cos\phi}$, Equation 2.28 can be transformed into Equation 2.29.

$$d = x\left(\frac{\cos\alpha}{\sin\alpha} + \frac{\cos\beta}{\sin\beta}\right) \tag{2.29}$$

Equation 2.29 can be rewritten in Equation 2.30, using the trigonometric identity $\sin(\phi+\psi) = sin(\phi)\cos(\psi) + \cos(\phi)\sin(\psi)$.

$$d = x \cdot \frac{\sin(\alpha+\beta)}{\sin\alpha\sin\beta} \tag{2.30}$$

Finally, the x is calculated in Equation 2.31.

$$x = d \cdot \frac{\sin\alpha\sin\beta}{\sin(\alpha+\beta)}. \tag{2.31}$$

Triangulation is based in AoA measurements, as seen in Section 2.3.5.

2.4.3 Trilateration

Trilateration is the technique that determines the location on an object by measuring the distance to this object from known points. It is often confused with triangulation, although their working principles are different.

Trilateration relies on the calculation of spheres of location (circumferences in 2D). As it can be seen in Figure 2.13, at least three measurements are required to calculate a unique position. In the left graph, only the distance to one reference point, A, is known, meaning that the object to locate could be anywhere on the circumference. In the middle graph, two distance measurements are available, given a total of two possible locations, P1 and P2. In the right graph, a third measurement to a reference point C is added, and the location determination is unique.

The positioning based on trilateration can be performed with several of the positioning measurements seen in the previous section. It can be done either

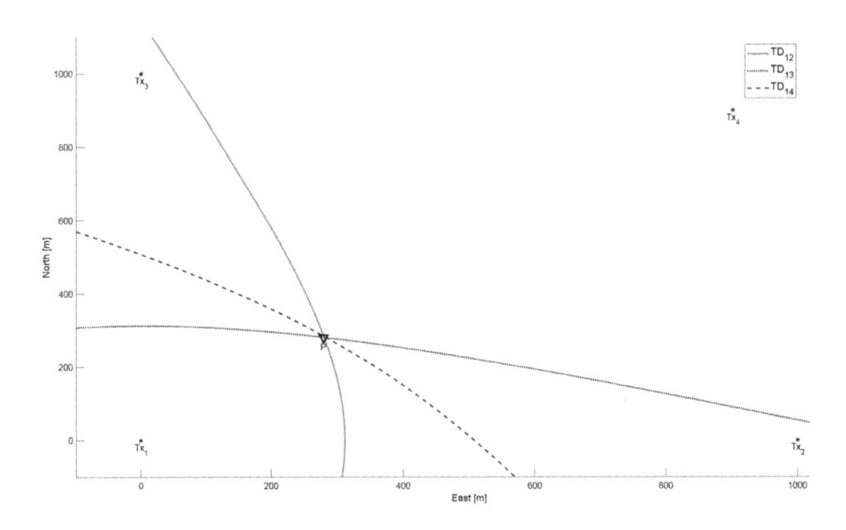

Figure 2.14 Multilateration.

with ToA (Section 2.3.1), with RTT (Section 2.3.3), or with RSS (Section 2.3.6) measurements. However, the accuracy of the results will depend on the type of measurements used and will in general be better with time measurements (ToA or RTT) than with power measurements.

2.4.4 Multilateration

Multilateration is the name given to the positioning algorithm based on the difference in distances from the object to locate to two reference points of known location. It relies on TDoA measurements, as seen in Section 2.3.2.

The positioning using multilateration is depicted in Figure 2.14. The object to locate is placed in point P. There are three TDoA measurements available, from transmitters Tx_2, Tx_3, and Tx_4 to the reference transmitter Tx_1. In general at least two measurements (three transmitters) are needed to obtain a 2D position, and at least three measurements to calculate a 3D position.

2.4.5 RF Fingerprinting

Radiofrequency (RF) fingerprinting is a positioning technique based on the analysis of the specific radio frequencies and properties of the signal originated by each

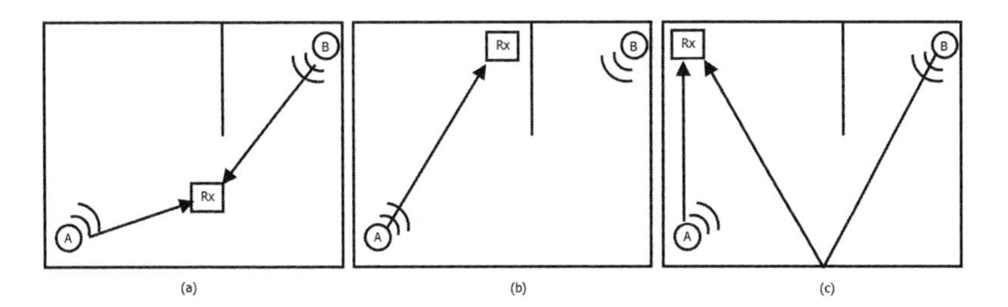

Figure 2.15 Examples of RF fingerprinting.

radio transmitter. Each transmitter generates a unique RF transmission. Based on the location of the receiver, the number of different signals seen and their respective powers will vary. RF fingerprinting generates a database with this information (the different signals and signal powers seen at each location). When the receiver reports the RSS seen for each transmitter in range, the measurements are compared against the database to find the best match for the location of the receiver [11].

A better understanding of the RF fingerprinting technique can be achieved by considering the three scenarios in Figure 2.15. In (a), the receiver is able to receive LoS signals from both transmitters, A and B. On the other hand, in (b), the receiver is next to a wall that blocks the signal from B, so it is only able to see transmitter A. In scenario (c), there is no LoS to B. However, the receiver is able to detect a reflection of the signal transmitted by B. In this case, the RSS for B will be much lower than in scenario (a). The example in Figure 2.15 depicts relatively simple scenarios. Nonetheless, the same approach can be followed for more complex scenarios with multiple transmitters and accurate power measurements. Furthermore, it can be combined with direction of arrival techniques to obtain more unique fingerprints [12].

RF fingerprinting is used predominantly with Wi-Fi and other short-range networks. It will be explained in more detail in Chapter 9.

2.5 POSITIONING CONCEPTS

This section will define some concepts related to the positioning systems.

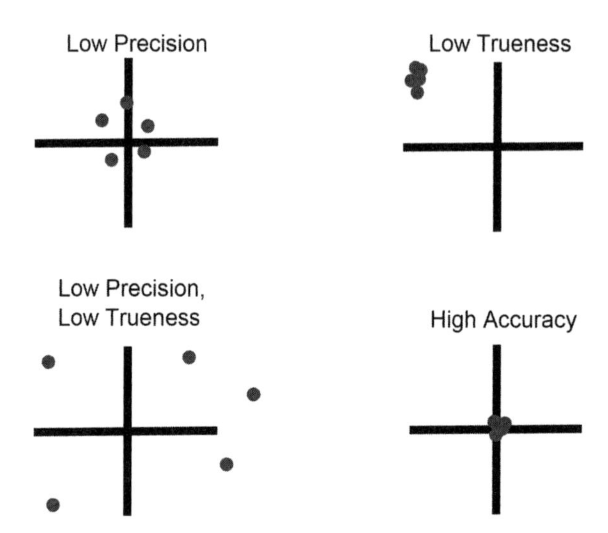

Figure 2.16 Comparison between accuracy, precision, and trueness.

2.5.1 Accuracy and Precision

When it comes to positioning measurements, it is important to understand the difference between accuracy and precision. According to the ISO 5725 [13], accuracy is "the closeness of a measurement to the true value," while precision is "the closeness of agreement among a set of results." The ISO also adds a third definition, trueness, or "the closeness of the mean of a set of measurements to the actual (true) value." Hence, if the accuracy is seen as a statistical function, the mean will be related to the trueness, while the variance will be related to the precision.

The difference between these three concepts can be seen in Figure 2.16. The true value is the point marked by the cross and the dots represent different measurements. If the measurements have low precision, the dispersion of the results is large, as in the upper left image. If the measurements have low trueness, the distance between each of the measurements and the true value is large, as in the upper right image. However, if the measurements have high trueness and high precision, all of them are close to the true value and they are repeatable and reproducible. This is defined as high accuracy, shown in the bottom right image.

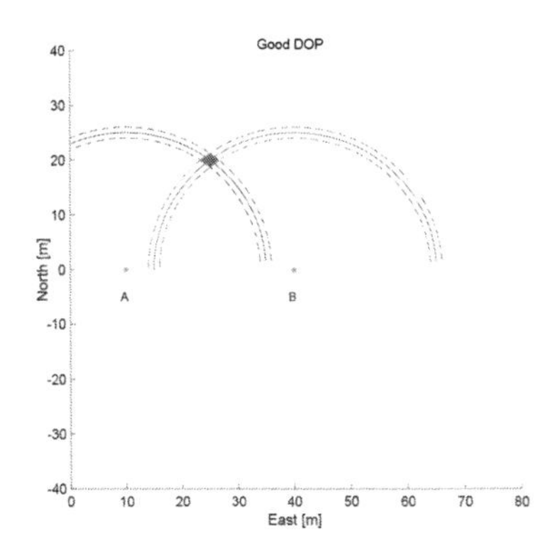

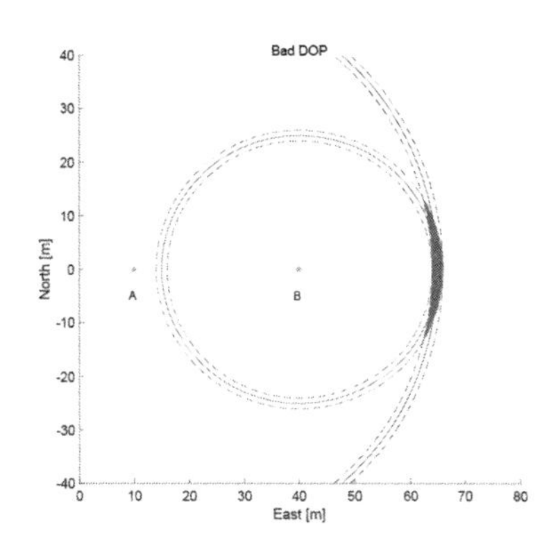

Figure 2.17 Definition of dilution of precision.

2.5.2 Dilution of Precision

Dilution of precision (DOP), also called geometrical dilution of precision (GDOP) is a term used in radio-localization systems to refer to the influence of the system geometry in the accuracy of the calculated position [14, 15]. It gives an idea of how much the distribution of the different location reference points (e.g., the satellites for GNSS systems) with respect to the object to locate will influence in the accuracy of the positioning.

In all the positioning methods explained in this chapter, the positioning measurements have been represented by a single line, as for instance the ToA measurement in Figure 2.3. However, in reality these measurements will have a certain measurement uncertainty (i.e. the measurement will be a certain range rather than just a line). This is represented in Figure 2.17. The measurement value is represented by solid lines. However, due to the uncertainty, the measurement can take any value between the two dashed lines. The reasons for this measurement uncertainty are manifold: noise, multipath, and other distortions of the signal, measurement quantification, and so forth.

The DOP of the positioning system tells how much the measurement uncertainty can affect the final results. If the DOP is low, the system geometry is good for positioning and the measurement uncertainty will not have a big impact on the calculated position. This is seen in the left part of Figure 2.17, where the object to locate is found within the filled area. However, if the DOP is high, the system geometry is not appropriate for positioning and the measurement uncertainty will greatly affect the accuracy of the calculated position. In the right part of Figure 2.17,

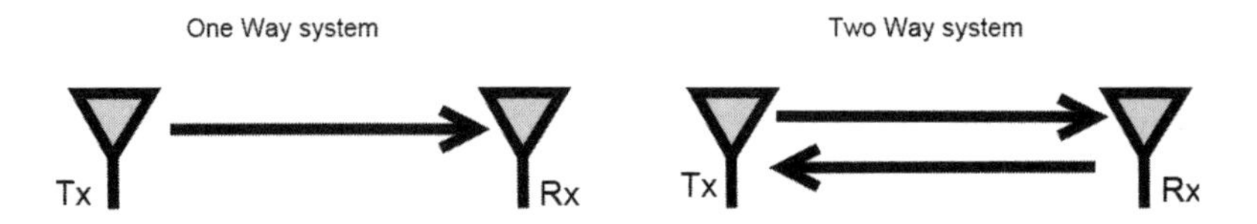

Figure 2.18 One-way and two-way positioning measurements.

it can be seen that the filled area is much bigger. The DOP is calculated for the whole positioning system, including as well the object to locate. In Figure 2.17, the coordinates of the two transmitters A and B remain unchanged in both cases. However, the DOP varies from good DOP to a bad DOP depending on where the object to locate is relative to A and B.

The concept of DOP will be further explained in Chapter 8 for cellular network positioning.

2.5.3 One-Way and Two-Way Positioning Systems

Positioning technologies can be divided between one-way and two-way systems, depending on whether the measured signal travels only in one direction, from the transmitter to the receiver, or in both directions from the transmitter to the receiver and back.

This is illustrated in Figure 2.18. In the left part of the figure, a one-way positioning system is represented. The transmitter sends a positioning signal to the receiver, who performs the positioning measurement. The right part of the figure shows a two-way system. The transmitter sends a positioning signal to the receiver. Upon receiving the signal, the receiver sends a response. The transmitter performs the positioning measurement.

In two-way systems, the initial transmission and the measurement are done at the same end. Thus, they eliminate part of the error sources of the positioning measurement (e.g., the clock biases). An example of one-way positioning measurement is ToA, while RTT is a two-way positioning measurement.

2.6 CONCLUSION

This chapter has laid out the basis for understanding how positioning works. The different types of positioning measurements have been explained and some useful positioning concepts have been introduced. The contents of this chapter will be the building blocks that form cellular positioning technologies in the chapters to come.

References

[1] NIMA Technical Report TR8350.2, *Department of Defense World Geodetic System 1984, its Definition and Relationships with Local Geodetic Systems*, Third Edition, 4 July 1997.

[2] Cardalda García, A. *RailSLAM: Simultaneous Localization and Mapping for Railways*, University of Oviedo, 2012.

[3] FCC, *Wireless E911 Location Accuracy Requirements Fourth Report and Order*, Tech. Rep. 07-114, Federal Communications Commission, Washington D.C., February 2015.

[4] Cardalda García, A. *Hybrid Localization Algorithm for LTE*, University of Oviedo, 2015.

[5] Bartlett, M.S. *Periodogramm Analysis and Continuous Spectra*, Biometrika, Vol. 37, Jun., 1950, pp.1-16.

[6] Tuncer, E., and Friedlander, B., *Classical and Modern Direction-of-Arrival Estimation*, Acamedic Press, 2009.

[7] Chen, Z., Gokeda, G.K., and Yu, Y., *Introduction to Direction-of-Arrival Estimation*, Artech House, 2010.

[8] Schmidt, R.O., *A Signal Subspace Approach to Multiple Emitter Location and Spectral Estimation*, Ph.D. thesis, Stanford University, 1981.

[9] Paulraj, A., Roy, R., and Kailath, T., "Estimation of Signal Parameters via Rotational Invariant Techniques - ESPRIT," *Proceedings 19th Asilomar Conference on Circuits, Systems and Computers*, San Jose, CA, 1985, pp. 83-89.

[10] Friis, H.T., "A Note on a Simple Transmission Formula," *Proceedings of the IRE*, Volume 34, 1946, pp. 254-256.

[11] Saraiva Campos, R., and Lovisolo, L., *RF Fingerprinting Location Techniques*, ch. 15 in *Handbook of Position Location: Theory, Practice, and Advances*, John Wiley & Sons, 2011.

[12] Guzmán-Quirós, R., et al. "Integration of Directional Antennas in an RSS Fingerprinting-Based Indoor Localization System," *Sensors*, 2016, Vol. 16, N. 4.

[13] ISO 5725-1, *Accuracy (Trueness and Precision) of Measurement Methods and Results - Part 1: General Principles and Definitions*, 1994.

[14] Langley, R.B., "Dilution of Precision," *GPS World*, 1999.

[15] Blewitt, G., "Basics of the GPS Technique," *Geodetic Applications of GPS*, Swedish Land Survey, 1997.

Chapter 3

Regulatory Positioning Requirements

3.1 INTRODUCTION

The ever-increasing number of emergency calls made through wireless phones have helped accurate localization to become a crucial part of cellular networks, as the survival chance of an emergency caller increases dramatically with every second saved in providing assistance. This has been the main driving force for increasing the reliability and accuracy of caller location determination. This chapter provides a comprehensive overview of how positioning technology has evolved along with the requirements for emergency services. It highlights the two major markets, the United States and the European Union, where many technologies have developed in parallel to cater for emergency services. This chapter provides an architectural description on the emergency services provided through mobile networks and their evolution towards IP based networks.

3.2 POSITIONING TECHNOLOGY IN THE UNITED STATES

The FCC [1] is a United States government agency that regulates communication, radio, television, and cable services for all States and territories in the United States. It creates and enforces rules and regulations related to these services. Under the guidance of the FCC, the Wireless Communications and Public Safety Act of 1999 (911 Act) took effect [2]. The purpose of this act was to improve public safety by setting up a communication infrastructure for emergency services for mobile phone calls. The 911 number has been used to call fire brigades since 1968, but there were no universal numbers used for contacting emergency services (not only fire brigades, but also police and ambulance services). The E911 act aims to have one number through which an emergency caller can get centralized access to all these emergency services promptly.

Following the FCC regulation, wireless operators must provide 911 services and route calls to the public safety answering point (PSAP). In 2003, the FCC also created a PSAP registry for all states and territories of the United States. Each PSAP is provided a unique identification and is registered with its state, county, and city information. This registry provides a comprehensive list of PSAPs and their readiness in handling different types of emergency calls. It also regulates the deployment of technologies, which enable delivery of location information during an emergency call. This deployment is covered under the FCC's Wireless Enhanced 911 (E911) rules.

The E911 program is divided into two phases. Phase I requires the operator to report the telephone number and associated location of the base station transmitting the call. Considering that cells can have ranges of multiple kilometers, this is not a very effective method for locating the caller. Phase II requires that the latitude, longitude, and altitude of the caller be reported within 50 to 300 m of the actual location. The FCC also encourages the development of new technologies that can further increase the accuracy of this reported location [4].

With the increased availability and improved latency of internet connections, there has been steady growth of VoIP calls. Next Generation 911 (NG911) is an evolutionary step aimed to support emergency calls over IP networks and enhance location accuracy performance for voice calls. VOIP calls will start coexisting with traditional circuit switched calls going forward and may ultimately replace them. They will provide not only voice but also other multimedia services such as video to the PSAPs.

3.3 FCC LOCATION ACCURACY REQUIREMENTS

The FCC has divided the United States into cellular market areas (CMAs), comprising metropolitan statistical areas (MSAs) and rural service areas (RSAs). For example, MSA 8 is Washington DC and RSA 336 is the region referred to as "California 1 - Del Norte." A complete list of the CMAs can be seen in [3]. Collecting data of all the CMAs, the FCC has estimated that around 240 million emergency calls were made in the United States in 2017 [5]. Depending on the area, approximately 80% of the calls were made from a wireless device. Estimates also show that 23% of the households use a wireless connection as the primary connection.

Accurately locating the caller is extremely important and determines the quality of service provided for the emergency call.

3.3.1 FCC Phase I Requirements

Before the eruption of positioning technologies, a wireless caller had to communicate his or her own location over a voice connection, often resulting in an unreliable position determination. Phase I of the FCC E911 regulations required wireless operators to report the caller number and location automatically to the PSAP. Initially, the cell tower servicing the call was automatically reported as the location of the caller, using the cell ID as a key to query a database of cell sites. This initial step was widely deployed by 2005.

3.3.2 FCC Phase II Requirements

The second phase of the FCC requirements introduced a series of targets with increasing location accuracy. Initially, wireless operators needed to provide the location information of the caller with an accuracy of 300 m under 6 minutes from the beginning of the call. This was limited usually to the outdoors in areas where a GPS signal was available. After that, there have been multiple updates to the location accuracy and reporting time requirements for phase II, including the addition of indoor location requirements in 2012 under the Next Generation 911 act of 2012 [4].

The FCC regulations apply to all nationwide (NW) operators and non-nationwide (NNW) operators. NW operators serve all states and territories of the United States and an estimated 80% of all wireless users, while NNW operators operate typically in smaller regions and a few states. The latest update of the FCC location accuracy requirements [4] has set the following goals:

- Since 2017, all NW operators must provide a horizontal location with 50 m horizontal accuracy or a dispatchable location for 40% of all wireless calls.
- Since 2018, both NW and NNW operators must provide a horizontal location with 50 m horizontal accuracy or a dispatchable location for 50% of all wireless calls.
- Since 2020, the requirement increases to 70% of all wireless calls. NNW operators can delay this and next deadlines based on the evolution of the VoLTE deployment on the networks.
- By 2021, the requirement increases to 80% of all wireless calls. Additionally, either dispatchable location or vertical positioning systems must be deployed in

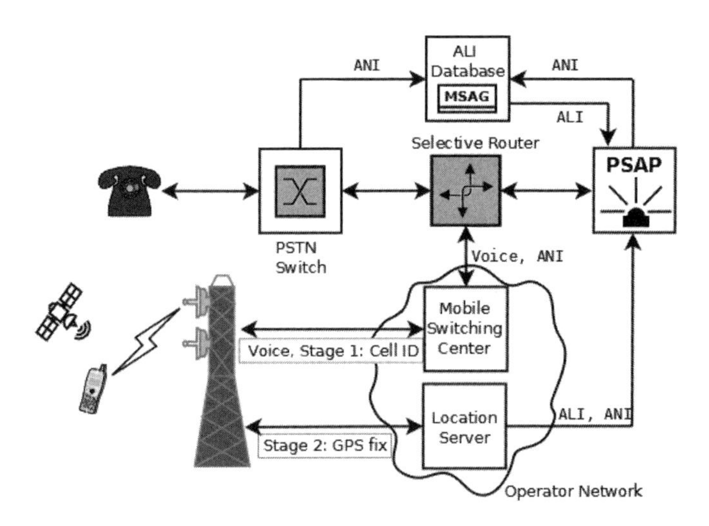

Figure 3.1 The E911 system architecture.

the top 25 CMAs. The vertical location should be reported within the approved vertical accuracy limits (which are still to be decided). For dispatchable location, the National Emergency Address Database (NEAD) must be populated with a density of 25% of the total population of the CMA. NNW operators have an additional year to meet this deadline.

- By 2023, the vertical accuracy and dispatchable location requirements expand to the top 50 CMAs. NNW operators have an additional year to meet this deadline.

A dispatchable location, as seen in Chapter 2, is defined as the civic address of the calling party, including additional details such as floor, apartment, or other information that could help to locate an apartment easily. A dispatchable location is considered a more accurate and suitable solution for emergency services when compared to latitude and longitude information.

3.4 THE E911 SYSTEM

The E911 network consists of databases that help routing the emergency call to the appropriate PSAP and guiding the emergency dispatchers to the caller location. The E911 system architecture is described in Figure 3.1. The automatic location information (ALI) database contains exhaustive information on county, region, or city and provides the mapping between a telephone number and the street location. The master street address guide (MSAG) database maps detailed street information and the corresponding PSAP emergency service number (ESN) associated with that location.

The key for querying this database is the telephone number associated with the call or a 10-digit unique identifier called automatic number identifier (ANI). When a 911 call is made on the public switched telephone network (PSTN), the call is directed to the selective router, which queries the MSAG database to determine the correct PSAP and then routes the call to that entity. The PSAP queries the ALI database to get the location information displayed on the screen of the operator. This enables accurate location information transfer even when the caller is unable to provide location information during the call.

The routing of wireless 911 calls follows a different scheme compared to the fixed line numbers, as the wireless caller could be located anywhere and the call requires routing to the PSAP responsible for the area in which the call is originated. The location of the serving cell provides a rough location based on FCC phase I requirements. This location is used by the selective router to direct the call to the appropriate PSAP. Afterward, a stage 2 location is calculated based on GPS, A-GPS, or other mobile and network-based techniques. This location is stored in the location server of the network, which can be queried by the PSAP using the ANI.

The exact architecture of the wireless emergency services network vary depending on the underlying RAT, the most popular being the 3GPP standardized networks like GSM, UMTS, LTE, and the 3GPP2 based networks such as IS-95 and CDMA2000. The diagram in Figure 3.1 presents a generalized view on how voice calls and location information are routed to a PSAP. However, each of these networks have their own architecture, which is discussed in the following sections.

3.4.1 Positioning on GSM, UMTS, and LTE Networks

The 3GPP is responsible for standardizing the LCS for GSM (2G), UMTS (3G), LTE (4G), and NR (5G) networks [6]. The LCS architecture in these networks is formed by base stations, core network elements, and location servers. The base station is the entity that connects with a mobile device using a wireless link and serves certain geographical areas ranging from hundreds of meters to a few kilometers. The core network elements interconnect the base stations and the other different elements of the network, including the location server. The location server

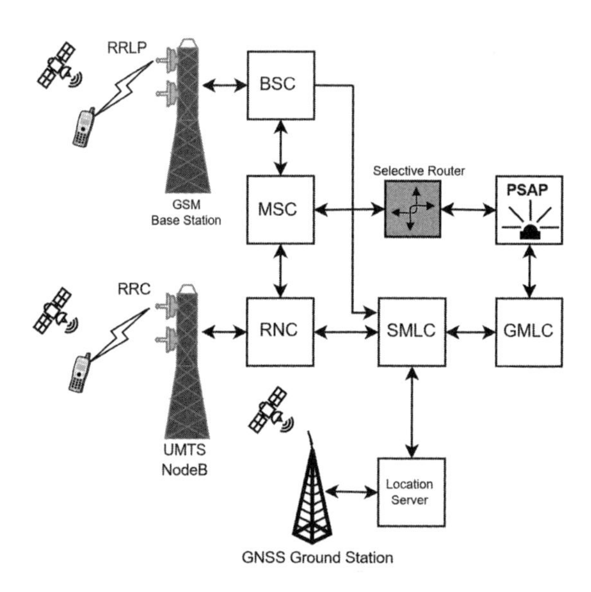

Figure 3.2 GMS and UMTS E911 architecture.

is the entity responsible for calculating and handling the mobile device's positioning information.

In GSM, the mobile is connected to the base transceiver station (BTS), which in turn is connected to the base station controller (BSC) and then to the core network consisting of mobile switching centers (MSCs). The MSC is responsible for routing the call to the appropriate destination. This network scheme is shown in Figure 3.2 [6]. In the case of E911, the call is routed to the selective router of the emergency services network responsible for routing the call to the corresponding PSAP. Typically, the base station location determines to which PSAP the emergency call is routed. The serving mobile location center (SMLC) is responsible for determining the stage 2 location of the user, allocating and coordinating resources for an accurate location determination [7]. The SMLC is the endpoint for many LCS protocols and it connects to servers that provide assistance data, such as GNSS navigation messages, in order to improve the localization process. The contact point between the SMLC and the PSAP is the gateway mobile location center (GMLC). In UMTS, the only major differences with respect to the GSM architecture is that the base station is known as the NodeB instead of BTS and that the radio network controller (RNC) replaces the BSC [8].

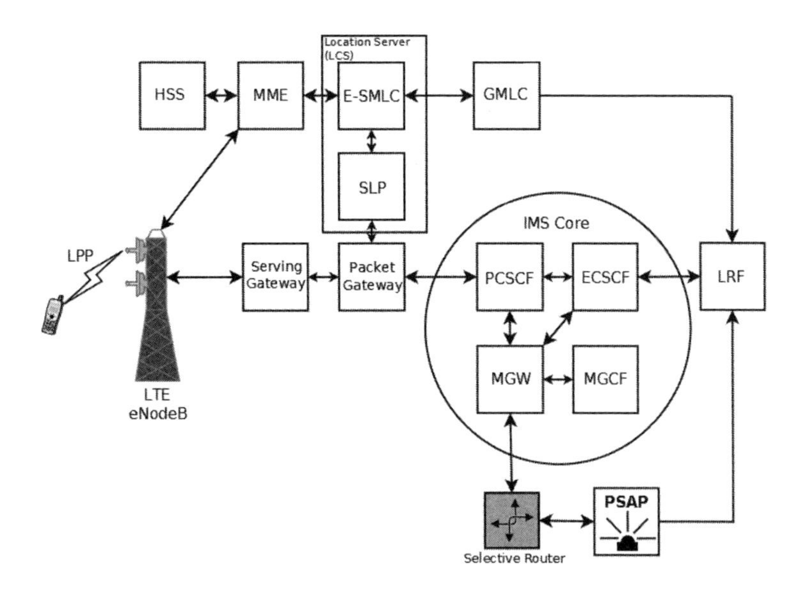

Figure 3.3 VoLTE E911 architecture.

On the other hand, LTE is a pure packet switched network and does not support the circuit switched voice calls from GSM and UMTS. In most network deployments, the LTE serving areas overlap with GSM, UMTS, or CDMA2000 networks. In such cases, circuit switched fallback (CSFB) methods can be used to switch the call over to a circuit switched network (GSM, UMTS, or CDMA2000). In other words, a voice call attempted via LTE networks causes the mobile to reselect to a circuit switched network and continue the call using legacy technologies. In such deployments, the rest of the emergency call follows the same path as in Figure 3.2.

However, advanced LTE networks implement voice over LTE (VoLTE) technology. VoLTE carries voice traffic through the LTE network and connects to an external IP Multimedia Subsystem (IMS), which provides a standardized framework for voice, text, and other multimedia traffic. This architecture is shown in Figure 3.3 [9]. The LTE network consists of a base station, also called eNodeB. The eNB connects to the mobility management entity (MME) and to a serving gateway (S-GW), which in turn manages the connection between different eNodeBs in the network. The packet data network gateway (PDN-GW) provides an IP address to each of the mobile devices in the LTE network, manages the QoS of the different connections, and serves as connection point to the IMS network. The enhanced serving mobile location center (E-SLMC) performs functions similar to the SMLC in GSM and WCDMA networks and is the terminating point for the LPP [10]. The E-SMLC may also support IP-based connections to obtain the location using Secure User Plane Location (SUPL) via the SUPL location platform (SLP). The E-SMLC and the SLP are logical entities that can potentially be located in the same physical server. Both together form the location server (LS) for LTE networks.

Inside the IMS core network, the primary call session control function (P-CSCF) is responsible for the session management of the multimedia connection and provides routing information for media servers. It provides the routing information to the destination multimedia gateway (MGW). In case of an emergency call, the session management is taken over by the emergency call session control function (E-CSCF), which then routes the call to the appropriate MGW. The MGW connects to the PSAP over PSTN connections. The location retrieval function (LRF) is the entity responsible for obtaining the location of the user from the E-SMLC through the GMLC and providing this location to the PSAP.

3.4.2 LCS Positioning Protocols

LCS protocols are broadly divided into control plane (C-Plane) and user plane (U-Plane) protocols. As the name suggests, the C-Plane protocols transmit the LBS messages for providing assistance data and obtaining location information on the control channels of the network. They were the default choice for networks based primarily on circuit switched connections. The different C-Pane protocols are:

- Radio Resource Location Protocol (RRLP) in GSM.
- Radio resource control (RRC) protocol in UMTS.
- TIA-801 in CDMA2000.
- LPP in LTE and 5G NR.

U-Plane protocols carry the location messaging over the data connection with the mobile and do not require extensive changes to the network elements. The Open Mobile Alliance (OMA) developed a SUPL standard to enable U-Plane positioning in a secure way [11]. The SUPL standard can be used in mobile networks such as GPRS, EDGE, HSDPA, LTE, HRPD, or 5G NR, with the ability to provide IP connections. The location protocols are exchanged between the SUPL client in the mobile and the SLP in the network, also referred to as SUPL Server. SUPL supports the same positioning protocols as C-Plane, RRLP, RRC, TIA-801, and LPP, plus an OMA enhancement to the LPP known as LPP enhancements (LPPe).

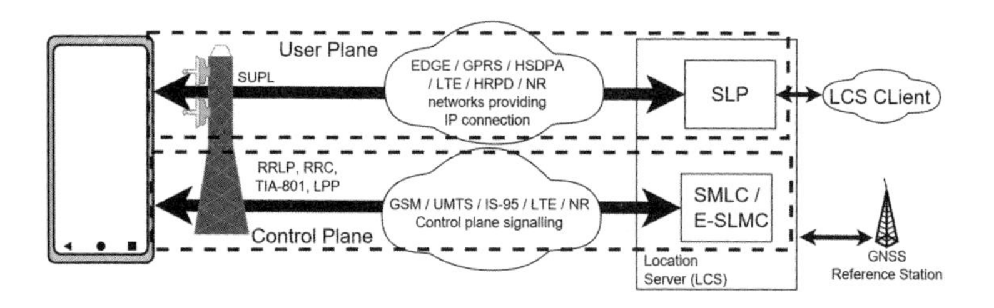

Figure 3.4 Comparison between SUPL and C-PL positioning.

In early deployments of GSM and UMTS networks, the C-plane method was preferred because data connections were not broadly available. However, as the availability of mobile data connections increased, SUPL offered an inherent advantage, being able to run transparently over all underlying data bearers such as GSM networks using GPRS, UMTS, LTE, or NR. The U-Plane protocols have also become increasingly relevant for commercial services (maps, messaging, etc.), which use positioning to enhance user experience. SUPL allows commercial usage of the location information enabling external location servers to be used for localization purposes. The comparison between SUPL- and C-PL-based positioning is shown in Figure 3.4. Table 3.1 gives an overview of the different protocols supported for SUPL and CPL for the different radio access technologies.

Table 3.1

Positioning Protocols Supported for the Different RAT over SUPL and C-PL

	GSM	*UMTS*	*LTE*	*CDMA*
C-Plane	RRLP	RRC	LPP	TIA-801
SUPL	RRLP	RRLP	RRLP, LPP	TIA-801

3.4.3 Indoor Localization

Meeting the FCC regulations for indoor location accuracy poses a challenge that is difficult to overcome with the existing cellular network and satellite-based technologies due to lack of availability of the signal, multipath effects, and other effects that will be further explained in Chapters 6 to 9. Furthermore, an additional problem is the requirement to provide accurate floor-level information. Thus, noncellular network based positioning technologies have been introduced in the OMA LPPe [12]

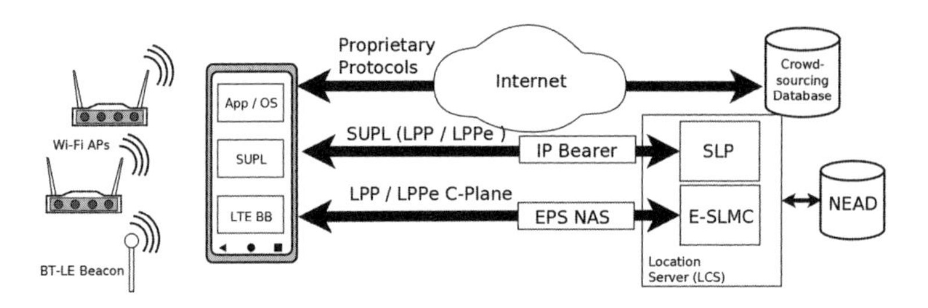

Figure 3.5 Indoor positioning architecture.

and 3GPP LPP (as of LPP Release 13 [10]). Both protocols benefit from the large-scale deployment of Wi-Fi routers across urban areas and other technologies such as Bluetooth Low Energy (BTLE), also known as BLE, for crowd-sourced location methods.

The location session based on these technologies follows a different scheme than satellite or cellular positioning technologies, as shown in Figure 3.5. When an E911 call is placed, the measured Wi-Fi access points and BTLE beacons information is sent to the E-SMLC over LPPe or LPP Release 13 using either SUPL or C-Plane. The E-SMLC sends the measurements to the location server, which uses this information to calculate the location. The E-SMLC also connects to a database, such as the NEAD in the United States, to obtain a dispatchable address. Such databases hold a relation of the Wi-Fi and BTLE MAC addresses and their respective physical location, and use this information to estimate the current address of the caller.

Both LPPe and LPP-R13 protocols also provide an interface for transmitting locations derived from techniques outside the scope of the network operators. These are provided as best effort location information and are typically used as a fallback location in case other methods are unsuccessful. Most mobile operative systems have proprietary positioning techniques for location determination, typically based on crowd-sourcing servers to estimate the mobile device's position from the list of visible Wi-Fi access points or Bluetooth beacons.

3.4.3.1 Floor-Level Information

The accurate vertical accuracy requirement is one of the toughest points of the FCC regulations, especially for indoor locations. One of the technologies that the FCC has considered for this is comparing measured barometric pressure by the mobile device with the pressure measured at a reference location in order to calculate the difference in altitude between both points. This measurement is also supported in both LPPe and LPP Release 13 protocols. The working principle of altitude calculation based on barometric measurements will be explained in detail in Chapter 12.

3.5 REGULATION IN THE EU RELATED TO EMERGENCY LOCATION SERVICES

3.5.1 Brief History of the E112

As early as 1991, the European parliament passed directive 91/396/EEC, defining 112 as the universal emergency number for calls across all EU member states. In 2002, this regulation was enhanced with a universal service directive that required member states to also provide location information for all 112 calls. During 2009, the 112 support was also extended to roaming mobiles and other technologies like VOIP and location accuracy requirements were mandated. Additionally, this directive required member states to ensure that 112 calls are provided service similar to any other national emergency call number [13].

The European Union, which inherits a conglomerate of different national emergency services, faces the problem of not having a uniform architecture for providing emergency services and location information across all member states. The legacy emergency services in most individual countries are primarily based on voice calls without automatic delivery of location information. The European parliament resolution in 2011 found that the availability and accuracy of location information varied across member states. Hence, it issued a recommendation for the different member states to upgrade their technology in order to deliver automatic and accurate location information for emergency calls. The European Commission aims at establishing directives and legislation, which then need to be implemented by individual national governments, authorities, and emergency services in that country. Each individual national telecommunication regulator should then ensure that the regulations are implemented fully as required by the legislation. As part of this task, the EC decided to fund initiatives and technologies to establish the Next Generation 112 (NG112), providing a service independent of network infrastructure. This activity has been especially pushed forward by the European Emergency Number Association (EENA), which is a nongovernmental organization comprising numerous members of European union, emergency services' representatives, and solution providers [13]. They provide a platform that brings together all stakeholders to discuss operations and technical aspects related to emergency services. They work along with EU officials to introduce legislation related to 112 and provide both operational and technical expertise in defining the legislation. Along mobile phone emergency calls, the EU also mandates the usage of the eCall service for all motorized individual vehicles.

In comparison to the USA, the location services traditionally offered by European operators were very basic: they typically relied on cell ID and lacked accurate centralized location reporting mechanisms for mobile-based emergency calls. Operators in the United States were mandated to update their networks to meet FCC stage 2 requirements. In contrast, in the EU there has not been any homogeneous standard towards that purpose. However, with the advent of smartphones, the ability to calculate a precise position improved greatly due to the usage of IP-based location services that utilize internet servers for GNSS (mostly SUPL), Wi-Fi, and cell ID localization. Consequently, there has been an influx of applications that can deliver highly accurate location information to the emergency services. Unfortunately, these applications are highly localized to a region or country. The EENA initiative called Pan-European Mobile Emergency APPs (PEMEA) aims at providing a framework that allows usage of emergency APPs across all countries in the European Union [14]. Another initiative from EENA is the promotion of AML technology, a capability directly built into the operating system of the mobile device. This has an advantage over PEMEA as an AML-enabled mobile does not need any specific installations. The two most dominating mobile phone operating systems (Android from Google Inc. and iOS from Apple) support variants of AML. The Android AML flavor Emergency Location Service (ELS) has also found acceptance outside the EU in America.

The following sections will explain in detail the different emergency location technologies present in the EU.

3.5.2 eCall

In 2015, the European parliament introduced regulation 2015/758/EC, mandating that all new cars sold from 31 March 2018 in the European Union shall be equipped with eCall technology. The eCall technology is part of the E112 initiative from the European Union to automatically report car crashes to the emergency services. Sensors in the car automatically trigger a 112 call when they detect an accident. Important data, called minimum set of data (MSD) is transmitted to the PSAP, including location information, vehicle identification, vehicle type, and direction.

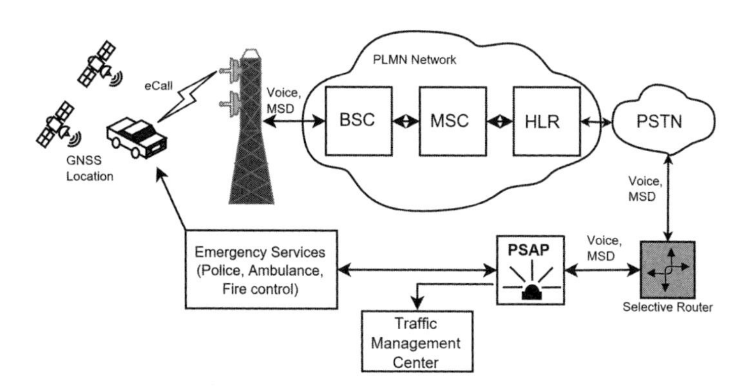

Figure 3.6 eCall system architecture.

This technology also enables the vehicle passengers to manually trigger a 112 call using an SOS button in the vehicle.

Based on the EU eCall requirements, the 3GPP along with ETSI have standardized eCall [15]. The eCall implementation establishes a voice call between the vehicle and the PSAP, carrying the MSD data in the same channel. This is achieved using a GSM in-band modem especially designed for eCall, called the eCall Inband Modem (eIM), which is present in the in-vehicle system (IVS). The eIM establishes an eCall with the public land mobile network (PLMN) using a GSM channel. The PLMN directs the eCall data through the PSTN network to the PSAP, as seen in Figure 3.6. The PSAP to which the call is routed should be updated to detect IVS transmissions over the PSTN channels and handle automatic emergency calls. Additionally to the MSD data, the operator can communicate directly with the passengers and provide assistance using the established voice channel. Once the location and nature of the accident is determined, the PSAP operator notifies the appropriate rescue services and provides them instructions on how to reach the vehicle.

An eCall-enabled system uses the same signaling as regular 112 calls, enabling the eCall to be generated independently of the roaming agreement related to the SIM card and benefiting from the existing PSAP and its associated network infrastructure. At call setup, the IVS shall indicate whether the call was automatically or manually initiated and it shall also terminate any other ongoing connections, as all the resources including microphones must be fully dedicated to the emergency session. The MSD data can be transmitted before, during, or after the voice call and has a maximum length of 140 bytes. The information elements have been standardized, consisting of mandatory and optional elements. In case not all data fields are filled, the rest of the bytes will be transmitted with padded data. The elements of the MSD are:

- Message identifier: MSD format version to enable backward compatibility.
- Activation: indicates if the eCall has been manually or automatically generated.
- Call type: whether the eCall is a real emergency or test call.
- Vehicle type: describes the kind of vehicle (such as passenger vehicle, buses and coaches, light commercial vehicles, motorcycles, etc.).
- Vehicle identification number (VIN): unique number for all vehicles sold in Europe after 1981. This number can be used to obtain all relevant information for a certain vehicle.
- Vehicle propulsion storage type: information on the type of motor type, important to determine risks related to fire and electricity (e.g., gasoline tank, diesel tank, compressed natural gas, or electrical motor).
- Time stamp: the time instant at which the message is generated.
- Vehicle location: the last known position of the vehicle (in latitude and longitude).
- Confidence in position: This element provides the assessment of the accuracy of the position as determined at the time of transmission. This item is to be set to "Low confidence in position" if the position accuracy is not within the limits of +/-150 m with 95% confidence.
- Direction: specifies the direction of the movement of the vehicle.
- Recent vehicle location (optional): last known positions of the vehicle.
- Number of passengers (optional): number of fastened seat belts.

The consistency of the transmitted data is checked with an ACK/NACK mechanism by the IVS and PSAP. In case an error is detected in the MSD transmission, the PSAP can transmit back a NACK to request a retransmission of the data. Should the MSD component not be included in an eCall, or if it is corrupted or lost for any reason, this shall not affect the associated 112 emergency call speech functionality.

One of the main challenges for the eCall system is the multiplexing of eCall MSD on the voice channel, as the codecs are optimized for voice transmissions and low bandwidths. The eCall In-band Modem (eIM) is designed to transmit data through the voice codec without significant losses and still with the ability to transmit the MSD reasonably fast. Once the connection between the IVS and the PSAP is established, the MSD data is transmitted by either using a push or pull

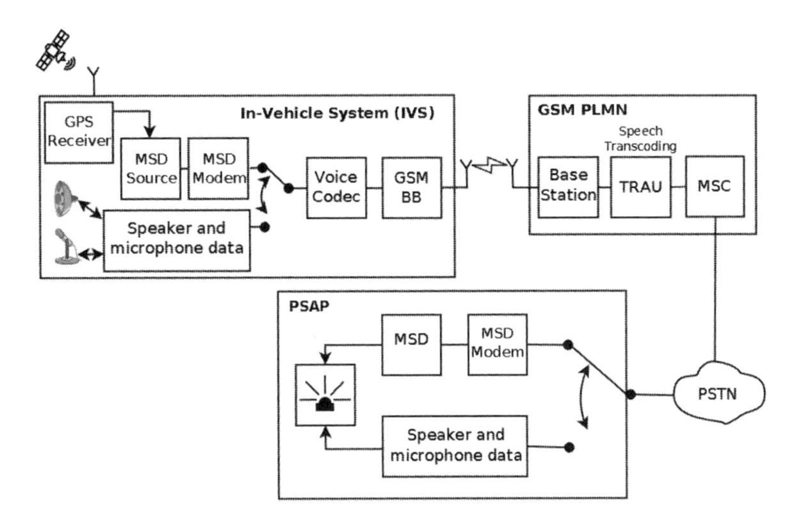

Figure 3.7 IVS and PSAP in-band modem.

method. In the push method, the PSAP requests the IVS to transmit the MSD. In this case, the IVS switches the voice codec connection from the microphone and speaker to the IVS data modem. The MSD data is transmitted on the channel and the IVS receiver switches the incoming stream to the PSAP data modem and displays the MSD data. The pull method is followed when the IVS indicates the PSAP to pull the MSD data. This process is shown in Figure 3.7.

3.5.3 Next Generation eCall

The eCall system is designed to send MSD information over the voice channel on circuit switched networks such as GSM. However, it is not efficient when used over packet networks, as voice packets encounter significantly higher jitter on PS connections compared to CS connections. In order to improve the voice transmission, packet switched networks implement techniques such as dynamic time warping, dejitter buffering, and sample removal. Unfortunately, these techniques cause synchronization issues between the IVS and PSAP, making them unsuitable for eCall.

The Next Generation eCall (NG-eCall) is a redesign of the eCall system to work on packet switched networks. Similar to VoLTE, it relies on IMS as a preferred protocol to handle data [16]. The IMS protocol has been updated to carry MSD information using the SIP Invite message that is sent at the start of the call, together

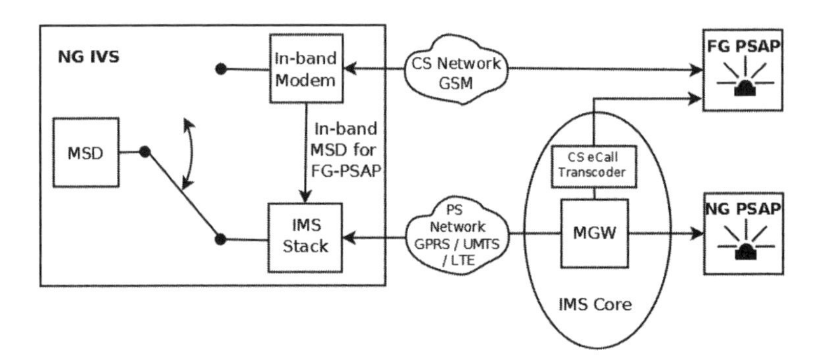

Figure 3.8 NG eCall system architecture.

with an eCall Flag to differentiate this call from other emergency calls. This ensures the availability of the MSD information at the start of the call. If the MSD needs to be updated during the call, the SIP Info message is used. This message also supports bidirectional transmission of ACK/NACK information for message reception. The network support of NG-eCall is broadcasted using system information blocks (SIB).

The next generation in-vehicle system (NG-IVS) system, shown in Figure 3.8, supports both the traditional in-band MSD transmission and the IMS-based MSD transmission. Hence, the same NG-IVS can be used in both packet switched networks supporting NG-eCall (NG PSAP in the figure) and circuit switched networks supporting the legacy eCall (FG PSAP in the figure). The following scenarios are possible when using a NG-IVS:

1. The network supports PS and CS connections and connects to NG-PSAP: in this case, the NG-eCall is used.
2. The network supports PS and CS connections and connects to FG-PSAP: in this case, the NG-IVS uses the CS fallback mechanism and completes the session using eCall.
3. The network is PS only but connects to a FG-PSAP: in this case the in-band MSD is added to the voice channel on the IMS call and the IMS MGW transcodes the voice data to a CS eCall. This is not a preferred deployment due to the associated problems on the voice transmissions.

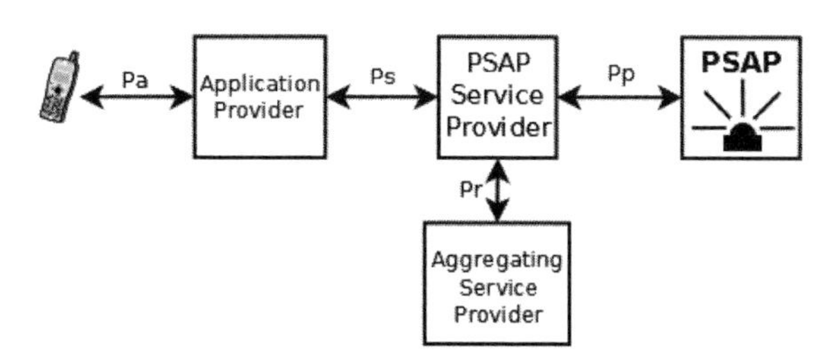

Figure 3.9 PEMEA system architecture.

3.5.4 PEMEA

The PEMEA [14] is a standard for smartphone applications to deliver location information for emergency services [17]. However, most of these applications are only of a country or regional scope [18], despite the EU efforts to make this service available to all citizens. PEMEA has the ability to enhance the service by additionally providing video, chat services, and other value-added information when compared to techniques such as the regular 911/112 calls or AML. On the other hand, PEMEA uses the regular network without emergency services priority, having access to a limited number of PSAPs, and it may not work reliably under roaming conditions. PEMEA provides an architecture, a protocol for the data exchange with the PSAP service provider (PSP) and a security model for protection of user data [19]. However, the aggregated service provider (ASP) for each PEMEA application maintains its own routing table for PSAPs and has a propriety format for the delivery of emergency data.

The system architecture in PEMEA follows the scheme in Figure 3.9. The PEMEA app runs on a smartphone, wearable, laptop, or similar device and is able to make an emergency call. It connects to the application provider over an IP connection obtained from a 3G, LTE, or Wi-Fi link. The app is required to authenticate itself with the application provider in order to transfer the emergency information. The app can use GNSS, Wi-Fi, and crowd-sourcing servers, as well as the services provided by the device's operating system to calculate accurately the location of the caller. Additionally, the app can try to acquire additional information such as cell ID of a serving cell or the Wi-Fi basic service set identifier (BSSID) if the call is made over a Wi-Fi link.

The application provider hosts servers where the users of a certain PEMEA app are registered. It communicates over the Pa interface, whose implementation is not in the scope of PEMEA. The application provider should authenticate the user and format the received information, combine this information with any other information stored in its server, and forward it to the PSP. Every application provider is connected to one PSP for the region or country to which it should forward the information. In case an application provider is connected to multiple PSPs, (for example, for multiple regions), it will transfer the information only to one of them in order to avoid multiple instances of the same call in the PEMEA network.

The PSP has a domain name and a certificate from a registered authority in the PEMEA network. It communicates with the application provider over the Ps interface specified by the PEMEA framework. Its role is to decide whether it is able to service the call using any of the PSAPs it is connected to and then forward the call to the selected PSAP. In case it is unable to service the call, it forwards the information to the ASP to service the call. The PSP authenticates itself with the PSAP and the ASP. A PSP may also handle incoming requests from an ASP in cases like a roaming user where the host PSP is not able to service the call.

The ASP also has a domain name and a certificate within the PEMEA network. It provides routing functionality between PSPs, especially in roaming cases where an originating PSP does not have direct information to which PSAP the call should be addressed. It has European wide access to all registered PSPs in the PEMEA network.

The main PEMEA messages are the *emergencyDataSend*, for sending the emergency information, and the *emergencyDataReceived*, to notify that the emergency information has been received and will be handled. Both are formatted as XML messages and transmitted using the HTTPS post method. The emergency information contains the following parameters:

- Time To Live: provides the number of hops allowed before the message will not be transmitted any further.
- Route: provides the nodes in the network, including the originating node, that have been used to transmit the message.
- Caller Identities: provides information on the caller identification such as an IMSI value. There may be multiple elements in this field if multiple SIM cards are present in the device.
- Location Information: this is the location calculated by the device.
- Application Provider Information: provides information on the contact details of the AP including telephone numbers, email, etc.
- Caller Information: contains personal information of the caller.
- Receive Error Post: when transmitted, it will prompt the receiving entity to transmit an error element to a certain address.

- AP Capabilities: optional element that lists down additional capabilities. For example, a URI for obtaining additional information from the AP.
- Receive Capability Support: provided in case AP Capabilities element is also present. The terminating node send its capability information to the address provided in this element.

When the PSP receives the emergency information message, it looks at the received location and determines whether the local PSAP can deal with the emergency request or not. In case the local PSAP is suitable, the PSP sends the emergency information to the PSAP for further processing. It also sends the response message back to the AP, containing the following elements:

- Time Stamp: accurate time stamp when the message was transmitted from the receiving node to the originating node.
- Route: contains a list of nodes that were used for the message, and the identifier of the receiving node will be appended to the list.
- Delivery: information on the node to which the emergency information has been passed on to.

In case the PSP determines that the location of the caller is not handled by its associated PSAP, it forwards the request to the ASP. The PSP duplicates the emergency information message, decrements the "Time To Live" element, appends its own identification to the "Route" element, opens a secure connection, and transmits the message to the ASP, which in turn forwards it to the appropriate PSP in the region of the call.

3.5.5 Advanced Mobile Location

In contrast to to the PEMEA service, which required an app installed in the mobile device, the AML system relies on functionality embedded in the device's own operating system. This enables mobiles to be preinstalled with AML capability and do not require any special installations from the user. When 112 is dialed, the mobile device provides automatic location information via SMS transfer to the emergency services. The use of AML enables a harmonized approach to location determination across different nations and network configurations without changing the network infrastructure significantly, since it can work directly on GSM supported on almost all networks across the EU. This technology was first developed by British Telecom in response to the lack of accurate location information during emergency calls. It was later adopted by many nations across the EU and standardized by the ETSI [20].

The AML architecture is explained in Figure 3.10. An emergency voice call follows the usual call-flow through the base station and RNC to the MSC that routes

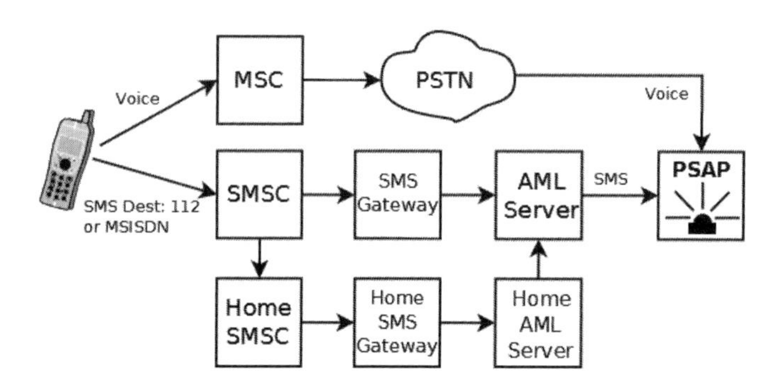

Figure 3.10 AML architecture.

the call through the PSTN network to the PSAP. As usual, the GMLC provides information on the serving cell location that is used to select the serving PSAP. The mobile triggers an SMS to the corresponding emergency call destination, in Europe 112. The short message service center (SMSC) receives this SMS and identifies it as an emergency SMS. In case the unified 112 number is not available in the network, the mobile can also use a known mobile station international subscriber directory number (MSISDN) for emergency services. The SMSC then forwards the SMS to the SMS gateway to be directed to the AML server, usually implemented as a part of the PSAP. The AML server compares the location information received by SMS with the one provided by the network using the MSISDN as the unique identifier. Once the consistency of the location information has been verified, the location is displayed on the PSAP operator's console.

In the case of a user roaming outside their home country, AML implements multiple techniques to route the location information to the correct PSAP. For instance, the mobile device could implement a built-in database with MSISDN numbers for emergency assistance in each individual country based on the mobile country code (MCC) present in the cell ID of the serving cell. In such a case, the visited network SMSC forwards the message directly to the AML server in the visited country. In case no MSISDN number is available for the visited country, the home country SMSC forwards the information to the home AML server, which in turn routes the information to the AML server of the visited country based on the MCC information provided in the SMS.

The AML technology allows the mobile device to determine its location based on any combination of all available methods, including GNSS, network-based positioning, or Wi-Fi. Furthermore, AML can automatically turn on any location settings in case of an emergency call, even if these were turned off before. AML specifies a maximum timeout value T1 (typically 20 s), within which the SMS containing the caller's location must be transferred from the mobile. In the ideal case, the mobile is able to estimate the caller's location using GNSS before T1 expires and it transmits this information. Otherwise, the device may try to calculate the location using Wi-Fi crowd-sourcing, cell ID, or any hybrid solution and send the best estimated location by the end of T1. In case of a T1 timeout without any successful location calculation, the last known location or cached location will be sent to the network. In case no location can be determined, a "No Data" message will be sent.

The AML SMS contains the following information elements:

- Header: provides the version of AML used, containing a double quote symbol as the second character of the header, as in AML=2.
- Latitude and longitude: the estimated location in WGS84 format, with a resolution of 1.1 m.
- Radius: the radius in meters of the uncertainty circle centered at the latitude and longitude provided in the message. Present only if its value can be determined.
- Altitude: altitude value in meters with respect to the mean sea level. Present only if its value can be determined.
- Floor number: estimated floor number for a high-rise building. Present only if its value can be determined.
- Time of positioning: time of the position fix specified in Coordinated Universal Time (UTC) format. The field format is YYYYMMDDhhmmss, where YYYY is the year, MM is month, DD is day, hh is hour, mm minute, and ss second.
- Cell ID: serving cell identifier. Present only if its value can be determined.
- Positioning method: specifies the method used by the mobile to determine the location. It can be one of the following methods: G for GNSS (GPS, Galileo, GLONASS, etc.), W for Wi-Fi measurements, C for cell ID, N if no position can be determined.
- International Mobile Subscriber Identity (IMSI): unique subscriber identifier present in the SIM card of the mobile device that has made the call.
- International Mobile Equipment Identity (IMEI): unique equipment identifier used to identify the make and type of the mobile device.
- Mobile Country Code (MCC): identifies the country where the emergency call is made.
- mobile network code (MNC): identifies the mobile network used to make the call.

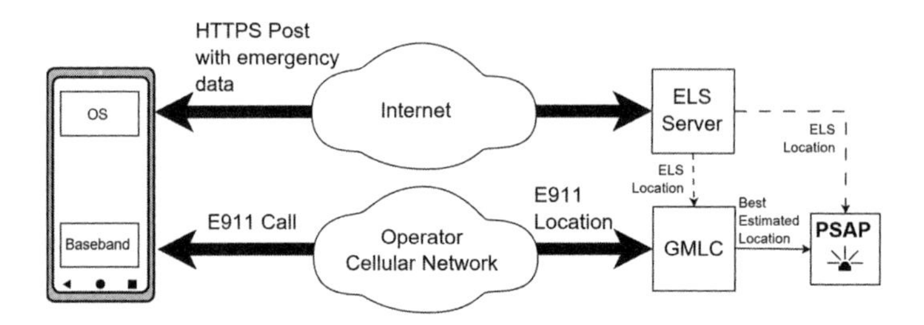

Figure 3.11 ELS architecture.

- Message length: length in bytes of the entire SMS message.

3.5.6 Emergency Location Service and other AML Enhancements

As data connections are getting ubiquitous in mobile phones and the PSAP implementation is moving increasingly to an IP-based network, the size-limited SMS transmissions from the AML system are becoming obsolete in comparison with data push methods. The HTTPS post method can used to transmit data push information to a centralized server in a certain country and this can be routed to a national-level PSAP that takes care of the coordination with the emergency authorities. The most popular smartphone operating systems, Android and iOS, have already implemented these services. An example is the ELS, supported by Android [21]. The HTTPS post message contains information similar to the contents defined for the AML SMS in the previous section. This technology is gaining popularity among the network operators, especially in the United States to supplement their E911 implementation. It can be used as a fallback mechanism when the network operator cannot obtain an accurate position using traditional techniques.

The ELS architecture differs from the AML architecture mainly in the mean used to send the emergency data message, as shown in Figure 3.11. The HTTPS post containing the emergency data is sent to an ELS server, which can be connected to the GMLC of the network or directly to the PSAP. The GMLC also gets location estimates using traditional E911 techniques and makes a decision on one of the two locations. The selected location is then transmitted to the PSAP.

With IMS being used widely in the next generation networks, the SIP protocol header can also be used to carry caller location, if this location is known at the start of the call. This method could become the reference in the future, as it allows the caller's location to be routed along with the voice call.

3.6 CONCLUSION

This chapter has explained the reasons why accurate location determination is a crucial part of emergency services, giving an overview of the status of emergency call regulations both in Europe, via the European Commission, and in the United States, through the FCC. The chapter has also detailed the different protocols and network architectures involved for both circuit switched and packet switched networks and the new tendency toward IMS implementations, eCall, and AML enhancements.

The ability to accurately locate the origin of an emergency call has the potential to save lives, and the accuracy requirements for the positioning systems are getting tougher and tougher. The FCC phase II requirements have set very ambitious targets for the next few years, including outdoor and indoor positioning and altitude determination. These regulatory requirements are the main driver for the mobile device positioning technologies that will be seen later in this book. However, regulatory bodies and emergency services are not the only entities demanding accurate positioning. In the copilot seat, also pushing for the improvement of location-based services, are the different commercial applications and the IoT, which rely on LTE and 5G networks. These commercial applications are described in detail in the next two chapters.

References

[1] https://fcc.gov, accessed on February 2020.

[2] *Wireles Communications and Public Safety Act of 1999*, Public Law 106-81-Oct. 26, 1999.

[3] https://www.fcc.gov/oet/maps/areas, accessed on February 2020.

[4] FCC *Wireless E911 Location Accuracy Requirements*, Fourth Report and Order, 2015.

[5] FCC *20th Mobile Wireles Competition Report*, 2017.

[6] 3GPP TS 23.271, *Functional Stage 2 Description of Location Services (LCS)*, V9.4.0, 2010.

[7] 3GPP TS 44.031, *LCS; MS - SMLC; Radio Resource LCS Protocol (RRLC)*, V15.0.0, 2018.

[8] 3GPP TS 25.331, *RRC Protocol Specification*, V15.4.0, 2018.

[9] 3GPP TS 23.167, *UMTS; LTE; IP Multimedia Subsystem (IMS) Emergency Sessions*, V9.4.0, 2010.

[10] 3GPP TS 36.355, *E-UTRA; LTE Positioning Protocol (LPP)*, V15.1.0, 2018.

[11] OMA-AD-SUPL, *Secure User Plane Location Architecture*, V2.0, 2008.

[12] OMA-TS-LPPe, *LPP Extensions Specification*, V2.0, 2014.

[13] EENA Operations Document, *112 and the EU Legislative Framework*, V1.0, 2013.

[14] EENA Operations Document, *PEMEA Requirements and Functional Architecture*, V7.0, 2015.

[15] 3GPP TS 26.267, *eCall Data Transfer; In-Band Modem Solution; General Description*, V15.0.0, 2018.

[16] 3GPP TS 22.101, *Service Aspects; Service Principles*, V15.6.0, 2018.
[17] EENA Operations Document, *112 Smartphone Apps*, V1.0, 2014.
[18] EENA Operations Document, *112 Apps Strategy*, V1.0, 2014.
[19] EENA Operations Document, *PEMEA Protocol and Procedures Specification*, V5.0, 2016.
[20] ETSI TR 103.393, *EMTEL; Advanced Mobile Location for Emergency Calls*, V1.1.1, 2016.
[21] https://crisisresponse.google/emergencylocationservice/, accessed on February 2020.

Chapter 4

Commercial Location-Based Services in LTE

4.1 INTRODUCTION

As introduced in Chapter 3, the main reason for the adoption of positioning technologies by mobile networks were the regulatory requirements for accurate positioning in emergency calls. However, mobile phones supporting location services also led to a wide range of commercial applications such as maps, navigation, and fitness trackers. This created many new business models for application service providers and made location awareness a very important aspect for future industry needs. This chapter focuses on the different LTE commercial applications using location technology and the design of a modern smartphone to support these requirements. Apart from the traditional commercial use cases (navigation, location-based advertisement, etc.), LTE defined two new applications demanding positioning services: D2D communications, including more specifically vehicle-to-vehicle (V2V) and V2X, and the IoT.

The ability of a device to conduct point-to-point communications was introduced with the D2D technology by adding a sidelink channel to communicate between devices independent of the wireless connection with the network. D2D technology did not have large commercial success but rather laid the foundation for the cellular V2X (C-V2X) technology, which is an important building block for enhanced driver safety and autonomous driving. This chapter will present an overview on how the car of the future will be designed in terms of wireless connectivity with a focus on C-V2X. A brief history followed by an overview of this technology should provide the user a good basis for understanding the details of vehicular autonomy.

The IoT has also seen a major adoption worldwide with applications such as smart cities, smart meters, wearables, smart watches, and tracking devices. A multitude of technologies exist that support IoT device connectivity depending on the distance and scope of operation. Nevertheless, the 3GPP offers a comprehensive solution based on its MTC and NB-IoT based technologies for large-scale deployment of IoT devices. This chapter will briefly look at the different technologies existing for IoT devices and focus on the 3GPP technologies.

4.2 LTE COMMERCIAL LBS APPLICATIONS

Although the first adoption of location technology was to serve safety and emergency services, its availability in smartphones opened the way for a multitude of applications, which directly and indirectly use location information for commercial purposes. The new business fields and revenue streams that they generate has led to development of many new applications and services that use location information. One can in fact argue that location determination accuracy has been accelerated due to the commercial interest in location-sensitive applications. Examples of such commercial applications are the following:

- Mapping and navigation applications: these applications use current user location to provide instructions on how to arrive to a planned destination. Google Maps is an example of such applications.
- Advertising based on location: advertising based on geofencing applications enable businesses to send advertisements to users who are in areas close to shops. It is a very powerful tool in addressing potential customers by using very accurate locations such as a building block, conference, or shopping location. Sephora is an example of a business that makes use of such an app.
- Social networking applications: location information is an enabler for new features for social networks, dating apps, and so on, allowing people to connect with each other in case they are geographically close. Advertising based on location on social networks enables combining personal preferences with targeted location advertising.
- Service and efficiency improvements: shopping apps that prepare the goods for delivery based on arrival time enable efficient logistics and an enhanced customer experience. An example of such an application is the Target app.

Another popular app is the Detour app that provides tourist information based on user location.

- Traffic management apps: navigation combined along with crowd-sourcing-based techniques enable apps to provide information on traffic congestion, traffic accidents, and detour information.
- Security applications: these provide safety features such as tracking your children. An example of such an app is Family Locator.
- Ride-sharing and car-sharing apps: these services provide a network of cars or other vehicles placed around the city and allow the user to rent them based on the user's location. They have become extremely popular in urban areas where shared cars and other shared resources are not only cost effective but also reduce congestion on roads. Similar applications also exist for bike or e-scooter sharing.
- Gaming and augmented reality: the ability to correlate real-life elements with virtual elements has created a new category of gaming applications, tourist information applications, and reality enhancement apps. The service utilizes location as well as other information such as the mobile phone camera to identify the surroundings of the user and provide information based on that.

As it can be seen, the above list offers a wide variety of different services and applications. Commercial location-based services are a very fast growing field and new applications are constantly emerging. During the early stages of LTE deployment, LBS was oriented primarily toward smartphones. However, with the adoption of V2V and V2X applications, the focus of LBS technology moved into cars and other types of vehicles. Finally, the IoT world opened LBS to any type of device, from wearables to shared bikes or even drones. The following sections will briefly introduce the typical architecture of mobile phones and vehicles to take full advantage of the positioning technologies.

4.2.1 Mobile Phone Architecture for LBS

Figure 4.1 shows the typical design of a mobile phone highlighting the components needed for location calculation. The mobile phone has a modem chip, which connects to the cellular network. The modem provides location services over C-Plane and/or U-Plane (using an IP connection to run SUPL clients for location services). Additionally, it uses measurements based on LTE positioning technologies

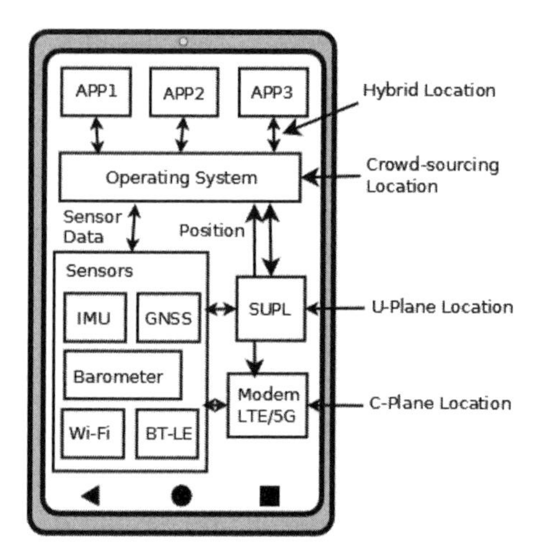

Figure 4.1 Example of mobile phone architecture for positioning.

like OTDOA or ECID. Mobile phones have also a number of sensors, typically a GNSS receiver, Wi-Fi and Bluetooth receivers, a barometer and an IMU. All these measurements can be combined to obtain location estimates. The operating system manages the device and provides services to the applications. Most of the mobile operating systems additionally connect to their own positioning servers to obtain crowd-sourcing-based location estimates. The applications in the mobile typically use the services provided by the operating system obtain the location estimates.

The methods for calculating the mobile device's positioning from the available measurements will be explained in detail in Chapters 6 to 12 of this book.

4.2.2 Automotive Applications

Out of all other LBS commercial applications for LTE, automotive applications deserve special consideration. The automotive environment is becoming increasingly connected and vehicles are equipped with a large number of sensors, GNSS receivers, and IP connectivity through the mobile network. The automotive industry is moving toward autonomous driving, which requires highly reliable car-to-car communications and where positioning accuracy is critical to ensure the safety of both pedestrians and passengers.

In order to address these requirements, LTE has developed C-V2X technology. However, before explaining C-V2X, it is worth having a look at modern vehicle

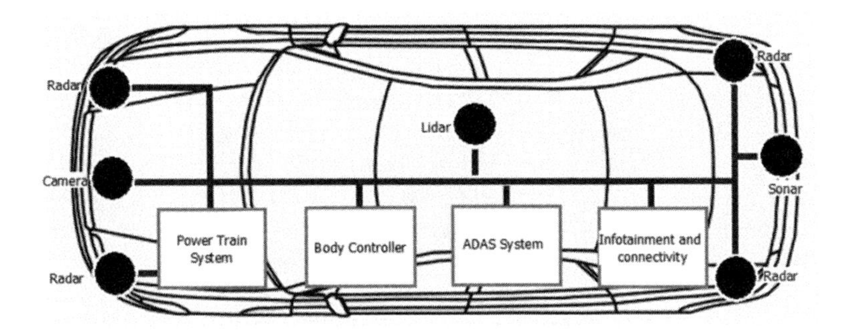

Figure 4.2 Example of car archicture for modern automotive applications.

architecture and the different components that can be used for positioning and communication aspects.

4.2.3 Vehicle Architecture

A modern-day car, although similar on the outside to earlier car models, has evolved over generations and is vastly modernized in terms of connectivity, sensors, control systems, and safety features. This section will focus on the different electronic systems present in the car and their interaction in order to support the driving experience. The car of the future will be increasingly connected, able to combine inputs from different sources, and make driving decisions. Driving decisions can be either in assistance mode (i.e., helping the driver) or autonomously (i.e., taking complete control). Such systems need to be extremely safe, secure, and resistant to jamming and spoofing attacks. In order to implement such a car, a number of sensors, connectivity modules, and other technologies are needed. Figure 4.2 shows some of these modules, in particular the ones that are needed for navigation and autonomous driving.

The sensors commonly used to detect obstacles and moving objects are radar, lidar, cameras, and sonars. They are the eyes of the car and are typically mounted on the car to enable a 360° view of the immediate surroundings. Each type of sensor differentiates itself from other sensor types based on the range of measurement, objects that can be measured, resolution, and reliability under different weather conditions. These sensors and their characteristics will be explained in detail in Chapter 12 of this book. Nonetheless, they will be briefly introduced here.

Radars transmit pulsed or continuous radio waves that impact on the surrounding objects and are reflected back to the receiver. Based on the time of flight and the Doppler shift of the signal, the radar can determine the distance, relative velocity, and approaching angle of different objects near the vehicle. They have a range of approximately 200 to 300 m. Sonars work on a very similar principle, but they use sound waves and typically have much shorter range, of a few meters.

Lidar functions much like radar, but uses laser beams instead of radio waves. It consists of a rotating head that transmits millions of pulses per second in all directions and generates a three-dimensional model of the surroundings. A lidar is usually placed on the roof of the car and allows a 360°view. It provides a higher resolution of the environment than a radar. However, it is adversely affected in foggy, rainy, snowy, and dusty conditions.

Camera systems in a car are used to identify artifacts on the road based on image-recognition capabilities similar to the human brain. They are used to detect road signs, traffic signals, lane markings, and other vehicles on the road. Multiple cameras are placed to cover all directions, including stereo cameras to get 3D images with depth resolution. Compared to lidar, camera systems are a much cheaper option and more robust in rainy and snowy conditions. They are also more useful in capturing road signs and detecting colors. Nonetheless, cameras are adversely affected by strong sunlight, reflections, and image saturation.

Apart from these sensors, a modern-day car consists of different control systems. Of special importance for autonomous driving is the advanced driving assistance system (ADAS). The ADAS is the brain of the car, where all sensory information related to the movement and surroundings is merged to offer driving assistance or control the motion of the car. As shown in Figure 4.3, the radar, lidar, camera, and other sensors supply information that is combined in a sensor fusion module of the ADAS. Combining such data requires highly sophisticated techniques involving artificial intelligence (AI) to create a 3D model of the surroundings of the car. AI algorithms are very powerful in processing lidar and camera inputs to perform image recognition, object detection, pedestrian recognition, and road sign detection, but they require high computational processing.

In addition to the sensor inputs, the ADAS also has access to connectivity inputs such as navigation and mapping data, routing information and other traffic management information, and access to the LTE network and GNSS measurements. Using all the available measurements, the positioning server on the ADAS can implement a sensor fusion algorithm to calculate the position of the vehicle in every instant, both in absolute terms (e.g., in latitude, longitude, and altitude coordinates) and relative to the road and the surroundings of the vehicle. Accurate knowledge

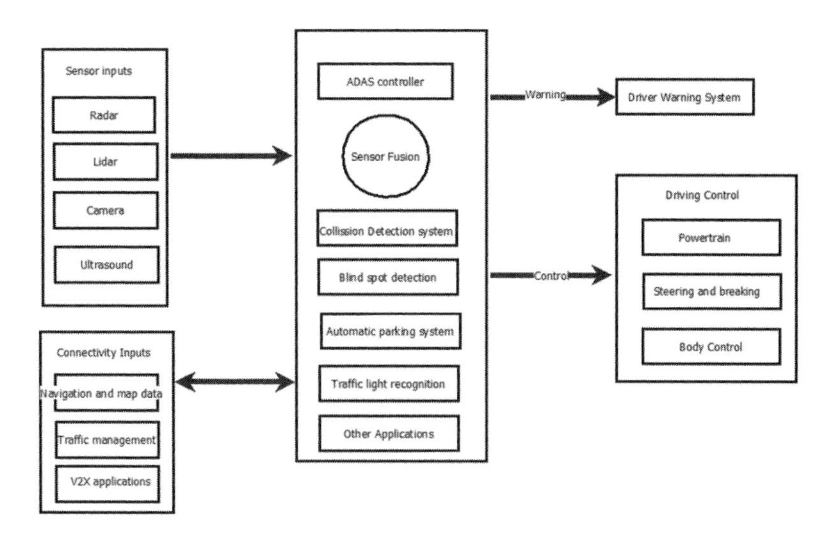

Figure 4.3 Vehicle sensors and interconnectivy to the ADAS.

of the vehicle's position is critical in autonomous driving to prevent collisions with other vehicles, pedestrians, or the infrastructure.

4.2.4 Autonomous Driving

The ultimate goal of the automotive industry is to get highly autonomous vehicles with low to no human interaction required to control them. Full autonomy is a very challenging problem to solve and needs different intermediate steps, starting from basic warning systems to driver assistance and finally full autonomy. The Society of Automotive Engineers (SAE) has classified these different steps in six levels [1]. The driving automation ranges from level 0 (no automation) to level 5 (full driving automation), as explained in Figure 4.4. In the first three levels, from level 0 to level 2, a human driver is still in control of the vehicle, but he or she receives assistance such as lane detection features or emergency breaking. In levels 3 to 5, the vehicle is self-driving. However, level 3 requires the human driver to take control of the car under certain conditions.

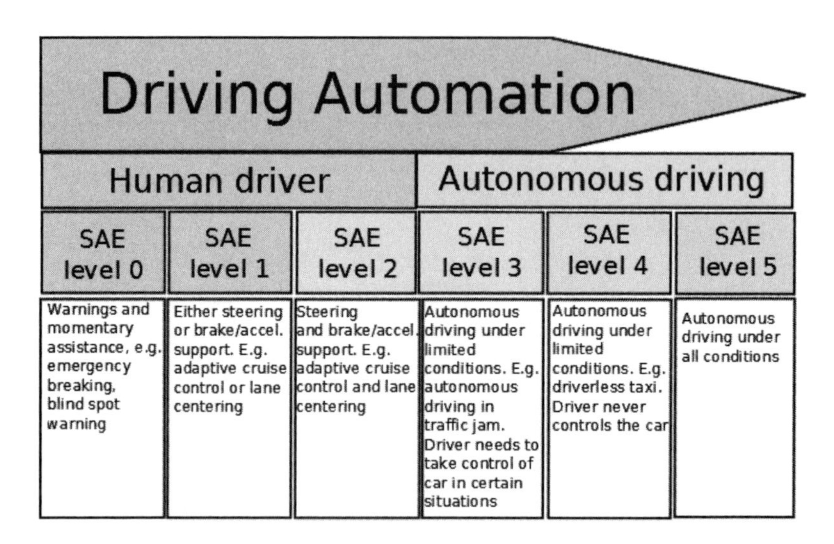

Figure 4.4 Autonomous driving levels as defined by the SAE.

4.3 D2D

D2D communications were introduced in LTE as a direct communication link between two different mobile devices without crossing the core network or the eNB. The communication link between devices is known as the sidelink (SL). The main advantages of the SL with respect to the usual uplink (UL) and downlink (DL) are the reduction of the network congestion, the energy efficiency, the spectral efficiency, and the reduced delay. D2D communications require proximity between the two devices. They can be used for commercial services such as the ones described in the previous section, but also for emergency communications in the case of natural disasters, for example by establishing an adhoc network to send alert notifications from device to device.

There are two main interfaces involved in D2D communication: the LTE-Uu (henceforth referred to as Uu) and the PC5. The Uu is formed by the traditional downlink and uplink channels, while the PC5 is the interface of the sidelink. The PC5 interface is part of the proximity services (ProSe) and defines mechanisms for nearby devices to communicate with one another on a point-to-point basis without using network resources. Communication using PC5 can work either on pre-configured resources or on resources allocated by the network.

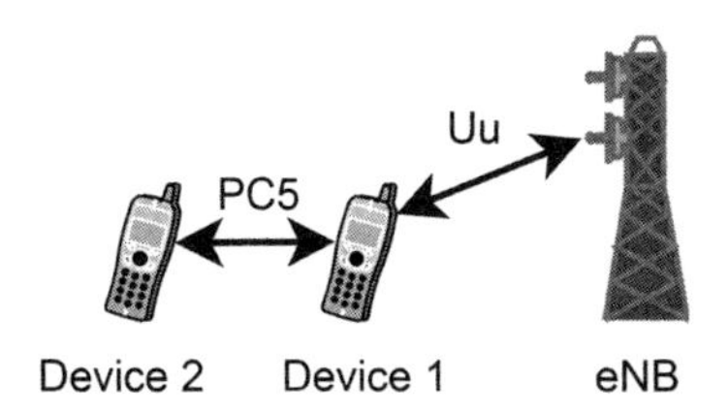

Figure 4.5 D2D interfaces.

Figure 4.5 shows an example of a D2D scenario with both interfaces, the Uu and the PC5. Device 1 is within the network coverage and has an active connection to an eNB via the Uu. It is also in the proximity of Device 2, and they maintain a D2D communication over the PC5. Device 2 in turn is out of network coverage and only has the PC5 connection.

4.4 V2X

V2X is the general term used to define any kind of communication between a vehicle and any entity in its surroundings, including other vehicles, road infrastructure, pedestrians, or the network. This section focuses on the 3GPP standardized V2X technology, commonly known as Cellular-V2X (C-V2X). However, there are also different implementations based on wireless local area networks (WLANs) such as the Wireless Access for Vehicular Environment (WAVE) program developed by the IEEE in 802.11p [2, 3].

4.4.1 Brief History

The U.S. Department of Transportation supports multiple programs for communication and safety on the road through the Intelligent Transportation Systems (ITS). One of these programs, known as Dedicated Short Range Communication (DSRC), has been developed by the SAE in the United States as a communication protocol and was adopted by the FCC in 2003. DSRC is defined to operate in the frequency range 5.850-5.925-GHz band with a maximum bandwidth of 75 MHz [4]. The European Union also standardized a similar technology in 2008, known as cooperative Intelligent Transport Systems (C-ITS) [5]. C-ITS has a frequency allocation of 30 MHz in the 5.9-GHz band. Both systems, C-ITS and DSRC, are similar in many aspects.

In order to develop these programs and support WLAN in the vehicular environment, the IEEE standardized the 802.11p standard under the WAVE. 802.11p provides the physical layer and MAC layer protocol stacks to be used along with DSRC and C-ITS. The standard provides the messaging protocols and interfaces to applications used for V2X communication and it was developed well before the 3GPP standard. However, it has experienced a slow adoption and little commercial success to date, giving 3GPP's C-V2X a competitive advantage.

C-V2X was standardized by 3GPP in E-UTRA Release 14 [6, 7], which was completed in 2016, and it builds on the existing LTE infrastructure already used in the automotive industry for E-Call and connectivity purposes. It is defined to work in the same band as 802.11p (i.e., 5.9-GHz). The C-V2X specification has also partly leveraged the safety requirements and applications well established for DSRC and C-ITS. It builds on the PC5 interface (also referred to as sidelink) developed by 3GPP to support D2D communications in previous E-UTRA releases.

Further development of the V2X standard in 3GPP is planned for Release 16 with 5G, mainly to add higher bandwidth and lower latency applications. When compared to IEEE 802.11p, C-V2X was developed later and is relatively untested in terms of latency and high-density usage. However, its underlying physical layer offers a more promising upgrade path and a harmonized worldwide standard. Thus, it is expected that C-V2X will be the predominant technology in the long run, although both will coexist. As this book is related to cellular network positioning technologies, it will focus solely on C-V2X and its use cases. For simplicity, the term V2X will be used instead of C-V2X.

4.4.2 V2X Technology

The abbreviation V2X implies the communication mechanism connection on one side a V ("vehicle") and on the other side X ("anything"). As shown in Figure 4.6, this can be other vehicles, implying a V2V connection; a pedestrian, implying V2P; traffic infrastructure, implying V2I, or the network operator, implying V2N. As it will be seen later, the possibility of having these interfaces enables different categories of applications not just limited to vehicle safety but also pedestrian safety, traffic optimization techniques, infotainment applications, and so forth.

The V2X technology builds on the sidelink channel (PC5 interface) defined for D2D communication, which was standardized by 3GPP in Release 12 and Release 13. The main purpose of the D2D technology was to allow devices close to one another to communicate without using network resources. Thus, the V2X technology needed additional modifications to this technology to meet the vehicular

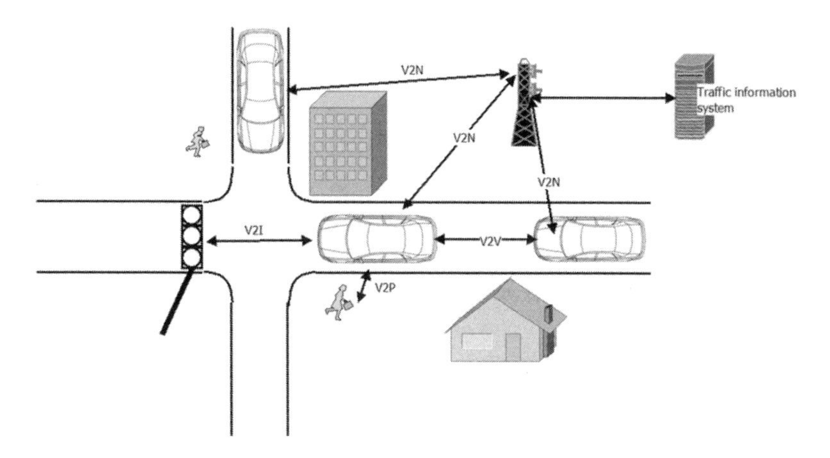

Figure 4.6 Examples of V2X communications.

environmental needs. These required changes have been standardized in Release 14.

Multicast: D2D was defined as a unicast service (i.e., the messages transmitted by a D2D device were directed to a specific user). The V2X transmissions are specified to work in multicast mode, where vehicles broadcast their V2X messages and any receiver may choose to use this information or not. Hence, the device discovery ability, defined in the ProSe specification, is no longer required and not supported in V2X operation.

Multicarrier: The ProSe specification allowed PC5 transmissions to use the uplink resources of the LTE cells. However, V2X operation requires more resources, and using the uplink resources allocated for normal LTE operation is not efficient. Thus, V2X has been defined over multiple carriers, with the Uu operation typically in an LTE operating band and the PC5 operation allocated in a different band. The band currently used for PC5 is TDD 47, around the 5.9-GHz frequency, which is reserved for ITS operation.

Channel enhancements: Latency for message transmissions over the PC5 interface has been reduced. Depending on the type of messages, the latency values vary from 20 ms to 100 ms. In addition, the PC5 interface has been defined to support relative speeds of 500 Km/h and range extension has been added to allow the driver up to 4 seconds of reaction time. The system has been defined

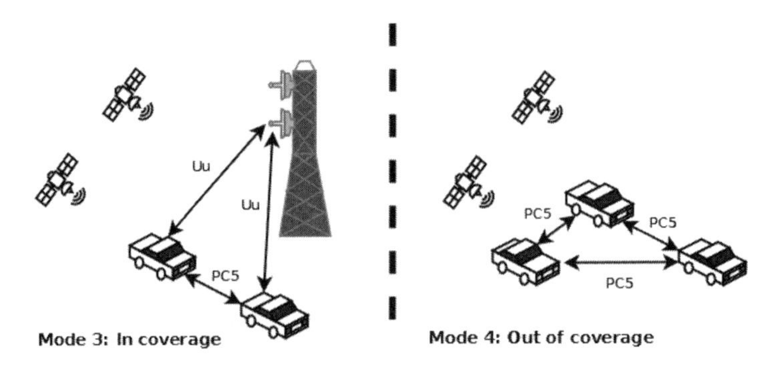

Figure 4.7 Sidelink modes 3 and 4 for V2X operation.

to support typical message transmissions of 10 Hz per car with peaks rates of up to 50 Hz.

Mode 3 and mode 4: These modes of operation have been introduced Release 14 to support V2X operation, as shown in Figure 4.7. The sidelink mode 3 can also be referred to as "in coverage" operation. Here, the vehicles are within the range of an eNB and the LTE network is responsible for resource allocation and congestion control. Each V2X UE is connected to the network using the Uu interface and communicates with other cars over the sidelink channel via PC5. The time synchronization on the sidelink is achieved using the LTE frame time. In the sidelink mode 4, also referred to "out of coverage" operation, the vehicles are not in range of any LTE eNB. Thus, the PC5 interface is the only connection available and the synchronization relies on the GNSS signal (i.e., all vehicles transmit the sidelink signal aligned to the GNSS frame timing). Additionally, this technology also allows alignment of the signal to the PC5 signals from another vehicle, considered as the reference vehicle. In this mode, the resources for sidelink operation are preconfigured per geographical area.

Congestion control: 3GPP introduced congestion control mechanisms based on semi-persistent scheduling (SPS) to reduce the occurrence of packet collisions. In contrast, the 802.11p standard uses carrier sense techniques for this purpose. SPS has already been used in VoLTE deployments for efficient resource utilization. This is especially useful for the sidelink mode 4 scenario, since no network coordination is available for resource management. SPS is suitable for V2X because the messages are periodic in nature and deterministic to a large extent. This enables resource usage decisions to be implemented directly on the V2X UE. It additionally makes use of the partial carrier sensing concept, where resource elements are monitored over a period of time by the UE. This monitoring allows to make decisions on which transmission resources to use based on prediction algorithms.

4.4.3 V2X Frequency Allocation

As it has been mentioned already, V2X uses the allocated ITS spectrum in the 5.9-GHz band coexisting with 802.11p. The coexistence of these technologies is ensured by allocating separate channels for V2X and 802.11p operation. Depending on the region, different parts of the spectrum are allocated to be used for V2X operation.

Table 4.1

C-ITS V2X Frequency Allocation in the European Union [8]

ITS-G5B	5.855 GHz to 5.875 GHz	ITS nonsafety applications
ITS-G5A	5.875 GHz to 5.905 GHz	ITS road safety
ITS-G5C	5.905 GHz to 5.925 GHz	Future ITS applications

In the European Union, the ITS band has been divided in subbands depending on the application. The safety related messaging such as collision avoidance are supported in the ITS-G5A subband. The nonsafety related messaging must use the ITS-G5B subband. Finally, the ITS-G5C subband is reserved for future ITS applications.

The 802.11 group in the United States has divided the ITS band into multiple channels of 10- or 20-MHz channel bandwidth. The channel numbering is shown in Figure 4.8. Some of the channels are reserved for specific DSRC applications.

The 3GPP has defined the ITS band as TDD 47, from 5.855 to 5.925 GHz, supporting channel bandwidths of 10 or 20 MHz.

4.4.4 V2X Network Architecture

This section describes the V2X network architecture including the PC5 and Uu interfaces. V2X builds upon the LTE architecture and the ProSe architecture, as it is shown in Figure 4.9. The different components and interfaces will be described briefly.

The LTE Uu interface is the standard interface used by mobile phones to connect to the LTE network. This interface is reused for V2X. The V2X UE

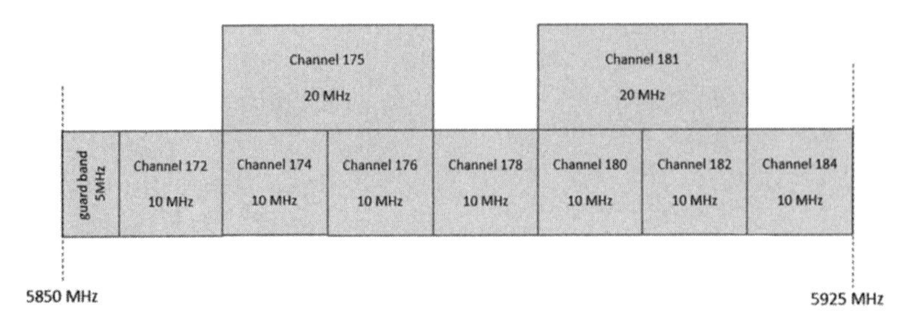

Figure 4.8 Channel numbering of the ITS band for 802.11p in the United States.

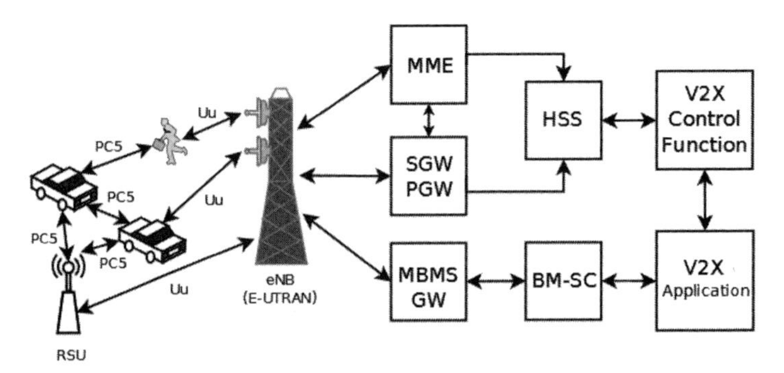

Figure 4.9 V2X network architecture.

connects to the eNB using resources allocated by the E-UTRAN and supports transmission of V2X unicast and broadcast messages. Unicast transmissions use dedicated data bearers defined in LTE, while multicast messages use the evolved multimedia broadcast/multicast services (eMBMS). This interface carries control messaging for the operation of the PC5 interface and the allocation of resources as configured by the V2X control function. This link is used by roadside units (RSUs) to transmit information relevant to large geographical locations to the V2X application servers.

The V2X UE generates safety messages in coordination with the ADAS unit and transmits this information to other vehicles, roadside units, the V2X control function, and V2X application servers. It also receives unicast and multicast messages and forwards these to the ADAS and head unit display. The V2X UE supports both the PC5 and Uu interfaces. The resources used for PC5 communication are provisioned by the V2X control entity of the PLMN for mode 3 operation or are preconfigured in the V2X module for mode 4 operation. There are two communication configurations:

- V2X communication over PC5 and eMBMS reception on Uu interface: the V2X module uses the PC5 interface for sending and receiving messages, and uses the eMBMS reception only to receive broadcast V2X messages. Traffic-related messages relevant to certain geographical regions are transmitted to the corresponding eMBMS service areas. Using eMBMS services is advantageous also in locations with a high density of V2X UEs, where the signal-to-noise ratio (SNR) of the PC5 link is degraded and the eMBMS link offers a viable option to transmit the information to a large number of UEs.
- V2X communication over PC5 and non-eMBMS-based transmission on Uu interface: in this case the V2X messages can also be transmitted to the V2X server by the UE. The application server then may transmit this information over the broadcast eMBMS channels.

The RSU is a stationary V2X UE that supports both the PC5 interface for communication between vehicles and Uu interface without eMBMS for connecting to the LTE core. This module can also extend the LTE coverage for V2X functionality along roadways. It connects to the vehicles over the PC5 interface and, depending on the relevance of the message, it may forward the message to the V2X application server, acting as a relay for V2X UEs that are not in range of the server. The RSU may also support other applications such as toll collection and other payment applications relevant to some road stretches.

The V2X control function is typically part of the PLMN for supporting V2X operation and manages the V2X UEs in that PLMN. It provisions the necessary radio resources for V2X communication through the MME for scheduling data over the Uu interface. It also is responsible for the allocation of the radio resources for the operation of the PC5 interface for the authorization of the UE and for ensuring the necessary QoS parameters.

The V2X application server hosts a variety of applications offered for V2X services in the network. There may be multiple servers present in the network depending on the region they serve and what services they offer. The application server takes care of receiving and transmitting information to and from the V2X UEs over unicast channels or eMBMS. It is responsible for determining the geographical scope of the received information and forwarding the message to the eMBMS service areas or the appropriate RSU. They also connect to other servers and existing traffic management systems to exchange and distribute information. The V2X application server supports most of the applications developed for DSRC and C-ITS as well as proprietary applications developed by vehicle manufacturers or toll management systems. There are numerous applications supported, which can be broadly divided into the following categories:

- Safety-related applications: these services include high-priority and low-latency messages related to safety applications such as queue warning systems, curve speed warning, and bad weather warning.
- Traffic management applications: these are services such as rerouting services, congestion reduction, toll payment services, and similar services to improve driving conditions.
- Auxiliary services: these services typically require higher bandwidth and are used to enhance user experience. Example of such applications are infotainment services, and mapping services.

4.4.5 V2X Protocol

The 3GPP V2X protocol makes use of some of the layers that were developed previously as part of DSRC and C-ITS. The V2X communication stack is a hybrid stack where the upper layers consist of DSRC and/or C-ITS components and the access layer is taken over by the ProSe stack and LTE stack components. Some of the layers have been adapted to the LTE environment.

Figure 4.10 depicts the V2X stack, where the physical layer and access control is based on the LTE and ProSe stacks. The LTE stack comprises the legacy PHY, MAC, RLC, and PDCP layers, while the ProSe stack manages the connection on the PC5 interface for V2X message transmission. The ProSe stack is configured by the V2X control function in the PLMN with security parameters, multicast IP addresses, and radio resource parameters to be used during operating mode 4 (out of coverage). For V2X operation, the ProSe is defined in 3GPP TS 23.303 [9]. It has minor modifications with respect to D2D, for example, procedures such as ProSe Direct Discovery are not supported, only multicast transmissions are supported.

The upper layers are beyond the scope of 3GPP and they depend on the deployment region. In the United States, the upper layers are based on the DSRC protocol and in the European Union they are based on the C-ITS protocol. Both these stacks have similar layers and functionality. The DSRC architecture is based

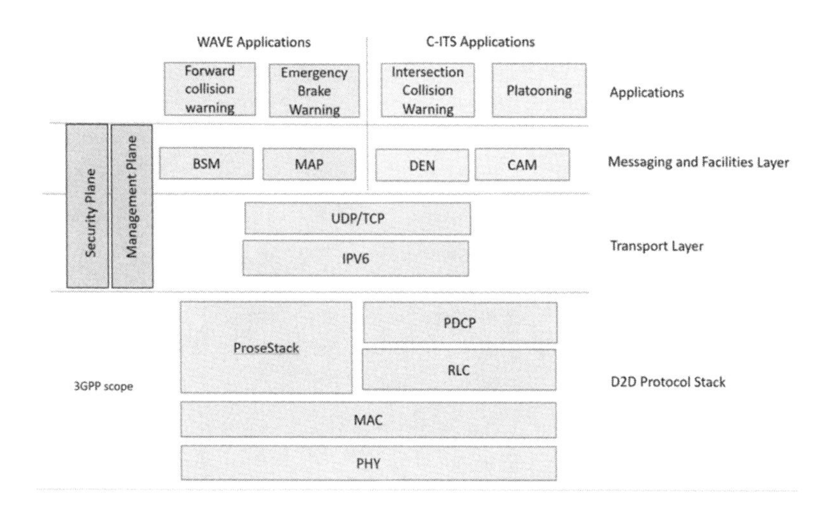

Figure 4.10 V2X protocol stack.

on the IEEE WAVE standard 1609.0-2013 [10] and the C-ITS is based on the ETSI EN 302 665 specification [11]. They consist of the following layers:

- Transport layer: based on IPV6, UDP and TCP protocols, the transport layer is used to establish the IP connection to the V2X application servers and other V2X clients. The V2X protocol stack supports only IPV6; IPV4 is not supported. UDP is used for port-based addressing, applying check sums to support best effort delivery of messages. TCP is used for acknowledged message transmissions.

- Messaging and facility layers: they represent the messaging layer to support the V2X applications on the layer above. Their main functionality is the maintenance of the message formats, message dictionaries, and data elements used for V2X communication. These layers are responsible, among other things, for the session management, including connection establishment, maintaining the session and addressing; message management (i.e., the generation of the messages); mapping information, combining static map information with dynamic information generated from V2X communications (e.g., traffic conditions or weather information), and position and time support. These layers provide the high accuracy positioning information essential for correct functioning of the V2X system.

- Management plane: consists of a collection of services to support the communication functions provided by the application and the messaging and facilities layers. However, it does not get directly involved in processing the application data. The services provided include time synchronization for channel coordination and processing service requests and advertisements, congestion control, and layer management.
- Security plane: this layer is responsible of ensuring the reliability and resilience of the V2X system. Among other functions, it takes care of the confidentiality, authentication, integrity, and privacy of the connection. It also validates the relevance of the information and prevents the reception of duplicated messages.
- Application layer: this layer consists of multiple entities, each serving a specific purpose. It makes use of the messaging and facilities layer to connect with peer entities. Some of the applications are specified under the DSRC and C-ITS initiatives, but there are also proprietary applications. They can be broadly classified into three categories: safety applications (e.g., collision warning or early braking warning), traffic management applications (e.g., congestion alerts or traffic light information), and other applications (e.g., eco-friendly driving or infotainment).

This section has described the principles of the V2X in LTE. In Chapter 5, the V2X enhancements in 5G will be explained. V2X aims at improving the automotive world and is seen as a key enabler for autonomous driving. Thus, high-accuracy positioning methods as the ones described later in this book are a mandatory feature for the success of V2X.

4.5 INTERNET OF THINGS

Wireless devices have found applications in machine communication and connecting systems since the early days of GSM. A wide range of applications benefit from such connections. One of the most popular early applications consisted of tracking devices installed on trucks and vehicles for logistical purposes. Wireless networks offer cheap connections over large areas with large-scale installation available worldwide due to the success of mobile phones. These early applications just reused the underlying wireless technology without significant modifications to support machine type communications.

The IoT is a vision where household devices such as fridges, washing machines, street lights, heating systems, or metering systems can be connected and controlled remotely. Industrial applications using drones delivery systems, security cameras, or robots have increasingly adopted wireless networks to connect to one another. Many applications that require point-to-point connection between devices such as headsets, health trackers, or smart watches benefit from IoT connections. The expectation is that the number of wireless connections allocated to devices will far exceed smartphones in the near future. This can be attributed to the development of low-cost technologies, longer battery life, and an explosion in consumer devices using machine-to-machine (M2M) communications. This section will provide an overview of different application areas for IoT devices and the different types of IoT technologies with the focus on the 3GPP standardized low power wide area (LPWA) technologies in combination with positioning.

4.5.1 IoT Applications

The applications in this area are numerous and developing at a very fast pace. Some examples of applications for IoT devices are the following:

- Smart home applications that connect various appliances such as smart fridges, alarm systems, air conditioning systems, or smart lighting.
- Wearables and sports devices, aimed at fitness and health tracking applications such as a heart rate monitor, fitness tracker, or navigation.
- Health care devices such as sugar-level tracker for patients with diabetes, or patient monitoring systems.
- Smart city services, such as waste management systems, traffic management, or lighting systems.
- Industrial applications such as inventory tracking and control, or drone-based delivery.
- Farming applications such as livestock management or irrigation control.

While some of these applications do not seem to need positioning features (e.g., the smart home applications), others clearly need location estimation capabilities, as for instance navigation applications using a wearable or livestock tracking.

4.5.2 Comparison of the Different IoT Technologies

Many of the above-listed applications have varied needs regarding the underlying wireless technology in terms of bandwidth and range of operation. Hence, multiple

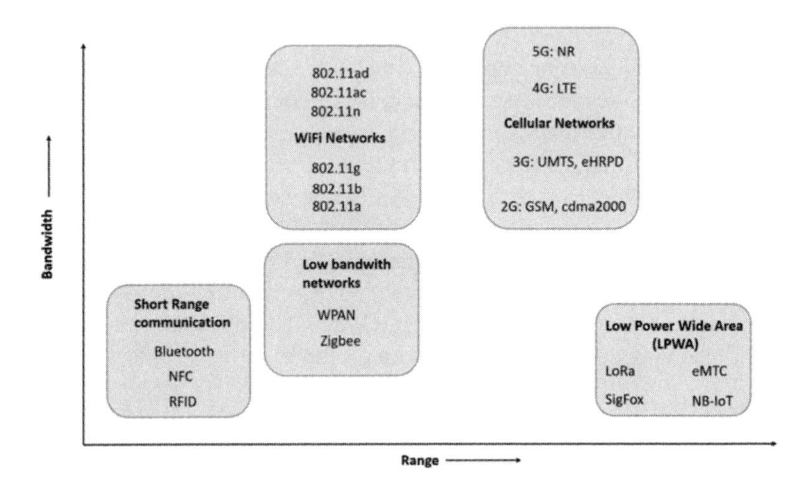

Figure 4.11 Comparison of the different IoT technologies.

IoT technologies have arisen in order to support them. Figure 4.11 details these different configurations.

Short-range communications, such as Bluetooth, near-field communications (NFC), or RFID, typically have a range of a few meters to connect devices close to each other. Longer range is supported by Wi-Fi networks and also other low-bandwidth alternatives such as Zigbee. Traditional cellular networks, such as GSM or LTE, can also be used for IoT. However, due to the high bandwidth of cellular network technologies, this is typically not efficient for most IoT use cases. Finally, LPWA technologies are designed specifically to target IoT applications with large range and low bandwidth.

Some LPWA technologies are based on non-3GPP standards, such as Long Range (LoRa) or SigFox. However, this book will focus on the 3GPP-based standards [12, 13], that is MTC and its evolution, enhanced machine type communications (eMTC) and further enhanced machine type communications (feMTC); NB-IoT, and briefly also Extended Coverage GSM IoT (EC-GSM-IoT). For many years, 3GPP focused their efforts on addressing UE with higher data rates and increased complexity. However, as of Release 12, they also started to consider simplified UEs with low data rates and bandwidths and typically only one receiver antenna. These types of UEs laid the foundations for the IoT devices, low-cost devices with up to 10 years of battery life expectation. As of Release 13, the 3GPP came up with three major technologies to address the Cellular IoT requirements of efficient transmission of low data quantities, increased coverage needs, support of high density of devices per cell, low network impact per device, and location services support. These three technologies are:

- eMTC shares the LTE in-band resources for operation and has a maximum bandwidth on 1.4 MHz in Release 13, increased to up to 5 MHz as of Release 14. It can operate in half-duplex and full-duplex modes. eMTC devices have much lower costs than regular smartphones but they are still able to support VoLTE and E911 applications. These devices support positioning and mobility procedures and are typically used for high-end IoT applications. eMTC devices support two coverage modes, coverage enhancement (CE) Mode A and B. CE Mode A has a coverage similar to a normal LTE device, while CE Mode B has been designed to enhance the coverage by approximately 15 dB compared to normal LTE operation. eMTC supports positioning technologies such as GNSS and OTDOA.
- NB-IoT uses a new radio access technology supporting only 180-kHz bandwidth, which can be integrated into the LTE bands in in-band mode, in the guard-band or as stand-alone deployments. It is typically used in low-end IoT applications with very low data rates, very long battery life requirements, and high density of devices per cell. Release-13 NB-IOT has not been designed to support positioning. However, Release-14 NB-IOT added OTDOA support based on a new narrowband positioning reference signal (NPRS).
- EC-GSM-IoT is based on eGPRS. The primary motivation to develop this technology was to reutilize the extensive worldwide GSM availability. The optimization needed to support EC-GSM-IoT in existing GSM networks is minimal and offers cost-effective upgrades to existing GSM networks. EC-GSM-IoT and NB-IOT have similar features and support similar applications. However, this book will focus on the LTE and 5G-based applications, and will not discuss EC-GSM-IoT further.

In Table 4.2, the characteristics of eMTC and NB-IOT are compared. For positioning, eMTC supports the full LPP, including ECID, OTDOA, GNSS and other RAT-independent methods, explained in Chapters 6 to 9 of this book. In comparison, NB-IOT only added positioning support in Release 14. ECID is only partially supported, as measurements can only be done in Idle mode. NB-IOT devices are meant to be low cost and most will not have a GNSS receiver. Thus, NB-IOT devices rely mostly on OTDOA for positioning.

Table 4.2
Comparison between NB-IOT and eMTC

	eMTC	**NB-IOT**
Bandwidth	Rel-13: 1.4 MHz Rel-14: Up to 5 MHz	180 kHz
Coverage	155 dB	164 dB
Peak rate	1 Mbps	25 Kbps (UL) 60 Kbps (DL)
Duplex mode	FDD, HD-FDD, TDD	HD-FDD
Voice	VoLTE	No support
Positioning	Full LPP support GNSS, OTDOA, ECID	Rel-13: no support Rel-14: OTDOA, GNSS

4.5.3 C-IoT Network Architecture

The IoT network works on a modified LTE network architecture. Typically, this network consists of a simplified form of the same LTE components tailored for IoT operation.

Figure 4.12 shows the architecture of an IoT network. The connection to the network is done through eNBs for eMTC devices and C-IOT base stations for NB-IOT devices. Comparing the design to a traditional LTE network, the IoT network offers a lightweight core network consisting of entities such as MME, S-GW and PDN gateway (PGW). These components are simplified and tailored to support IoT features such as:

- Reduced signaling procedures.
- Efficient procedures to transfer small data packets.
- Simplified USIM and security procedures.
- Simplified non-access stratum (NAS) and session management (SM) procedures.
- Subscription and charging services efficiently for IoT devices.
- Paging optimization for wide coverage area support.

The IoT functionality is part of one entity called the C-IOT serving gateway node (C-SGN). There are additional changes added in the data transmission architecture to support IoT devices, which are important in understanding the C-IoT architecture:

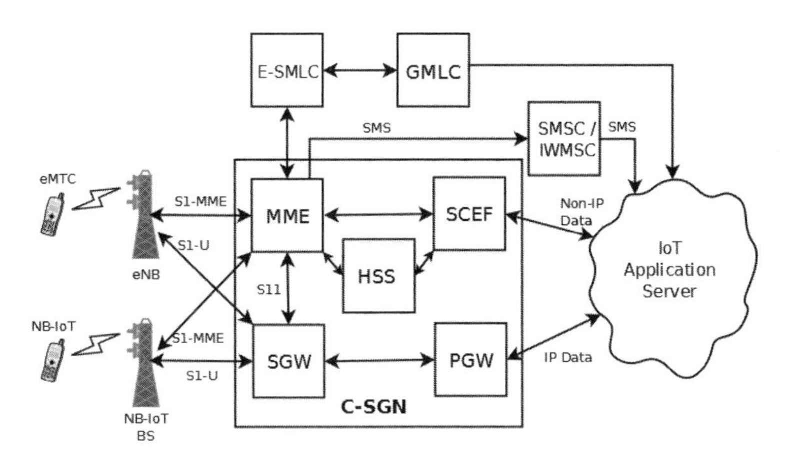

Figure 4.12 C-IoT network architecture.

Control Plane IP: uses data-over-NAS functionality to piggyback IP packets over the NAS signaling messages, thereby reducing the overhead of a dedicated IP bearer. The C-IoT RAN (eNB, C-BS) transfers the data over NAS on the S1-MME interface. The MME then transfers the IP data to the SGW over the modified S11 interface. High-end IoT devices requiring larger data transfers continue to use the S1-U interface to establish a dedicated bearer for IP data.

Non-IP data: C-IoT supports transfer of non-IP data between the IoT device and the application server, using a new component called the service capability exposure function (SCEF). The SCEF provides a set of interfaces that connect the services of the C-SGN with external application servers. It connects to the MME and HSS for IoT device and connection management purposes.

Data over SMS: one enhancement to the C-IoT network architecture is the support of SMS over PS networks, where the data is transferred by the MME to the SMSC/IWMSC and to the application server. Similar to the above two methods, data over SMS also uses the data over NAS feature for the SMS transfer in the RAN.

Much of the C-IoT deployment has been done by upgrading the LTE network installations to support IoT devices. Nonetheless, the C-IoT core network is a separate entity with dedicated elements, although still reusing some components from the LTE network. This allows the IoT network to benefit from cost reduction due to synergies with the LTE network, without deploying all the complexity required for regular LTE.

4.5.4 Positioning on IoT Devices

The LCS network architecture is similar to the LTE architecture explained in Chapter 3, where the E-SMLC connects through the MME using the LPP protocol. The location session is triggered by the application server though the GMLC.

The need for positioning for IoT devices is a very important aspect as a large number of IoT applications such as tracking devices, livestock management devices, bike sharing, and wearables need location information for functioning. LBS services are a very important enabler for large-scale deployment of such devices. However, there are many challenges when implementing LBS services for IoT devices due to cost and battery life considerations. The use of GNSS technology for positioning is not very optimal for many categories of IoT devices because GNSS always needs an extra receiver, which increases the cost of devices and results in higher battery consumption. Additionally, the challenging environments for IoT devices (e.g., indoors, basements, or garages) make the reception of GNSS very difficult.

One of the most promising technologies for position determination for IoT devices is OTDOA, because there is no need for an extra receiver and signal power is relatively high compared to GNSS. However, the narrow bandwidths used for eMTC and NB-IOT reduce the ranging and multipath resolution capability of OTDOA. Additionally, the need for such devices to work at much lower SNR environments compared to traditional mobile phones led 3GPP to define OTDOA enhancements to address IoT applications in Release 14. These enhancements are explained in Chapter 14 of this book.

4.6 CONCLUSION

This chapter has focused on the commercial applications of location-based services for LTE, as opposed to the regulatory and emergency applications discussed in the previous chapter. Setting aside traditional commercial use cases such as navigation applications or advertising based on location, this chapter has explained in detail two very important technologies for the future of LBS. Both C-V2X and C-IoT deployments will continue to grow in the future, setting positioning requirements that often are more challenging than the regulatory E911 and E112 requirements.

In the next chapter, it will be seen how, with the emergence of 5G, the automotive and IoT worlds will still play a determining role in positioning requirements.

Furthermore, 5G will add additional challenges such as UAV or health care applications (eHealth) such as patient tracking.

References

[1] SAE J3016_201401, *Taxonomy and Definitions for Terms Related to On-Road Motor Vehicle Automated Driving Systems*, January, 2014.

[2] U.S. Department of Transportation, *IEEE 1609 - Family of Standards for Wireless Access in Vehicular Environments (WAVE)*, 2009.

[3] IEEE 802.11p-2010, *IEEE Standard for Information technology– Local and Metropolitan Area Networks– Specific Requirements– Part 11: Wireless LAN MAC and PHY Specifications Amendment 6: Wireless Access in Vehicular Environments*, 2010.

[4] Federal Communication Commission, Report No ET 99-5 *FCC Allocates Spectrum in 5.9 GHz Range for Intelligent Transportation Uses*, October, 1999.

[5] European Commission, *C-ITS Platform, Final Report*, January, 2016.

[6] 3GPP TR 22.885, *Study on LTE Support for Vehicle-to-Everything (V2X) Services*, V14.0.0, December, 2015.

[7] 3GPP TR 36.885, *Study on LTE-Based V2X Services*, V14.0.0, June, 2016.

[8] European Commission, COM/2016/0766 final, *A European Strategy on Cooperative Intelligent Transport Systems, a Milestone Towards Cooperative, Connected and Automated Mobility*, 2016.

[9] 3GPP TR 23.303, *Proximity-Based Services (ProSe); Stage 2*, V15.1.0, June, 2018.

[10] IEEE WAVE 1609.0-2013, *IEEE Guide for Wireless Access in Vehicular Environments (WAVE) - Architecture*, December, 2013.

[11] ETSI EN 302 665, *Intelligent Transport Systems (ITS); Communications Architecture*, September, 2010.

[12] 3GPP TR 23.720, *Study on Architecture Enhancements for Cellular Internet of Things (CIoT)*, V13.0.0, March, 2016.

[13] 3GPP TS 36.300, *E-UTRA and E-UTRAN; Overall Description; Stage 2*, V15.4.0, January, 2019.

Chapter 5

The Evolution of LBS for 5G

5.1 INTRODUCTION

Chapter 3 introduced the positioning requirements for emergency services and the different regulations in regions such as the United States and Europe. Chapter 4 focused on the location-based services and commercial applications that developed with LTE and the IoT world. Furthermore, it presented the safety-critical positioning aspects needed for new fields such as V2X communication, one of the key technologies to enable autonomous driving.

The advent of 5G NR came with the promise of much higher bandwidths, very low latency, and very large capacities. This chapter gives an insight on how these milestones can be achieved and the main changes coming with 5G NR. It will also take a look at the 5G NR network architecture and deployment scenarios and how they impact the location session, providing as well a brief description of the access network (AN) and CN components, concepts that will be further clarified in later chapters of the book.

Regarding the positioning use cases, 5G NR must fulfill the same regulatory requirements for emergency services as LTE and previous cellular networks. Thus, the 5G NR emergency call architecture, the interactions with the IMS network and the different fallback techniques for emergency call routing are one of the key aspects handled in this chapter. Nevertheless, 5G NR also brings a plethora of new commercial use cases for location-based services, motivated by the increasing number of connected devices (massive IoT) and the advances in the autonomous driving area, with further improvements of V2X. New fields for positioning also include UAVs, eHealth or wearables. This chapter will introduce these use cases and their corresponding positioning requirements.

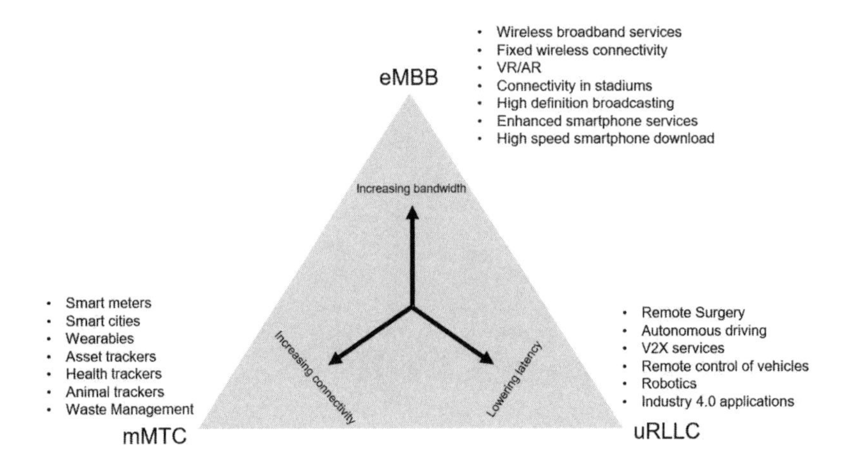

Figure 5.1 Use cases of the 5G NR network.

5.2 THE 5G SYSTEM

5.2.1 Motivation for 5G

Each of the cellular network generations was motivated by specific goals. The primary goal of 2G systems was to provide large-scale circuit switched voice connectivity. 3G systems introduced data on wireless devices along with circuit switched connectivity. 4G systems moved to fully packet switched networks with higher data rates, complemented later by additional technologies such as IoT or the sidelink communication for V2X. The requirements for 5G technology have been driven by the ITU in the International Mobile Telecommunication 2020 (IMT-2020) [1]. The IMT-2020 requirements crystallized in one of the key aspects of 5G, eMBB.

This aspect of 5G is the traditional cellular network business and the one that requires higher data rates, higher bandwidths and spectral efficiency, and the primary goal of the 5G NR specifications released as part of the 3GPP Release 15. However, the industry view on 5G NR compounded two other main use cases as well: massive IoT, also known as mMTC, and URLLC. These together form the typical triangle of 5G NR use cases, which is shown in Figure 5.1. As a result of combining all use cases, some of the main requirements are summarized in Table 5.1.

Table 5.1
5G System Requirements

Parameter	Target Requirement
Peak throughput	DL: 20 Gbps, UL: 10 Gbps
User experience throughput	DL: 100 Mbps, UL: 50 Mbps
Area traffic capacity	10 Mbps/m^2
Latency	eMBB: 4 ms, URLLC: 1 ms
Connection density	1 million devices per km^2
Mobility interruption time	0 ms
Minimum BW supported	100 MHz

Each of the use cases has its own specific requirements:

- **eMBB** aims to provide extremely high data rate to each user and high network device density for large-scale mobile broadband services. The goal is to enhance the user experience and increase service availability. The eMBB focuses as well on high data rate services to crowded places such as stadiums, shopping malls, and other public places. Furthermore, it attempts to provide an alternative to traditional internet connections for homes and enterprises, high-speed fixed wireless access (FWA) connections, private networks for companies, broadcast of high-definition television, and so forth. Such services are expected to require a peak throughput of up to 20 Gbps.
- **mMTC** has a primary goal to connect a very high number of different devices, also known as the massive IoT world. Applications for mMTC are very different in nature: smart home applications, such as remote-controlled heating or wireless security cameras; smart city applications, such as traffic control or smart parking; wearables and eHealth, or multiple industry verticals part of the Industry 4.0, such as warehouse organization. Typical mMTC applications may not require large bandwidths or high data rates, but benefit from enhanced battery life performance and low cost. Furthermore, they demand a network capable of supporting high device density, with up to a million devices per km^2.
- **URLLC**: aims to provide reliable, stable communication with very low latency. Applications for URLLC include V2X and autonomous driving, mission-critical applications, industrial automation, and remote control of medical procedures. Similar to mMTC, many of these applications do not benefit from large bandwidths and high data rates. Being mostly critical applications for safety-of-life

(SoL), aspects like battery saving and low cost are also secondary. Instead, availability, reliability, real-time communication, and robustness are key requirements for URLLC.

The heterogeneous nature of the requirements for the different 5G NR applications clearly shows the need for a radical change in the 5G NR network with respect to legacy cellular networks, such as LTE or WCDMA. Concepts like network slicing, cloud computing, virtualization, or high-speed fiber-based backbone play a very relevant role in achieving the goals of the 5G NR network.

5.2.2 Standardization Plan

Given the complexity of 5G and the aggressive timeline imposed by the IMT-2020 requirements, the standardization bodies, in particular the 3GPP, opted for a phased approach when defining the 5G NR specifications. The first version of the 5G NR network has been included in the 3GPP Release 15, and focused mainly on the necessary improvements to meet the IMT-2020 requirements and the eMBB use case. The RAN updates were prioritized over the CN updates, as the different 5G deployment scenarios offered the possibility to connect 5G base stations, known as next generation Node-B (gNB), to the LTE core network, the evolved packet core (EPC). The basic CN functionality of the 5G Core (5GC) was also specified in Release 15, but setting aside anything nonessential. At the time of the writing, 3GPP is about to conclude the Release 16, which builds upon the Release 15 and includes further updates to support the other two 5G NR use cases (mMTC and URLLC). Release 17 is already in the planning phase and the initial studies are started.

Table 5.2

5G Release Planning

Release 15	**Release 16**
NSA and SA operation	Industrial IoT
eLTE updates	URLLC enhancements
Service-based architecture	V2X phase 3
Network slicing: e2e logical networks	Commercial positioning requirements
V2X phase 2	RAT-dependent positioning
Regulatory positioning requirements based on RAT-independent technologies	

Table 5.2 gives a very high-level overview of what features are included in Release 15 and Release 16. There are of course many other features in both

releases, but Table 5.2 captures only those that may have an impact on location-based services.

5.2.3 5G Frequency Spectrum

One of the main novelties of 5G NR is the usage of the millimeter-wave (mm-wave) frequency spectrum, unprecedented so far in cellular network technologies. The 5G NR spectrum has been divided in two ranges:

- Frequency range 1 (FR1) refers to the low frequencies, comprising the traditional cellular network spectrum and increasing it up to 7.125 GHz. It is sometimes referred to as sub-6, because the initial 3GPP definition considered only frequencies up to 6 GHz, but it has been extended. Many of the frequency bands in the FR1 reuse some of the legacy 2G, 3G, and 4G frequencies. The available bandwidth is limited in this range, but offers many advantages in terms of signal range and network coverage.
- Frequency range 2 (FR2) refers to the higher frequencies, starting at 24 GHz and with frequency bands defined up to 40 GHz in Release 15. This range is also called mm-wave, due to the wavelength of the radio waves at these frequencies (although strictly speaking, 28-GHz signals are still in the centimeter wavelength range). The upper bound of the range is open and is likely to be extended with higher frequency bands in the near future, since at the time of the writing there are already auctions for frequencies around 50 GHz. The available spectrum in this range offers very large bandwidths and is necessary to achieve the very demanding data rates required by 5G. This range is particularly challenging in terms of signal propagation. The link losses at these frequencies are much higher than at the lower frequency bands, being the line of sight (LoS) attenuation between 16 and 24 dB higher than for FR1. They are also more susceptible to vegetation, rain, and attenuation due to oxygen, and the presence of any obstruction causes severe signal quality degradation.

At the time of the writing, the 3GPP has also started a study on the viability of using the range between 7.125 GHz and 24 GHz for 5G NR. However, whether this range can be used or not and its possible naming has not yet been concluded. The frequency ranges and the corresponding NR frequency bands are defined in [2, 3].

Due to the very high propagation losses at FR2, additional mechanisms are needed to improve the link budget and enable this frequency range for 5G cellular networks. In legacy systems, the base stations used primarily 120°wide beams

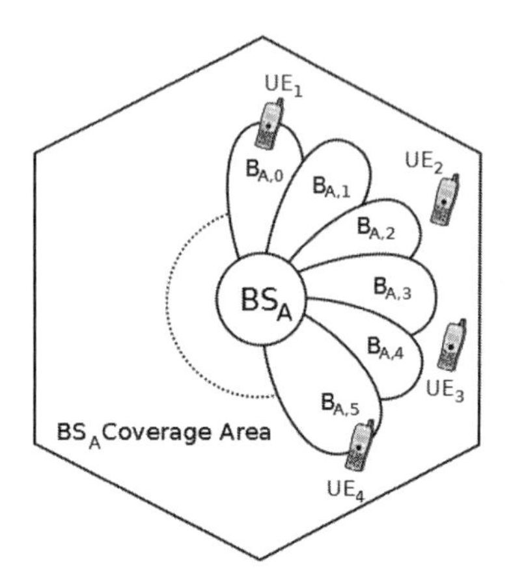

Figure 5.2 Example of a beamforming deployment.

covering large areas of the cell. 5G NR base stations rely on massive MIMO (large antenna arrays) and beamforming, using a larger number of narrow beams to cover the same spatial area as in legacy networks, as shown in Figure 5.2. Narrower beams have the advantage that they concentrate the transmitted power into a smaller region, improving the link budget.

The same concept is applied as well on the mobile device side. LTE devices were expected to have isotropic antennas. On the other hand, 5G devices, especially devices supporting FR2, will make use of multiple beams to receive and transmit signals in different directions in order to achieve a higher antenna gain. While base station antennas are rather likely to produce static beams, it is possible that mobile devices could use antennas that can be tuned to steer the beam in different predefined directions in order to perform a scan or sweep of the space.

Beamforming and massive MIMO open up new possibilities for positioning, since they enable angle-based techniques as seen in Chapter 2 of this book.

5.2.4 5G Network Deployment Scenarios

As mentioned earlier, the 3GPP defined multiple deployment scenarios for 5G NR [4] in order to allow a phased upgrade from the existing LTE / EPC networks. These scenarios are based on the decoupling between the RAN and the CN. The network deployment scenarios allow multiple combinations, including the next generation radio access network (NG-RAN) connected to the EPC network. Conversely, eLTE

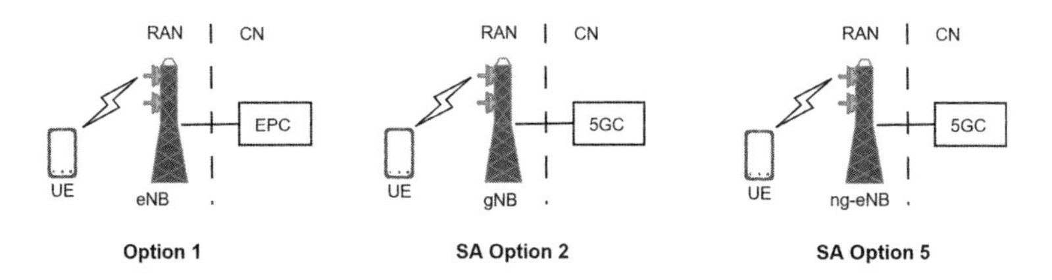

Figure 5.3 Standalone deployment options.

(LTE Release 15 onward) allows the LTE RAN to connect to the 5G core (5GC), which is also called 5GCN or next generation core (NGC), depending on the source. There are two main deployment types for combining 5G NR and LTE:

- Standalone (SA): as the name implies, in SA there is only one RAN, which can be NG-RAN or LTE, connected to the CN, which can also be either EPC or 5GC.
- Non-standalone (NSA): in this case, both RANs are available and connected in a dual connectivity (DC) fashion to either one of the CNs.

The NSA deployments have the advantage for the network operators to start deploying their NG-RAN on top on their existing LTE network without replacing the core. Thus, it allows a faster deployment compared to SA. On the other hand, the new 5G gNBs will be connected to the in many cases already-overloaded EPC network and consuming additional resources. Furthermore, many of the advantages in 5G come from key developments in the core network such as network slicing and core virtualization.

5.2.4.1 SA Deployment Options

There are three types of SA deployments depending on the combination of possible RAN and CN, as shown in Figure 5.3.

- Option 1: this is the legacy option, where the eNB is directly connected to the EPC network. This option does not include any 5G NR deployment and is often not listed in most of the sources when talking about the 3GPP deployment options.
- Option 2: this the option where both RAN and CN belong to the 5G NR system. It is, together with Option 3, one of the most popular deployment options among network operators. However, this option does not support legacy devices, since the UE must also be a 5G UE.

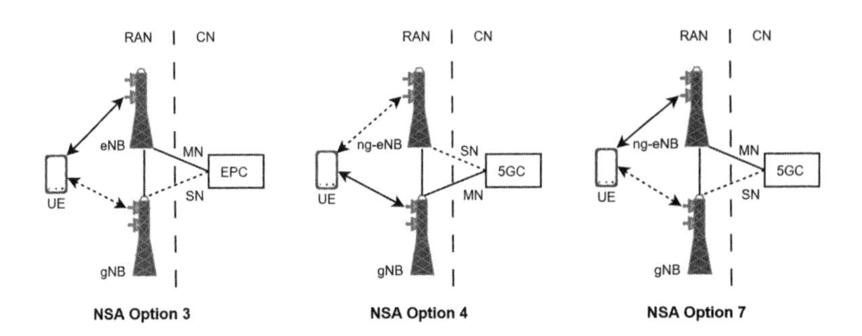

Figure 5.4 Non-standalone deployment options.

- Option 5: in this option, the network has already deployed 5GC, and the legacy LTE RAN has been upgraded to eLTE with next generation enhanced Node-B (ng-eNB) in order to be connected to the CN. The mobile device is connected to the RAN by LTE. However, the device must still be compatible with the 5GC.

5.2.4.2 NSA Deployment Options

In NSA deployments, both RANs coexist and the mobile device can be connected to both using DC. Thus, there is a master node (MN), which is the base station of the RAN to which the UE does the initial connection, and a secondary node (SN), which is added via DC. The 3GPP has defined three main options for NSA deployments, as shown in Figure 5.4:

- Option 3: the eNB acts as the master radio link and the gNB as the secondary radio link. Both RAN networks are connected to the EPC and the 5GC is nonexistent in this configuration. This type of deployment has been popular among network operators in order to speed up their 5G plans.
- Option 4: this deployment is the opposite of Option 3. The gNB acts as the MN and the ng-eNB is the SN. Both RANs are connected to the 5GC.
- Option 7: the connection between the UE and RAN is almost identical to Option 3, with the only difference that the LTE part has been upgraded to eLTE. However, the core network has been upgraded to 5GC.

The MN of the dual connectivity scenario is responsible for the connection establishment and for the RRC signaling using the signaling radio bearer (SRB).

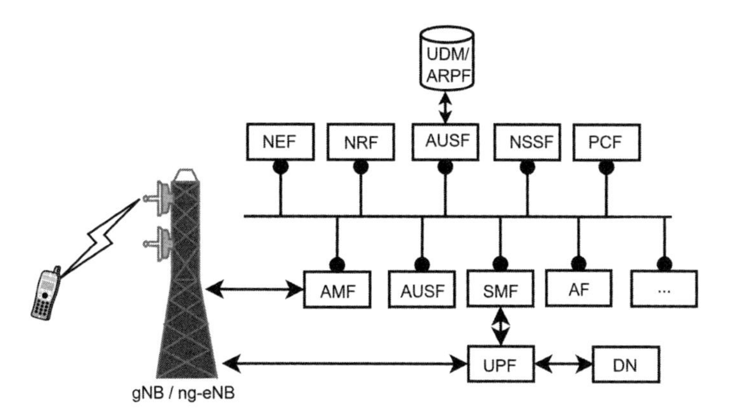

Figure 5.5 Architecture of the 5G NR network.

The routing of the SN RRC signaling is controlled by the MN. The routing of data transmissions follows a similar approach.

5.3 5G NR NETWORK ARCHITECTURE

The 5G NR network is, as in LTE, divided in a CN, called 5GC, and a RAN, called NG-RAN. 5G NR incorporated several new concepts, such as cloud computing, integrated access and backhaul (IAB), and edge computing, which enabled a major upgrade compared to previous RATs. Some of the major design considerations in the 5G NR architecture are the following [6]:

- Service-based architecture (SBA): each element of the network is defined by a set of services accessible through network functions (NFs). Each NF consists of an implementation and a service-based interface (SBI), which defines the way the NF communicates with other NFs. As represented by the single line in Figure 5.5, every two NFs can connect to each other without additional functions in between.

- Network virtualization. The NFs are not bound to a specific hardware. Instead, they are implemented completely in software and run on general-purpose processors. Thus, the NFs can be located centrally or distributed depending on the application needs. Virtualization simplifies network maintenance and allows rapid deployment of new services or scaling up the network capacity.

- Control and user plane separation (CUPS) can be used to centralize the C-Plane functionality while distributing the U-Plane functionality to the edge nodes for reduced latency. Data connections are not routed to a central hub. Instead, they are routed directly to the user, allowing an optimal use of the network resources.
- Network slicing divides the network in multiple parallel networks (slices) tailored for specific QoS needs.

The main 5G NRs NFs are shown in Figure 5.5 in a reference point representation. This book will only focus on those functions that can be relevant for positioning. For more detailed information on the rest of the functions, please refer to [7,8].

- The access and mobility function (AMF) is the 5G equivalent to the MME in LTE and serves, together with the gNB, as an entry point to the network for the user equipment (UE). The tasks of the AMF include connection management, mobility, and NAS signaling. See TS 29.518 [9] for further details.
- The session management function (SMF) is responsible for setting up data sessions and managing QoS flows. It covers part of the functionality managed by the MME, S-GW, and PGW in LTE. A UE can be connected to one SMF per network slice.
- As its name indicates, the user plane function (UPF) implements the functionality related to data handling. It is responsible for data packet routing and mapping of the protocol data units (PDUs) to the corresponding QoS flow. The UPF is located close to the network edge to enable low latency applications.
- The gNB is an evolution of the enhanced Node-B (eNB) in LTE and is responsible for maintaining the radio frequency (RF) link with the mobile and managing the radio resources for the link. The gNB tasks include connection establishment, resource scheduling, and mobility management.

Figure 5.6 shows the interconnection between the different layers of the UE and the 5G NR network. For most of the layers, the UE connects primarily to the gNB. This is true for Service Data Adaptation Protocol (SDAP), RRC, Packet Data Convergence Protocol (PDCP), radio link control (RLC), medium access control (MAC), and physical layer (PHY). The routing of the positioning messages, which will be further explained in Chapter 15 of this book, is different for C-Plane and U-Plane. In case of C-Plane positioning, the LPP messages are embedded in the NAS signaling; that is, between the UE and the AMF. NAS messages are encapsulated in

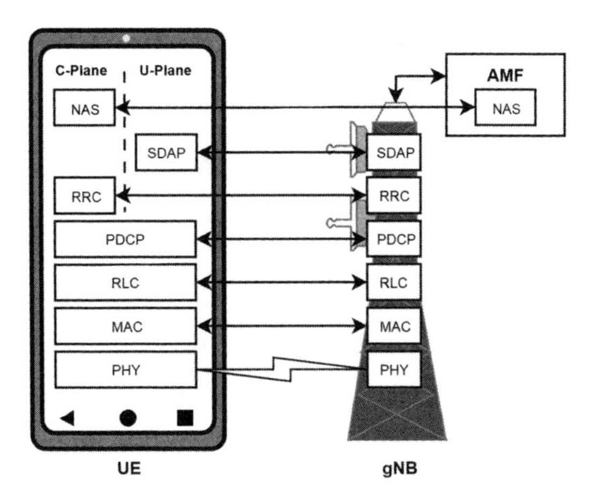

Figure 5.6 UE and gNB protocol stack.

the RRC signaling. For U-Plane, the SUPL messages are transmitted via the SDAP layer, as will be seen in the next section.

5.3.1 U-Plane Architecture

The U-plane is primarily responsible for sending and receiving data using an Internet Protocol (IP) connection. There may be one or more IP addresses assigned to the UE for different services, such as internet access or emergency calling. For the 5G NR system, the 3GPP has introduced new concepts that directly impact U-plane functionality.

5.3.1.1 PDU Session and QoS Flows

A PDU session refers to an end-to-end data session associated with an IP address. Each PDU session consists of one or more QoS flows (i.e., logical links carrying data packets with the same QoS needs). A QoS flow is uniquely identified by the QoS flow identifier (QFI).

Figure 5.7 shows an example of a PDU session with three QoS flows, carrying data between the UE, the gNB, and the UPF. A dedicated radio bearer (DRB) carries the QoS flows on the radio interface (between the UE and the gNB). Within 5GC, the QoS flow is carried by the GPRS Tunnelling Protocol (GTP).

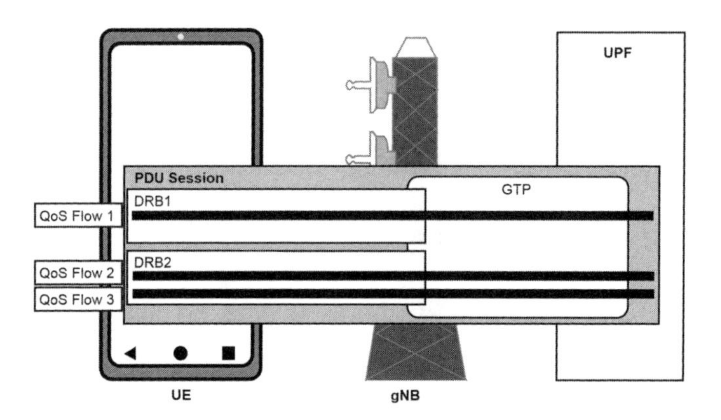

Figure 5.7 PDU session and QoS flows.

The U-Plane functionality in 5G NR is managed by a new layer, the SDAP, which resides above the PDCP. Each PDU session is associated with one SDAP entity, which manages all the QoS flows linked to that session.

5.3.1.2 Network Slicing

As it has been seen earlier in this chapter, each of the 5G NR use cases (eMBB, mMTC, and URLLC) has its own specific requirements in terms of latency, capacity, reliability, and security. In order to offer the best possible service tailored to each specific set of requirements, 5G NR has introduced the concept of network slicing. Network slicing refers to the partitioning of physical network resources into groups of logical network components optimized for the different service types. Each of these groups of logical components is considered a slice and is identified by a network slice instance identifier (NSI ID).

A network slice spans over multiple domains, such as the RAN and the CN. The 3GPP introduced network slicing for the 5GC in Release 15 and enhanced its functionality in Release 16. During the Release-17 time frame (starting at the beginning of 2020 until mid-2021), 3GPP plans to also add network slicing in the NG-RAN. The Global System for Mobile Communications Association (GSMA), a worldwide association of network operators, has provided a summary of the network slicing use cases and their respective requirements [10]. The GSMA defined the slices according to industry verticals, with their main requirements collected in Table 5.3.

Table 5.3

GSMA Identified Network Slices

Vertical	Applications	Requirements
Augmented and virtual reality (AR, VR)	Gaming, simulated environment for training, etc.	High bandwidth, low latency
Automotive	V2X, autonomous driving Infotainment	Low latency, reliability High bandwidth
Energy	Smart grid systems, smart meters, dynamic electricity provisioning	Low cost, long autonomy
Healthcare	Patient tracking, remote surgery	Reliability, low latency
Industry 4.0	Factory automation, logistics, asset management	Long autonomy, reliability
IoT	Waste management, smart city, etc.	Low cost, long autonomy
Public safety	Mission-critical applications	Reliability, low latency

The network slice selection is performed during registration procedure by the network slice selection function (NSSF) (see Figure 5.5). The 3GPP allows one UE to connect multiple network slices. However, a particular PDU session can be associated with one network slice only. The network slices can be flexibly organized, sharing the access network and core components as required by the network deployment. For example, two network slices could share the same frequency and physical layer resources but support different core networks.

5.4 POSITIONING IN THE 5G NETWORK

The 5G NR positioning and emergency call architecture builds on the 4G architecture that was covered in Chapter 3 of this book. Thus, this section will mainly focus on pointing out the differences from what was explained in the previous chapter.

The 5G NR deployments offer multiple choices to how the voice service (including emergency calls) accesses the IMS core. The LTE infrastructure, leveraged in NSA deployments, offers network operators a proven and reliable routing for emergency calls. Call routing can also take place directly over the 5G network. This section starts of by explaining a pure 5G deployment of IMS and then the alternative deployment scenarios based on the different deployment options discussed in Section 5.2.4.

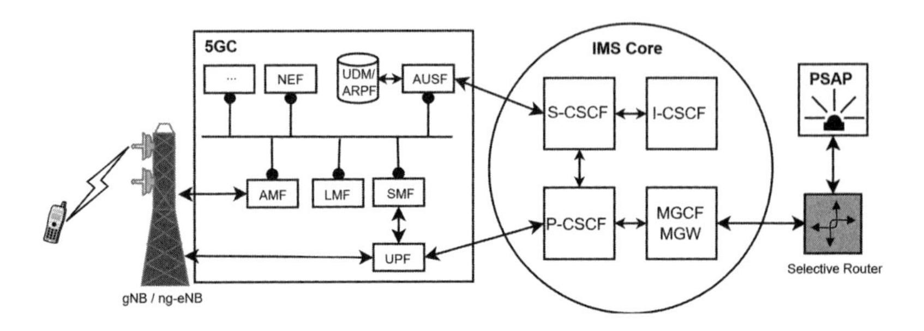

Figure 5.8 5G VoNR and IMS architecture.

5.4.1 SA Emergency Call

The voice services in 5G NR, called VoNR, leverage the IMS network already used for VoLTE services. The 5G VoNR and IMS architecture is shown in Figure 5.8. In the SA Option 2 deployment, the mobile device establishes a connection to the 5GC via the gNB. The voice call establishment in such deployment is explained in TS 23.502 [11]. The simplified procedure is listed below:

1. The UE establishes a RF connection to the gNB and performs a registration procedure with the AMF serving the preferred network slice.
2. The AMF selects the SMF for data session management. This completes the 5G NAS registration procedure.
3. The SMF establishes the PDU session, providing the UE an IP address and linking it to a proxy-call session control function (P-CSCF). This is followed by a registration to the IMS core using the Session Initiation Protocol (SIP). At least two QoS flows will be configured: one for the IMS signaling and one for the VoNR traffic.
4. Once the QoS flows have been established, the routing of the voice call is done through the UPF to the P-CSCF and M-GW of the IMS core. Routing through the different components of the IMS core to the PSAP is similar to the LTE procedure.

For SA Option 5, as seen in Figure 5.3, the UE is connected to the 5GC network via a ng-eNB. Apart from the initial RF connection establishment, the rest of the procedure is the same as for SA Option 2.

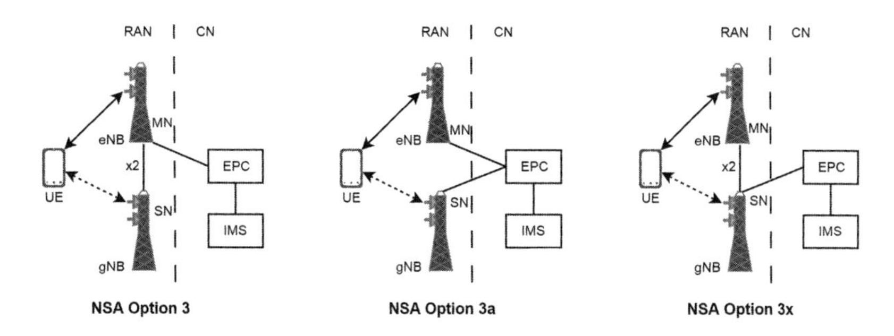

Figure 5.9 NSA Options 3, 3a, and 3x for IMS and eCall.

5.4.2 NSA Emergency Call

NSA Option 3 is typically the first step for network operators stepping from a 4G to a 5G network. Since the core network is still the EPC, the voice call routing from the EPC to the PSAP via the IMS core is the same as for LTE, which was already explained in Chapter 3 of this book. Nonetheless, there are different subdeployment options depending on the connections between the eNB/gNB and the EPC [13]:

- Option 3: in this deployment, the gNB is connected to the eNB via the x2 interface and not connected to the EPC. The eNB carries the signaling and is responsible for the data transmissions with the EPC.
- Option 3a: in this deployment, the EPC connects directly to both RANs and coordinates the transmission of the data over eNB and gNB. Lower bandwidth voice data is transmitted over the eNB and data sessions requiring higher bandwidth over the gNB. There is no direct connection between gNB and eNB.
- Option 3x: this deployment is the opposite of Option 3. There is no direct link between the eNB and the EPC. The gNB carries the signaling and the data connection. The gNB can route some traffic, as for instance a VoLTE call, to the eNB via the x2 interface.

The other NSA deployments (Option 4 and Option 7) are based on the 5GC. Thus, the communication between 5GC and IMS is the same as described for SA Option 2. The difference between the different subdeployments resides in how the RAN is connected to the CN, and also how the UE connects to the RAN. The possible permutations are shown in Figure 5.10 [13]. The differences between them are whether the UE is primarily connected to a gNB (Options 4 and 4a) or to a

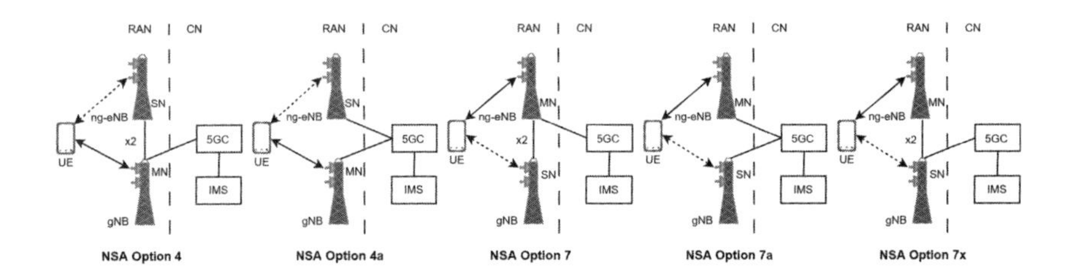

Figure 5.10 NSA Options 4, 4a, 7, 7a, and 7x for IMS and eCall.

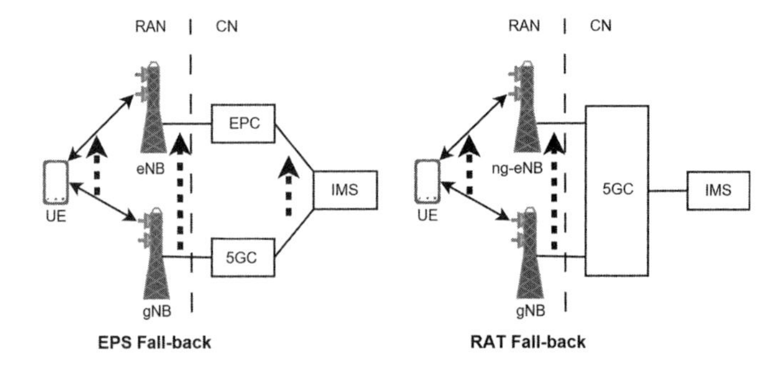

Figure 5.11 EPS and RAT fallbacks for IMS voice call routing.

ng-eNB (Options 7, 7a, and 7x) and which one of the nodes (master or secondary) holds the primary connection to the 5GC.

5.4.3 Emergency Call Fallback

Fallback procedures, for rerouting the emergency call to a different RAT in case of failures, were already introduced for LTE. They were useful for instance to relay voice and emergency calls over WCDMA and GSM circuit switched connections, in the time when LTE networks did not support IMS voice services. This technique was referred to as CSFB. Similarly, the 5G NR system supports fallback procedures to LTE networks. In Release 15, 5G NR does not support legacy CSFB procedures, since there is no interworking defined between 5G networks and legacy 2G and 3G networks. However, the 3GPP is discussing handovers and other procedures between 5G and 3G / 2G during Release 16. Hence, CSFB may also be supported in the future.

Figure 5.11 shows the two fallback techniques supported in 5G, which are:

- The evolved packet system (EPS) fallback provides an alternative to route voice traffic over the EPC. In case the voice data cannot be carried, the 5GC triggers a redirection or inter-RAT handover procedure to route the emergency call via the EPC. The EPS fallback requires the coexistence of the two core networks, EPC and 5GC.
- The RAT fallback, also referred to as ES fallback, is performed when the 5GC is able to handle voice data, but the gNB cannot. When the emergency call is dialed, the 5GC triggers a handover or reestablishment procedure to move the connection from the gNB to a ng-eNB. However, the connection to the IMS core is still done from the 5GC.

5.4.4 LCS Architecture

Same as for the IMS, the 5G NR LCS are based on the LTE LCS architecture. 5G NR LCS has the focus on improving critical key performance indicators, such as accuracy, availability, and privacy. A differentiating aspect between LTE and 5G LCS is the increasing importance of a wide range of commercial applications, in addition to the emergency location services' needs. This is triggered by the appearance of multiple commercial applications, which rely on the location services. Furthermore, since the LCS clients are often external to the network, the need for user privacy increases. Allowing only authorized LCS clients to obtain the UE location is an important design consideration in the 5G location architecture.

Similar to LTE, there are three main possibilities to retrieve the location of a UE in the 5G network:

- Network induced location request (NI-LR): the location request is generated internally in the network for services such as emergency calling or other regulatory needs.
- Mobile terminated location request (MT-LR): the location request is originated by an external application or another PLMN to the serving network to determine the UE location. For example, this is the method that the PSAPs can use to request the location of an emergency caller.
- Mobile originating location request (MO-LR): the location session establishment request comes from the mobile to the network. Such a request could be the result of a commercial application requesting the UE location.

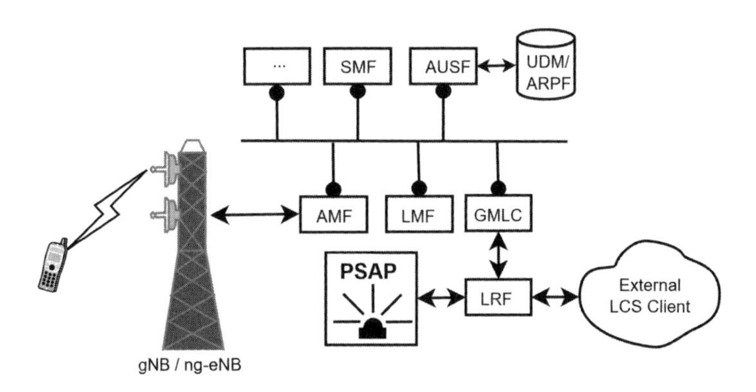

Figure 5.12 Service-based 5G location services architecture.

Figure 5.12 shows the LCS architecture of a service-based 5G network [14]. The location determination procedure and the routing within the 5G network is slightly different for commercial and emergency services.

5.4.4.1 Location Determination for Emergency Calls

The emergency call, as seen in Section 5.4.1, establishes an IMS emergency session to carry the voice data over to the external PSAP. The AMF may select the appropriate location management function (LMF) and start a NI-LR to determine the location of the caller. The LMF is analogous to the E-SLMC in LTE networks. The LMF is the terminating point of the 5G location protocols and provides the required assistance data (e.g. GNSS assistance data). Once the location is available, the GMLC forwards this information to the PSAP.

Alternatively, the PSAP can also trigger a MT-LR via the LRF to the GMLC to obtain the caller's location. The GMLC forwards the location request to the LMF, who starts the positioning session.

5.4.4.2 Location Determination for Commercial Applications

An external LCS client can use the MT-LR to trigger the LRF and the GMLC. However, an additional step is necessary in order to ensure the user privacy: the authentication of the application requesting the location. For that purpose, the GMLC connects to the authentication server function (AUSF), the 5G equivalent of the home subscriber server (HSS). The AUSF authenticates the application initiating

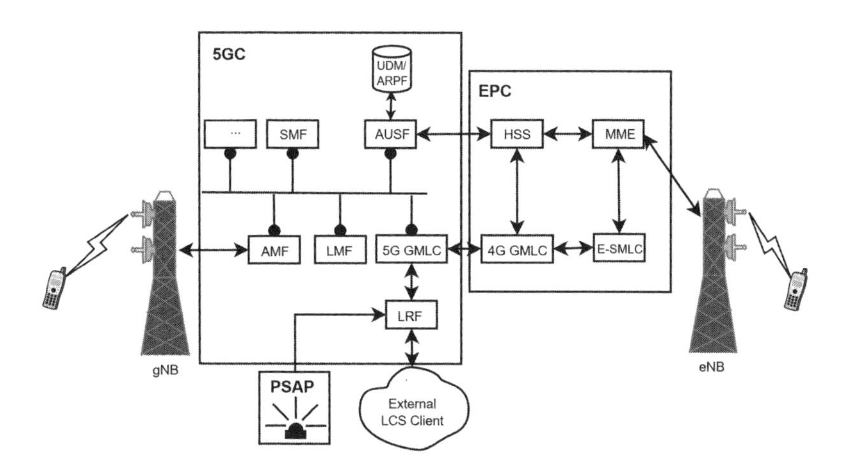

Figure 5.13 Combined 5G and 4G location services' architecture.

the location request with the help of the unified data management (UDM) and the authentication credential repository and processing function (ARPF).

5.4.4.3 Non-Standalone LCS Architecture

In case of non-standalone deployments, the combined positioning architecture of LTE and NR is shown in Figure 5.13, detailing the logical connection between EPC and 5GC. In newer implementations of the EPC, components such as the EPC GMLC and the 5GC GMLC or the E-SMLC and the LMF may coexist on the same physical unit. This architecture supports positioning for both 5G UEs and legacy LTE UEs, since the positioning session can be carried on LTE and EPC only, on NSA deployment on either core, or in SA deployment using 5GC.

The positioning session can be carried over U-Plane or C-Plane, similar to how it is done in the legacy networks, which already explained in Chapter 3 of this book. For C-Plane, the LPP protocol, already used for LTE, has been extended for NR and eLTE as of LPP Release 15. U-Plane allows, in addition to LPP, implementations based on legacy protocols, such as RRLP. The termination points of these protocols are the LMF in 5GC or the E-SMLC when used with the EPC.

5.5 POSITIONING USE CASES AND TECHNOLOGIES

5.5.1 Positioning Technologies Supported for NR

As in legacy networks, the positioning technologies supported in 5G NR can be categorized as RAT-independent (i.e., the position calculation is based on technologies external to the cellular network) and RAT-dependent (i.e., the position calculation uses the cellular signal).

Due to the tight timeline for concluding the first version of 5G NR, during the 3GPP Release 15 it was decided to postpone NR-based positioning technologies for Release 16. However, in order to be able to provide support for the regulatory emergency positioning requirements (seen in Chapter 3 of this book), the RAT-independent technologies (A-GNSS, Wi-Fi, Bluetooth, etc.) and the LTE-based positioning technologies (OTDOA, ECID, etc.) are supported and the positioning information can be exchanged in a 5G NR call flow.

For Release 16, the 3GPP has researched and agreed on several NR-based positioning technologies. Some of them, such as DL-TDOA, are an evolution of the legacy LTE positioning methods. On the other hand, 5G NR, thanks to the massive MIMO and the beamforming capabilities, opened up the door to precise angle-based technologies. Furthermore, the procedure for RTT measurements, part of LTE ECID, has been extended to allow RTT measurements to neighbor gNBs. This technology is called multi-RTT. All the 5G-based positioning technologies will be explained in Chapter 10 of this book.

The different positioning call flows, both for RAT-independent and RAT-dependent, and the extension of LPP will be explained in Chapter 15 of this book.

5.5.2 Positioning KPIs

Up to LTE, the positioning requirements were driven by the regulatory bodies and mobile phone industry. The requirements focused mainly on two performance indicators: the accuracy of the position fix (either two-dimensional or also including the altitude), and the time-to-first-fix (TTFF), which is the elapsed time until the positioning result is available.

However, with the advent of 5G systems and the new industry verticals, the positioning requirements evolved in two directions. On one hand, the typical positioning requirements for accuracy and TTFF became more stringent, driven by V2X and other safety-critical applications. On the other hand, the new application areas need additional positioning requirements that go further than accuracy and

TTFF. Gaming and virtual reality need to know the orientation, velocity, and heading values. Mission-critical applications need higher reliability and availability of the positioning fixes. Low-power IoT applications may be more relaxed in terms of accuracy but need very low cost and battery power consumption.

A complete list of all the positioning KPIs and their definitions can be found in TR 22.872 [15].

5.5.3 Commercial Positioning Use Cases

During the early stages of 5G NR, the 3GPP started a study on potential positioning use cases and their corresponding requirements [15]. The different use cases have been classified into six broad categories: LBS-related use cases, industry and eHealth use cases, emergency and mission-critical use cases, road-related use cases, rail and maritime use cases, and aerial-related use cases.

This section will briefly introduce the different use cases and give a summary table with the positioning requirements. For further details, the complete study can be found in [15].

5.5.3.1 LBS-Related Use Cases

This category focuses on giving services to users based on their location. The 3GPP has further divided this category into the following particular examples:

- **Bike sharing**: oriented to shared bike services in urban areas, which allow a rider to rent a bike using their smartphone and drop it off anywhere within the service area for the next user. These services currently exist in cities. They are not limited only to bikes but also include other means of transport such as electrical scooters.

- **Augmented reality**: the location information is used to enhance the user experience when using AR glasses or head-up displays (HUDs) for outdoor sports and leisure activities, as well as social networking.

- **Power-saving mechanism of wearable devices**: the intention is to use the location information to indicate to a wearable when continuous monitoring may not be required. Thus, the device could go into power-saving mode.

- **Location-based advertising push**: the location of users and their daily routines and interests are used in exhaustive data analyses in order to offer them personalized advertisements. This use case was already popular in LTE.

- **Flow management in large transportation hubs**: optimize the organization of places such as airports or metro/subway and rail stations by helping passengers to find their way, reducing the time of transit.

Table 5.4 summarizes the location requirements associated with these use cases.

Table 5.4

5G LBS-Related Positioning Requirements

Use cases	Service Area	Position Accuracy	Speed	Avail.	Update Rate	TTFF	Lat.
Bike sharing	Outdoor	2m hor.		90%		10s	1s
Augmented reality	Outdoor	1-3m hor. 0.1-3m ver.	2m/s 10°	80%	1-10 Hz	10s	1s
Wearables	Outdoor / indoor	2m hor. 1-3m ver.		90%	30-300s	10s	1s
	Outdoor / indoor	2m hor. 1-3m ver.		99%	1-30s	10s	1s
Advertisement push	Outdoor / indoor	3m hor. 3m ver.		99%			60s
Flow management	Outdoor / indoor	10m hor.		80%	10s	10s	
Note: from [15].							

5.5.3.2 Industry and eHealth Use Cases

This category is mainly oriented to improve logistics and introduce new features like patient location, among other factors:

- **Person and medical equipment location in hospitals**: wristbands or other location trackers can be used to locate patients that are at risk of running away, as for instance patients in a psychiatric hospital or patients with dementia and Alzheimer. Additionally, it may be necessary to locate caregivers and medical equipment, especially in emergency situations.

- **Patient location outside hospitals**: similar to the previous case, but focusing on locating patients outside of the hospital facilities in order to bring them back. In addition, it can be also be applied to ambulatory patients with a potentially critical condition who need to be located in case of emergency.

- **Trolley location in factories**: monitoring of the location and track of the movement of trolleys in big factories.
- **Waste management and collection**: bins and waste containers can periodically report their location and whether they are full or not, improving the efficiency of the waste collection services.
- **Support to accurate and reliable handling of containers**: track the location of the containers in ships, docks, and factories. Includes the use of sensors to place the containers correctly and prevent issues such as blocked doors.

Table 5.5 summarizes the location requirements associated with these use cases.

Table 5.5

5G Industry and eHealth Positioning Requirements

Use cases	Service Area	Position Accuracy	Avail.	Update Rate	TTFF	Lat.
Person and medical equipment location in hospital	Indoor	3m hor. 2m ver.	99%			60s
	Outdoor	0.2m hor.	99%		10s	1s
Patient location outside hospital	Outdoor / indoor	10m hor. 3m ver.	99%			
Trolley location in factories	Outdoor / indoor	0.5m hor. 1-3m ver.	99%			20ms
Waste management	Outdoor	3m hor.	99%	2h - 1 day		60s
Note: from [15].						

5.5.3.3 Emergency and Mission-Critical Use Cases

This category aims to improve the response time of the first responders and emergency services in case of emergency calls and mission-critical situations such as natural disasters.

- **Accurate positioning for emergency services**: such as the traditional regulatory positioning requirements in case of an emergency call.

- **Accurate positioning for first responders**: oriented to monitor the location of the first responders such as fire-fighters, for mission planning and organization as well as in order to rescue them if their situation becomes critical.
- **Alerting nearby emergency responders**: targets to locate and warn individuals qualified to provide urgent care in case of a medical emergency.
- **Emergency equipment location outside hospitals**: real-time location of life-saving medical equipment (e.g., defibrillators) deployed in public and private spaces to be used in case of emergency. The equipment would be connected to the 5G network and include a 5G positioning module, periodically sending information about its position to a mobile application.

Table 5.6 summarizes the location requirements associated with these use cases.

Table 5.6

5G Emergency and Mission-Critical Positioning Requirements

Use cases	Service Area	Position Accuracy	Avail.	TTFF	Lat.
Emergency call	Outdoor / indoor	50m hor. 3m ver.	95%	30s	60s
Accurate positioning for first responders	Outdoor	1m hor. 0.3m ver.	98%	10s	5s
	Indoor	1m hor. 2m ver.	95%	10s	1s
Alerting nearby emergency responders	Outdoor / indoor	50m hor. 3m ver.	99%	10s	
Emergency equipment location outside hospitals	Outdoor / indoor	10m hor. 3m ver.	95%	10s	
Note: from [15].					

5.5.3.4 Road-Related Use Cases

This category includes all positioning services related to traffic management as well as road usage charging. It is worth noting that 3GPP has decided to handle the V2X-related positioning requirements as a separate entity part of the V2X study. V2X will be discussed in a separate section.

- **Accurate positioning to support traffic monitoring, management, and control**. The Traffic Management Server (TMS) receives real-time information from all the vehicles connected to the 5G network. This information is used for flow management, congestion avoidance, and other services such as smart parking.
- **Road-user charging**: monitors the vehicle position and movement in order to collect a charge or certain tax based on the usage of the road infrastructure. This use case is not limited to toll collection in highways, and could potentially include pay-as-you-drive services. For instance, car insurance could be more or less expensive depending on the amount one drives.

Table 5.7 summarizes the location requirements associated with these use cases.

Table 5.7

5G Road-Related Positioning Requirements

Use cases	Service Area	Position Accuracy	Speed	Avail.	Update Rate	TTFF	Lat.
Traffic monitoring and control	Outdoor	1-3m hor. 2.5m ver.		95%	10Hz	10s	30ms
Road-user charging	Outdoor / tunnels	<1m (across) 3m (along)	2m/s	99%	1 Hz	10s	
Note: from [15].							

5.5.3.5 Rail and Marine Use Cases

This category focuses mainly on the tracking of cargo loaded onto a train or ship. This feature will improve the worldwide logistics by optimizing the overall transportation efficiency and improving end-to-end traceability. Table 5.8 summarizes the positioning requirements for asset and freight tracking.

5.5.3.6 Aerial Use Cases

UAVs are becoming increasingly popular for different purposes, from surveillance missions to delivery services:

- **Accurate positioning to support UAV missions and operations**. High-precision geolocalization information can be used to combine and superimpose images sent

Table 5.8

5G Rail and Marine Positioning Requirements

Use cases	Service Area	Position Accuracy	Speed	Avail.	Update Rate
Asset tracking and management	Outdoor	1-30m hor.	5 m/s	99%	300s-1day
Note: from [15].					

by a UAV onto a ground digital map. The accurate 3D geolocalization is also needed during UAV automatic landing.

- **Transport by drones for medical purposes**. Use of drones to transport medical equipment between locations in a secure and time-controlled process.

Table 5.9 summarizes the location requirements associated with these use cases.

Table 5.9

5G Aerial Positioning Requirements

Use cases	Service Area	Position Accuracy	Speed	Avail.	Update Rate	TTFF	Lat.
UAV data analysis	Outdoor	0.1m hor. 0.1m ver.	0.5 m/s 2 deg	99%		10s	
UAV remote control	Outdoor	0.5m hor. 0.3m ver.		99%			150ms
Note: from [15].							

5.5.3.7 Summary

Taking a look at all the requirements collected in the previous sections, it can be seen that although very different in nature, the performance requirements are fairly similar. The most demanding use cases need a position accuracy in the centimeter-level for horizontal as well as vertical positioning. The desired latency is smaller than 1 second and the availability up to 99% of the time.

5.5.4 3GPP Positioning Requirements

The first official set of the 5G NR standards, published during the 3GPP Release-15 time frame, did not have any additional positioning requirements. The only positioning requirements came dictated by the emergency service regulations, detailed in Chapter 3 of this book. The regulatory horizontal accuracy requirement is 50 m, with a TTFF of 30 s. The vertical accuracy requirement will only be enforced after 2021.

During the Release-16 time frame, the 3GPP conducted a study item on the achievable positioning accuracy using 5G NR technology [16]. The outcome of this study item was that the current industry expectations, captured in [15], were not realizable with the current technologies. Nonetheless, TR 38.855 managed to define a set of positioning requirements, tighter than the regulatory requirements, which could cover at least some of the desired positioning use cases:

- **Regulatory cases**
  - Horizontal positioning error $\leq$ 50 m for 80% of UEs
  - Vertical positioning error $<$ 5 m for 80% of UEs
  - End-to-end latency and TTFF $<$ 30 s
- **Commercial use cases**
  - **Indoor deployment scenarios**
    * Horizontal positioning error $<$ 3 m for 80% of UEs
    * Vertical positioning error $<$ 3 m for 80% of UEs
  - **Outdoor deployment scenarios**
    * Horizontal positioning error $<$ 10 m for 80% of UEs
    * Vertical positioning error $<$ 3 m for 80% of UEs
  - End-to-end latency $<$ 1 s

There are several different positioning technologies that can be used in 5G NR, both RAT-independent and RAT-dependent. Given the diverse nature of the positioning use cases, it is understood that not all the technologies need to fulfill the tightest requirements. The requirements in TR 38.855 are considered minimum performance targets that must be guaranteed in any region with 5G coverage. More demanding applications can be limited to certain areas, where additional positioning services (for example, high-accuracy GNSS methods, as described in Chapter 7 of this book) can be deployed to enhance the positioning performance. A more detailed analysis of the performance of the different positioning methods with respect to the accuracy and other parameters, such as the reliability, is given in Chapter 11 of this book.

At the time of the writing, the 3GPP has started a follow-up study for Release 17 (expected to be completed by June 2021), in order to address more demanding positioning use cases. It is expected that in this study item, the 3GPP takes a step forward toward centimeter-level positioning accuracy.

5.5.5 V2X in 5G

Car-to-car communication has been around for a while. The first 3GPP technology focused mainly on the V2V communication, in Release 13, and it was enhanced with V2X in Release 14, as seen in Chapter 4 of the book. Nonetheless, the Release-14 V2X features were limited to safety-critical messages over the PC5 interface. This includes the basic safety message (BSM) for DSRC-based implementations, the cooperative awareness message (CAM), and the decentralized environmental notification message (DENM).

This section will briefly explore the upgrades to V2X during Release 15 (mainly for LTE) and Release 16.

5.5.5.1 Release-15 V2X

Phase 2 work for V2X started with Release 15, introducing additional capabilities into the LTE-based V2X system and increasing the data rate of the transmissions. The use cases aimed to be supported by V2X Phase 2 are vehicle platooning, extended sensors, advanced driving, and remote driving. The main features introduced are the following:

- Carrier aggregation on PC5 with up to 8 carriers allowed. The maximum aggregated bandwidth for communication using the PC5 interface has been defined as 60 MHz.

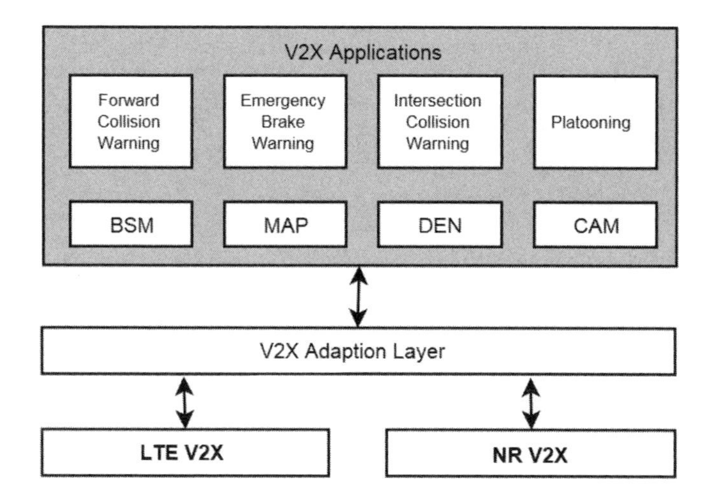

Figure 5.14 Coexistance of LTE and NR-based V2X technologies.

- Support for high modulation schemes of up to 64 QAM on the PC5 interface.
- Tx diversity support in PC5 transmissions to increase the redundancy.
- New power class devices on PC5 for higher transmission power and hence increased coverage.

As mentioned before, the Release-15 V2X implementation still used LTE for communication between vehicles and the network, and the LTE physical layer technologies as a basis for the PC5. More information on Release-15 V2X can be found in [17]. The NR-based implementation has been introduced in Release 16.

5.5.5.2 Release-16 V2X

Release 16 introduced the V2X standard for the first time over the NR link [18]. It also defined the Sidelink channel using the NR physical layer. NR V2X will coexist with Release 14 and Release 15 based LTE V2X technologies, as shown in Figure 5.14.

NR V2X will provide advanced and reliable V2X services with higher throughput, lower latency, wider bandwidths, and high accuracy positioning. Some of the enhancements planned for NR-V2X are:

- Addition of wider bandwidths, such as 80 MHz or 100 MHz.
- Support of modulation schemes up to 256 QAM.
- Compatibility with Release-14/15 LTE V2X implementations.
- Support for ultra-reliable low latency links.
- Support for very high mobility, with relative speeds above 500 Km/h.

An NR V2X device always has both the NR stack and the LTE stack. The LTE stack handles the messaging for basic safety features, while the NR stack handles V2X enhancement features. This ensures backward compatibility, especially for vehicles that support only LTE Release-14 safety messages.

5.6 CONCLUSION

This chapter introduced the basics about the 5G NR system and network architecture, with special emphasis on the architecture for LCS and emergency calls. It has pointed out the main differences between legacy networks and 5G, for example, service-based architecture or network slicing.

The second part of the chapter focused on the increasing number of applications and fields requiring positioning services based on 5G. The main groups of applications, also called industry verticals, have been explained briefly, paying special attention to the positioning performance requirements desired by the industry.

As seen in the last part of the chapter, the Release 16 of 5G has met the regulatory positioning requirements and part of the commercial requirements as well. However, 5G NR RAT-dependent positioning has not been able to meet some of the most demanding commercial requirements, for instance centimeter-level positioning accuracy. Nonetheless, 5G standardization is still ongoing, and Release 17 has put of the focus on improving positioning performance to meet the industry needs.

References

[1] ITU M.2410-0 (11/2017), *Minimum Requirements Related to Technical Performance for IMT-2020 Radio Interface(s)*, November, 2017, accessed on January 2020.

[2] 3GPP TS 38.101-1, *NR; UE Radio Transmission and Reception; Part 1: Range 1 Standalone*, V16.2.0, January, 2020.

[3] 3GPP TS 38.101-2, *NR; UE Radio Transmission and Reception; Part 2: Range 2 Standalone*, V16.2.0, December, 2019.

[4] 3GPP TS 38.300, *NR; Overall Description; Stage-2*, V16.0.0, January, 2020.

[5] 3GPP TS 37.340, *NR; Multi-Connectivity; Overall Description; Stage-2*, V16.0.0, January, 2020.

[6] 3GPP TS 23.501, *System Architecture for the 5G System (5GS)*, V16.3.0, December, 2019.

[7] Rohde & Schwarz, *5G New Radio: Fundamentals, Procedures, Testing Aspects*, 2019.

[8] Dahlman, E., Parkvall, S., and Skold, J., *5G NR The Next Generation Wireless Access Technology*, Academic Press, 2018.

[9] 3GPP TS 29.518, *5G System; Access and Mobility Management Services; Stage 3*, V16.2.0, December, 2019.

[10] GSMA, *Network Slicing: Use Case Requirements*, April, 2018.

[11] 3GPP TS 23.502, *Procedures for the 5G System (5GS)*, V16.3.0, December, 2019.

[12] NGMN Alliance, *5G Project Requirements and Architecture – Work Stream E2E Architecture*, V1.0.8, September, 2016.

[13] GSMA, *5G Implementation Guidelines*, July, 2019.

[14] 3GPP TS 23.273, *5G System; Location Services (LCS); Stage 2*, V16.2.0, December, 2019.

[15] 3GPP TR 22.872, *Study on Positioning Use Cases*, V16.1.0, September, 2018.

[16] 3GPP TR 38.855, *Study on NR Positioning Support*, V16.0.0, March, 2019.

[17] 3GPP TR 36.788, *Vehicle-to-Everything (V2X) Phase 2; User Equipment (UE) Radio Transmission and Reception*, V15.0.0, July, 2018.

[18] 3GPP TR 22.886, *Study on Enhancement of 3GPP Support for 5G V2X Services*, V16.2.0, December, 2018.

Part II

Positioning Technologies

Chapter 6

Assisted GNSS

6.1 INTRODUCTION

GNSS is the generalized term for positioning systems like GPS/Navstar (United States), Galileo (European Union), BeiDou (China), and GLONASS (Russia). Systems which cover only certain regions, like QZSS (Japan) and Navic (India), are summarized under the term regional navigation satellite system (RNSS). All these systems consist of a space segment (the satellite constellation), a user segment (e.g., a mobile phone with GNSS receiver), and a ground segment (control station). The position is calculated in the user segment by measuring the RF signal's travel time from four or more visible satellites to the receiver using the principle of trilateration as explained in Chapter 2. Naturally, the position of the satellites must be known to perform this calculation. While the fundamental principle of GNSS and RNSS can be explained in a few words, a thorough overview and explanation of all the systems' aspects can fill books on their own right. It is the intent of this chapter to give professionals and students with a telecommunication or computer science background profound insight into the topic. The interested reader may additionally refer to *Global Positioning System: Theory and Application* [1] by Bradford W. Parkinson and James J. Spilker Jr. and *A-GPS: Assisted GPS, GNSS and SBAS* [2] by Frank van Diggelen for more in-depth information.

This chapter is organized in an order similar to how a GNSS receiver acquires, tracks, calculates satellite positions, and consequently computes a final position. The basic terms and concepts necessary for GNSS and A-GNSS are introduced in the first part. Helpful equations to understand signal power calculation are introduced. It will also help the reader to understand the limitations of standalone receivers and the benefit of A-GNSS, which is described in detail in the second part. The evolution to multi-GNSS and multifrequency receivers is covered in the third part. The chapter concludes with the RF spectrum environment adjacent to the GNSS signals, integrity concepts, limitations of GNSS, and alternative low-Earth-orbit constellations that can be used for positioning as well. A-GNSS protocol aspects are treated in Chapters 13, 14, and 15.

6.2 GNSS BASICS

6.2.1 GNSS Signal Power

The first hurdle to overcome is the low received signal power. The satellite transmit power is in the order of only 44 dBm for GPS L1C/A ([1][Chapter 6.2 D.2, Table 2]) and the free-space path loss for GPS L1 coarse/acquisition (C/A) is around 174 dB. As a consequence, the signal power received on Earth is around -130 dBm under good conditions. This value needs to be compared against the noise floor, which depends on the receiver implementation, but is usually higher than this value. In [2][6.4.3, Table 6.3, C8], the noise floor in an exemplary receiver with a bandwidth of 3 MHz is -109.1 dBm.

The received GNSS signal power at the intermediate frequency (IF) is therefore below the noise floor, leading to an IF SNR of -20.9 dB in the example ([2][6.4.3, Table 6.3, C9]).

6.2.1.1 Coherent Integration and Spread Spectrum

The negative IF SNR is not a problem for the receiver, since coherent integration can be employed thanks to the special characteristics of the signal. To achieve a high time resolution, a signal with a broad bandwidth is preferable. All GNSS signals are designed to distribute the power over several MHz using the spread-spectrum technique. In the case of the L1 C/A signal, the bandwidth is 20 MHz, but most of the signal energy is concentrated in the inner part, which explains why a receiver with a bandwidth of only 3 MHz works fine (see Figure 6.1, [1][Chapter 3.1]). The spread-spectrum technique allows coherent addition, as the spreading sequences are a priori known. This results in a coherent gain: If the receiver uses 2 samples per chip and a coherence time of 1 ms is assumed, van Diggelen calculates the ideal coherent gain with 33.1 dB in [2][6.4.3, Table 6.3]. In our example, this results in a coherent SNR of 11.7 dB, which allows decoding the signal.

Almost all GNSS systems also use CDMA for multiplexing the different satellites. GLONASS' legacy signals are the only exception, which use FDMA (as

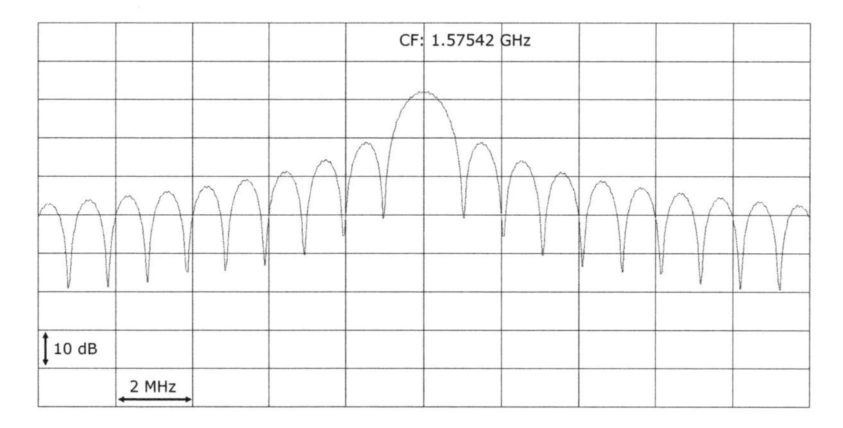

Figure 6.1 Spectrum of the GPS L1 C/A signal without noise.

can be seen in Figure 6.2). However, they still use spread spectrum for distributing the signal power over a larger bandwidth.

6.2.1.2 Converting Signal Strength: SNR, Carrier-to-Noise Density and Received Signal Strength in dBm

(A-)GNSS receivers often do not provide the received signal strength in dBm, but provide the SNR value or the related carrier-to-noise density (C/N_0) value instead. The SNR and C/N_0 can be converted, if the receiver's two-sided IF bandwidth (BW) is known (Equation 6.1).

$$C/N_0 = SNR + 10log_{10}(BW) \tag{6.1}$$

If the receiver's full front-end configuration is known, one can deduce the equivalent received signal strength from the C/N_0 value by a simple formula given in Equation 6.2 ([2][6.4.3]):

$$C/N_0(dB - Hz) = SS(dBm) - 30 - 10log_{10}(kT_{eff}) \tag{6.2}$$

with SS being the signal strength and $T_{eff} = T_A + (F - 1) * T_0$, where F is the receiver's noise figure calculated with Friis' formula and T_0 being the ambient temperature.

For a live signal, which is broadcast from the cold GPS satellite in space, $T_A = 130K$ is typically used according to [2][6.4.3].

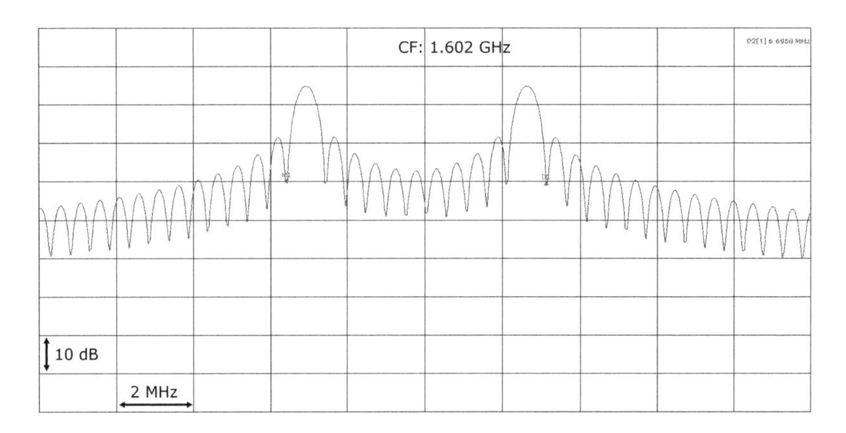

Figure 6.2 Spectrum of the GLONASS L1OF signals on channels -7 and 6.

For a signal simulator at room temperature, the transmitter's temperature is also ambient temperature: $T_A = T_0 = 290K$. The formula can be simplified to Equation 6.3:

$$C/N_0(dB - Hz) = SS_{sim}(dBm) + 174 - F_{dB} \tag{6.3}$$

6.2.2 The Ephemeris: Satellite Position

Another obstacle is the unknown satellite position on the receiver side. For the trilateration, the receiver must know from where the signal was broadcast. Orbits used by GNSS systems are depicted in Figure 6.3. As of today, GPS, GLONASS, and Galileo use only MEOs. Navic and QZSS use only IGSOs for their satellites. Beidou uses MEO for most of the satellites, but also employs a few GEO and IGSO orbits to improve coverage over Asia. In short, IGSOs and GEOs rotate synchronously with the Earth and allow regional coverage, while MEOs are ideal for global coverage.

When looking at the ground-tracks of the satellites as depicted in Figure 6.4, only the GEO satellite seems static in respect to the Earth's surface. The MEO and IGSO orbits result in a relative movement as a function of time. Since the position of the satellites needs to be known in the submeter range, even the GEO cannot be considered static anymore. In fact it also follows the characteristic "8" ground-track of an IGSO if one would zoom in on the image. This is due to minor orbit imperfections.

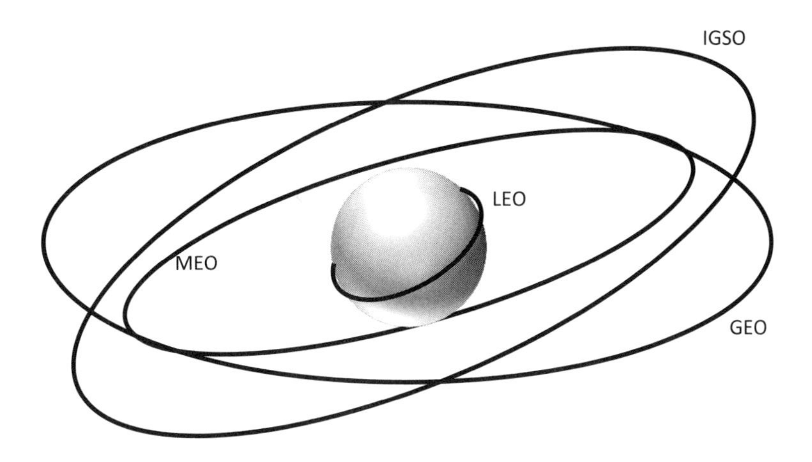

Figure 6.3 Visualization of common GNSS orbits.

Consequently, for all satellite types alike, the trajectory is a function over time. Fortunately, the satellite movement is quite predictable and an astronomer found a convenient representation for the parameters describing an orbiting body in space: Johannes Kepler. In his *Astronomia Nova* of 1609 he described the orbits of the planets, but his equations are valid for artificial satellites orbiting Earth as well. The definition of the Keplerian orbits can be found in Section 6.4.2.2.

For the planets, the sun, and the moon, these values can be predicted very precisely over a long time because of their large masses. Long before the age of radio navigation, sailors and aviators used the natural celestial bodies for navigation with a sextant. Orbital parameters have been published since the late 18th century. To this day, the astronomical almanac is available in printed format, for example, by the U.S. Government Publishing Office. It is updated every year and contains the predicted values for the next 12 months.

For an artificial satellite, it is not possible to predict the precise orbit over such long periods because of their small mass, the Earth's irregular mass distribution,

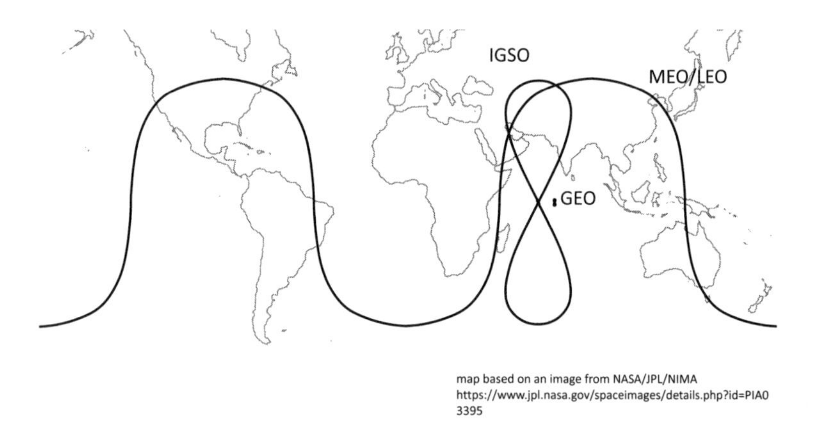

Figure 6.4 Ground tracks of common GNSS orbits.

the influence of the moon and sun's gravity (leading to a three-body-problem), the atmospheric drag, and so forth. Distributing a book (or an electronic file) once a year is not sufficient to achieve the accuracy required. This raises another problem: how to transmit the ephemeris to the user. The GPS's solution is to transmit the ephemeris along with the time marks directly from the satellites. With a bit rate of 50 bit/s it takes around 20 s for a satellite to transmit its own orbit (in bad signal conditions it might take much longer, however, to receive an error-free copy). To speed up the data delivery, A-GNSS solutions provide the ephemeris via mobile protocols as described in Chapter 14. The calculation of the ephemeris and its upload to the satellites is the task of the ground segment operated by the respective governments. Improvements in the orbit prediction at the ground segment are one reason why all GNSSs are getting more precise over time.

Once the satellites' positions are known, a famous mathematician comes to our aid to find our position: C.F. Gauss originally developed his least-squares method to trace down the orbit of the comet Ceres in 1801. This method is still used in many implementations to solve the equations of the trilateration.

6.2.3 Clocks

Using trilateration, it is also crucial that all transmitters are synchronized (or have at least a known offset). Clock offsets translate to position offset with the speed of light as a factor. Hence, clock errors need to be corrected. Similar to the orbits, the ground segment also monitors each satellite's clock and distributes clock correction parameters along with the orbit information to the users. These correction terms are also part of every A-GNSS protocol.

6.2.3.1 Relativity

Talking about clocks, high speeds, and gravity, it is virtually impossible to avoid one prominent physicist: Albert Einstein. Interestingly, both of his theories of relativity have an impact on GNSS operation: While the special theory of relativity predicts that clocks run slower if they have a high speed relative to the observer, his general theory of relativity predicts that clocks run faster relative to an observer if they experience less gravity. Both effects, however, do not cancel each other out and need to be compensated.

6.2.3.2 UTC Time

Since all GNSS systems except GLONASS use their own system time independent of UTC, a message carrying the time difference from each system to the UTC time is also broadcast in the navigation message (or sent via A-GNSS) and can be used to relate the various system times. Some systems also transmit direct GNSS-GNSS differences (e.g., Galileo transmitting the difference to GPS time).

6.2.4 Atmosphere and Ionosphere

On their way from space to Earth, the GNSS signals travel through the Earth's atmosphere, which is another major source of ranging errors. Namely, the troposphere and the ionosphere influence the signal travel time. Dual-frequency GNSS receivers can remove most of this error since the impact is frequency-dependent. However, a lot of the mass-market receivers as of 2020 are still single frequency and need to use error models to mitigate the effects. While the impact of the troposphere is usually compensated by a global model (NATO's Standardization Agreement (STANAG) Document 4294) and the influence of the weather (humidity) is neglected, the influence of the ionosphere is larger and also dependent on the sun's activity, which is not perfectly predictable. The designers of the GPS system decided to create a mathematical model of the ionosphere based on empirical data (Klobuchar model). The parameters of the model are again transmitted along with the ephemeris in the navigation message of the satellites. The BeiDou system uses the same global model, while GLONASS does not transmit any ionospheric correction terms at all. The designers of the Galileo system decided to use a more sophisticated ionospheric model. This model is also based on empirical data and called NeQuick G (see [3]). It is important to note that a ionospheric model of one system can also be used for other systems in a multi-GNSS receiver. For this reason, A-GNSS protocols may only allow to deliver one model, even in the case of multi-GNSS. An overview of the ionospheric models natively used in each system is given in Table 6.1. Further atmospheric error reduction methods are discussed in Section 6.6.2 and Chapter 7, Section 7.4.2.

Table 6.1

Native Ionospheric Models in Broadcast Navigation Messages

GNSS	Ionospheric Model
GPS	Klobuchar
GLONASS	None
Galileo	neQuick
BeiDou	Klobuchar
Navic	Klobuchar
QZSS	Klobuchar

6.2.5 The Almanac

Finally, there is one more element transmitted in all GNSS: the almanac. It contains once again the satellites' orbits, but with less resolution than in the ephemeris. Transmitting the same information twice may look like a waste of bandwidth, but there is a valid rationale behind this: while every satellite transmits its own orbit in high resolution in the ephemeris, it transmits the orbits of ALL satellites of its constellation in the almanac. As the bandwidth is very limited, it was decided to transmit the other satellites' orbits only in reduced resolution. The advantage is that the receiver gets information about the other satellites' coarse locations and their visibility above horizon. In A-GNSS mode, the almanac is normally less important, as the location server already indicates which satellites the receiver may see.

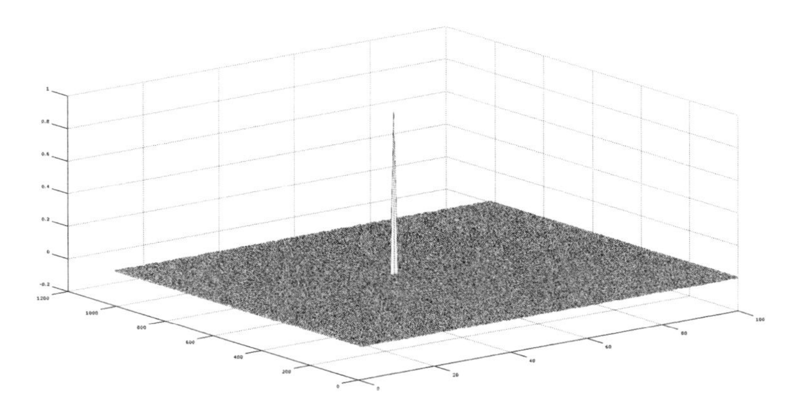

Figure 6.5 Correlation peak in a search space of 1,023 code positions and 100 bins in frequency range.

6.2.6 Acquisition and First Location Fix

The previous section introduced the CDMA system characteristics of GNSS systems and the payload of their navigation messages. This section will explain in a step-by-step approach how a receiver, which has no a priori knowledge of its location, obtains an initial position fix.

Since the received signal is deeply buried in thermal noise, the receiver needs to utilize the coherent integration to be able to decode a signal. This means that it needs to correlate a locally generated PRN sequence with the received signal. If the PRN sequence and the received signal's code are aligned, the correlation will result in a peak. However, not only is the received signal's code phase initially unknown, its exact frequency is also not known, because the satellite movement (and the receiver movement to a smaller part) lead to an unknown Doppler shift (see Diggelen A-GPS [2] 3.1.2). This results in a two-dimensional search space, with the axes code-phase and Doppler shift as shown in Figure 6.5. Without prior knowledge, the receiver must perform a brute-force search of this space for each PRN code until it finds the first signal.

Once the first correlation is found, the receiver can despread the signal and read the navigation message. The channel is then in tracking state. As soon as the first three satellites are found and 30 seconds of navigation message (including the ephemeris) are decoded for each, an initial 2D position can be calculated.

6.2.7 The Positioning Algorithm

The 3GPP conformance test cases provide a good example for a simple position algorithm. It is a variant of the Gauss' least-squares algorithm mentioned earlier, used to calculate the mobile's position from raw measurements. The weighted least squares (WLS) algorithm is specified in [4]. The interested reader is also referred to [1], Chapter 9.

Based on the ephemeris, the positions of the satellites are calculated according to Section 6.4.2. Using an initial position estimate (in the worst case, the center of the Earth), the Earth's rotation is compensated and clock, ionospheric, and tropospheric corrections (e.g., according to the GPS ICD) are applied to the geometrical distance from each satellite to the estimated user location. The resulting values constitute the initial estimated pseudoranges for each satellite's signal. Each estimated pseudorange is subtracted from the actually measured pseudorange, leading to the vector $\Delta\rho$.

The satellite positions, along with the initial location estimate, form the geometry matrix:

$$G = \begin{bmatrix} -\hat{\mathbf{1}}^{\mathsf{T}}_{GNSS_{1,1}} & 1 \\ -\hat{\mathbf{1}}^{\mathsf{T}}_{GNSS_{1,2}} & 1 \\ \vdots & \vdots \\ -\hat{\mathbf{1}}^{\mathsf{T}}_{GNSS_{1,n}} & 1 \\ \vdots & \vdots \\ -\hat{\mathbf{1}}^{\mathsf{T}}_{GNSS_{m,1}} & 1 \\ -\hat{\mathbf{1}}^{\mathsf{T}}_{GNSS_{m,2}} & 1 \\ \vdots & \vdots \\ -\hat{\mathbf{1}}^{\mathsf{T}}_{GNSS_{m,l}} & 1 \end{bmatrix} \tag{6.4}$$

$$-\hat{\mathbf{1}}^{\mathsf{T}}_{GNSS_{m,l}} = \frac{\mathbf{r}_{sGNSS_{m,i}} - \hat{\mathbf{r}}_u}{|\mathbf{r}_{sGNSS_{m,i}} - \hat{\mathbf{r}}_u|} \tag{6.5}$$

where $\mathbf{r}_{sGNSS_{m,i}}$ is the location of satellite i in the GNSS constellation m and $\hat{\mathbf{r}}_u$ is the estimated user location.

In Equation 6.4, n measurements are available for GNSS number 1, and l measurements are available for GNSS number m.

In a nutshell, all normalized distance vectors between each satellite and a estimated user location are stacked, leading to a matrix of size $k \times 4$ with k being the total number of satellite signals.

Each satellite's code-phase measurement is accompanied by an RMS error $w_{GNSS_{m,l}}$, expressing the quality of each measured signal. The inverses of the squared RMS errors are arranged in a diagonal matrix $\mathbf{W}$, the weighting matrix.

$$\mathbf{W} = diag\left(\frac{1}{w^2_{GNSS_{1,1}}}, \ldots, \frac{1}{w^2_{GNSS_{1,n}}}, \ldots, \frac{1}{w^2_{GNSS_{m,1}}}, \ldots, \frac{1}{w^2_{GNSS_{m,l}}}\right) \tag{6.6}$$

As the name suggests, it is provided to the least-squares algorithm to evaluate the measurements based on their quality.

The WLS algorithm is defined by the following equation:

$$\Delta\hat{\mathbf{x}} = \left(\mathbf{G}^{\intercal}\mathbf{W}\mathbf{G}\right)^{-1}\mathbf{G}^{\intercal}\mathbf{W}\Delta\rho, \tag{6.7}$$

where $\hat{\mathbf{x}}$ is a state vector containing the estimated user location and the clock bias of the receiver.

In an iterative process, the initial estimate is refined by the following formula:

$$\hat{\mathbf{x}} \rightarrow \hat{\mathbf{x}} + \Delta\hat{\mathbf{x}} \tag{6.8}$$

The result is an estimated position in a Cartesian frame (ECEF, seen in Chapter 2), which can be easily converted to the well-known geodetic latitude, longitude, and altitude.

6.2.8 GNSS Basics Summary

A GNSS receiver can obtain a position fix without any prior knowledge by using a brute-force search with short integration times (low sensitivity). Once the receiver operates for a longer time, it can use a more elegant search strategy leading to a short time-to-first-fix (warm-start) after short interruptions. Furthermore, the receiver can remove the data bits and achieve a higher sensitivity if it knows the whole navigation message (the data-bit wipe-off). These characteristics can be directly observed when using a hiking GNSS receiver (without a cellular modem) after winter-break: If switched on for the first time, it will only find satellites if there is a relatively unobstructed sky view. If the sky view is poor (e.g., in a forest), the receiver will show bad performance for a longer time. On the other hand, if the receiver is placed in open sky for several minutes it will subsequently have a much better performance.

6.3 A-GNSS

6.3.1 Motivation

The previous section explained how a GNSS receiver works without aiding. While it can obtain a position by using only GNSS satellites, provided it has a good sky view when starting, it has some clear limitations, which degrade the user experience. When it comes to integrating a GNSS receiver in a mobile phone, there are even further challenges: while a specialized GNSS receiver typically has a well-designed GNSS patch antenna, a mobile phone's GNSS receiver typically has to use a tiny metal strip as an antenna due to size constraints. Furthermore, the RF environment in a mobile phone with its RF transmissions on New Radio (NR), LTE, WLAN, Bluetooth, NFC, and other modules is anything but silent. Integrating a standalone GNSS receiver in a mobile phone is therefore not a good idea. A simple experiment similar to the previous one can demonstrate the problem: taking an old smartphone (which was unused for one week and has no WLAN or mobile network access) and starting a GPS-based application, one will observe the limitations as mentioned above. After inserting a SIM card with internet access it will obtain an almost instantaneous fix, even in challenging environments.

Besides the improved time-to-first-fix and the superior sensitivity compared to a standalone device, A-GNSS also saves energy in the mobile, since the receiver can be turned off after a few seconds once it has a fix.

6.3.2 A-GNSS Protocols

The delivery of the A-GNSS data over cellular networks can be achieved by various protocols, which are also supporting other location technologies besides A-GNSS. Therefore, they are handled in Chapters 13, 14, and 15 in detail.

6.3.3 Providing the Navigation Message via Cellular Networks

The major shortcoming of a standalone GNSS is the low data rate of the navigation message of typically 50 bit/s and the difficulty to receive it in obstructed environments. Having access to a high-speed data connection in a mobile phone, A-GNSS in its simplest form does nothing more than transmit the content of the navigation message over the cellular network. The data transmitted by a GPS satellite within 12.5 minutes at 50 bit/s is only 37.5 kbytes, an amount practically irrelevant for a 3G, 4G, or 5G connection.

6.3.4 Reference Location

Optimizing one step more, the cellular network can also provide a coarse position: based on the Cell ID, the receiver's location is known to be typically within a 3-km radius around the base station.

6.3.5 Reference Frequency

Furthermore, the receiver can use the network as a frequency standard, since base stations typically provide a highly accurate frequency. This reduces the Doppler uncertainty and hence the search area.

6.3.6 Reference Time

Yet another optimization can be achieved by using the network as a reference time source. However, not all cellular networks are accurately synced to GPS or other GNSSs, and therefore the network can signal the time uncertainty provided to the receiver. In case the time is delivered via IP protocols (SUPL), the uncertainty signaled is typically up to +-3s. If the reference time is provided over the network's lower layers, the message transfer time can be more accurately estimated. The most accurate time synchronization can be achieved if the network transmits the GPS or UTC time aligned to the physical layer frame structure, which is also called fine-time assistance [2][Chapter 7.1].

6.3.6.1 Fine-Time Assistance via System Information Blocks

A time relation to the LTE frame can be achieved by transmitting the GPS time or the UTC time in the system information blocks (SIBs). SIBs are broadcast by base stations and usually carry information about neighbor cells for handover procedures or other network parameters. In LTE, system information Block SIB8 or SIB16 carry GPS and UTC time, respectively. Typically an accuracy of 10 ms can be achieved.

6.3.6.2 Fine-Time Assistance via System-Frame Number Time Relation

The LPP protocol can also signal a time-relation between the cellular network's system frame number (SFN) and the GNSS time (see Figure 6.6). The SFN is a monotonously increasing number that rolls over after a certain value. The ambiguity

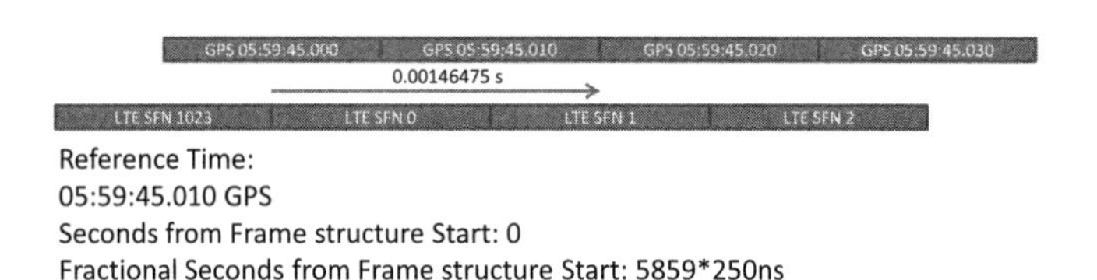

Figure 6.6 Fine-time assistance via time relation to SFN

due to the rollover depends on the actual RAT. It is 10.24 s for LTE/NR, 40.96 s for WCDMA, 12533.76 s for GSM, and 10485.76 s for NB-IOT. The resolution of this time relation is 250 ns and can optionally include a drift between the two time scales. The estimated signal travel time should be compensated by the location server. The LPP protocol allows to convey up to 16 of those time relations, which can be used in case the neighbor cells are not synchronized.

6.3.7 MS-Assisted vs. MS-Based

The method of A-GNSS described above is called MS-based (in 3GPP language) or UE-based (in OMA language), since the position calculation is performed within the user equipment. There is a second variant, called MS-assisted/UE-assisted, which works differently. Instead of transmitting the navigation message to the receiver over a cellular network and letting the receiver calculate the search window, performing the correlations and doing the trilateration, in this mode, the cellular network directly transmits the parameter of the search window (Doppler window and code phase search window; frequency is provided implicitly) [2][Chapter 3.6.2]. The receiver performs only the correlation and reports the raw measurements back to the network. The trilateration is performed in the cellular network. Originally, this mode was introduced to reduce complexity on the mobile phone. However due to advancements in GNSS receiver designs, all modern phones can perform the trilateration themselves and this mode became less common. Nevertheless, there are other advantages of this mode:

1. The raw measurements are processed in the network (an early form of cloud computing). Thus, improvements in the trilateration algorithm will benefit all UEs.

2. The raw measurements can be combined with other raw measurements to form a hybrid OTDOA/GNSS position fix as described in Chapter 8.
3. C/N_0 values (a measure related to SNR) are transmitted to the network. This feature is used for mobile phone GNSS antenna pattern testing during the development process.

Hence, it is still used by some network operators and supported by most phones.

6.3.8 A-GNSS Reference Station Network

This chapter so far has described the content of the assistance data that is delivered to a mobile phone and the meaning of each element. To obtain the broadcast navigation message from all GNSS satellites, it is necessary to operate stations distributed around the globe [2][Chapter 7.3], since there is no direct provision of broadcast navigation message from any of the GNSS operators (the coarse almanac information is provided, but the other information is not directly available from the official websites). Several companies specializing in the operation of these monitor networks and provide the navigation messages to the location server entities in a mobile network.

6.4 FROM GPS TO MULTI-GNSS

6.4.1 GPS, GLONASS, BeiDou, Galileo: Similarities and Differences

So far, the generic principles of GNSS and A-GNSS systems have been sketched mainly based on GPS. In general there are many similarities between the GNSS systems. Even though the various GNSS are portrayed as "competing" with each other in some publications, they actually rather augment each other from a civil user perspective. While there were concerns some years ago that later GNSS systems might not be added in mass-market receivers, the past few years clearly showed that most mass-market receivers implement all global GNSSs to the benefit of the end user.

Nevertheless, special care must be taken when measurements from different GNSSs are combined. This is mainly due to different time systems and different frequencies used in the L1 band. Since the GNSSs are operated by independent governments, they all rely on their own time source. The various countries try to

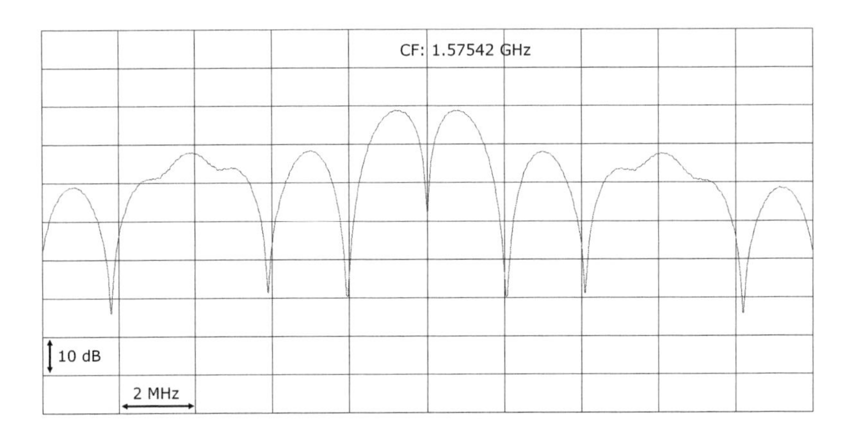

Figure 6.7 Spectrum of the E1 signal.

align their UTC times to each other, but slight differences remain, which can be handled in two ways: compensating the delay or solving for the time offsets. In the first variant, the GNSS-GNSS time offsets are transmitted in the navigation message of one or more systems (Galileo-GPS in Galileo, GLONASS-GPS in GLONASS, etc.). The receiver can treat all satellites as if they belong to one constellation. A 3D position fix can be obtained with one GPS + one GLONASS + one BeiDou + one Galileo satellite, for example.

However, there is a caveat: since BeiDou and GLONASS operate on slightly different L1 frequencies than GPS and Galileo, the signals might experience different group delays in the receiver's RF front-end leading to slight time offsets induced by the receiver. Either special care must be taken to compensate for these delays or Gauss' least-squares algorithm needs to be adapted: instead of solving for four unknowns (X, Y, Z, Time), the algorithm can be solved for seven unknowns in a quad-constellation receiver $(X, Y, Z, Time_{GPS}, Time_{GLO}, Time_{GAL}, Time_{BDS})$. Under the assumption that a sufficient number of satellites are visible (seven or more), this removes the intersystem time differences and even the time offsets due to filter delays. The price to pay for the latter variant is the higher number of satellites required for a fix.

Another hurdle for multi-GNSS receivers is the different representation of the orbit in the GLONASS system. The designers of the GLONASS system came up with a much different approach to convey the ephemeris information of their satellites. Instead of providing parameters for the Keplerian equations, they decided to transfer each satellite's position, velocity, and acceleration in ECEF coordinates

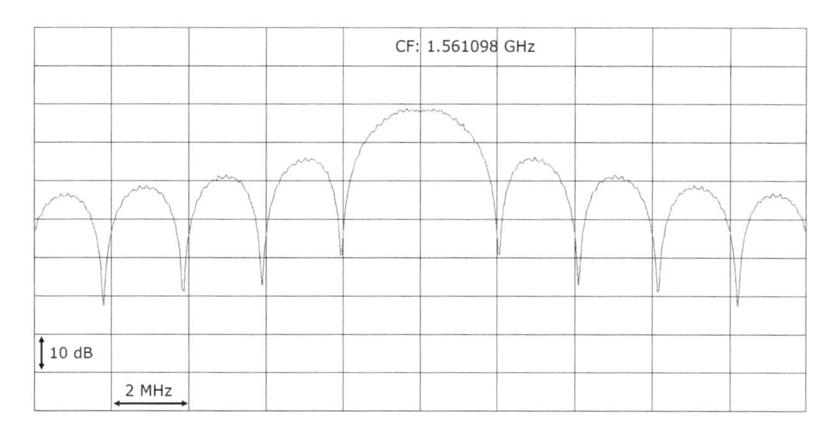

Figure 6.8 Spectrum of the B1I signal.

at a certain point in time. To derive the position at arbitrary time instances, the GLONASS interface control document (ICD) specifies a differential equation that needs to be solved using numerical integration (e.g., by using the Runge-Kutta approach).

Even though GLONASS differs in this aspect from the other systems, the end result of the calculation is again a satellite position in the ECEF frame, which can be seamlessly integrated in a multiconstellation fix. To better understand this difference, the following section will give an introduction to orbital mechanics used in GPS/BeiDou and Galileo and contrasts them with the approach of GLONASS.

6.4.2 Orbital Mechanics

6.4.2.1 ECEF and ECI frame

Before addressing the task of calculating a satellite's position, it is important to familiarize oneself with the various coordinate frames. The well-known position format of latitude/longitude/height (geographic coordinates) is an ECEF-polar coordinate frame. This format is traditionally chosen for terrestrial navigation. It has the convenient feature that a landmark on the Earth's surface does not change its coordinates over time (e.g., Munich will always be at 48N, 11E, 550 m). Calculating in a Cartesian coordinate system is easier in many cases and therefore GNSS receiver algorithms typically use the ECEF with Cartesian coordinates for the majority of the calculations. Using the position of Munich as an example, the Cartesian coordinates are still fixed over time, but defined as a 3D vector from the gravity center of the Earth (e.g., Munich is at X=4197522 m, Y=815916 m, Z=4717285 m). However, things are not that simple, as the Earth is not a perfect sphere, but rather an ellipsoid. To convert between LLH and Cartesian ECEF coordinates, it is necessary to make assumptions on the ellipsoid that approximates the Earth's shape. WGS84 is the most common ellipsoid and used in GPS. GLONASS (PZ-90.11), BeiDou, and Galileo use ellipsoids that currently are almost identical; hence, in many cases the differences are in the low cm range and may be ignored for standard precision applications.

While ECEF is convenient for terrestrial navigation and aviation, it is less appropriate for calculating orbital mechanics. Calculations using Newton's and Kepler's laws are typically performed in an inertial coordinate frame. The ECI coordinate frame introduced in Chapter 2, Section 2.2.

Since the average positioning problem requires a location in ECEF coordinates rather than ECI, the designers of GPS, Galileo, and BeiDou decided to use pseudo Keplerian parameters instead of the plain Keplerian parameters. This has the practical advantage that the result of the equations specified in the ICDs can be easily converted to the format typically required by end users.

In the case of GLONASS, it is required to convert the ECEF ephemeris parameters to ECI for the numerical integration.

6.4.2.2 Keplerian Orbit Description

The Keplerian orbit is a suitable model for a GNSS satellite orbiting Earth. The shape of the trajectory is defined by two parameters:

- $\sqrt{A}$ (square root of semi-major axis)
- e (eccentricity)

If the eccentricity has a value below 1 (which is the case for all GNSS orbits), the orbit has the shape of an ellipse.

The orientation of the ellipse versus the frame is specified by three angles (defined in rad for GNSS):

- Ω (angle between vernal equinox and ascending node)
- ω (angle between ascending node and perigee)
- i (inclination)

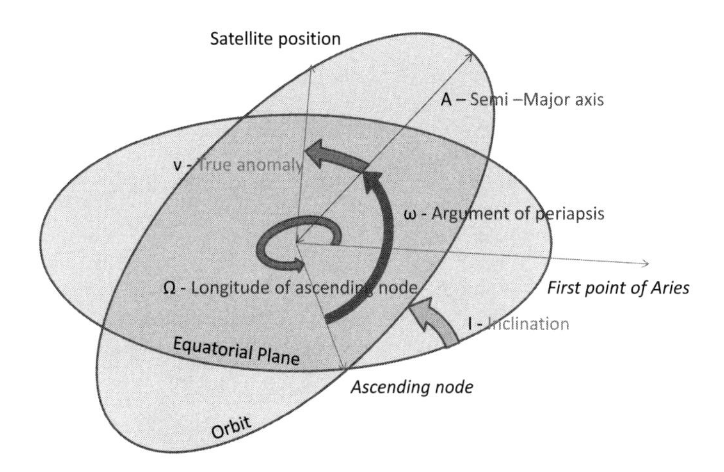

Figure 6.9 Keplerian parameters.

The vernal equinox is defined as the direction from Earth to Sun at the day-night equilibrium in spring.

Finally, the position of the satellite at a specific time is given by the angle of the mean anomaly:

- M_0 (mean anomaly at reference time)

Additionally, a time of ephemeris or t_{oe} is given (in the case of the almanac: time of almanac, t_{oa}):

- t_{oe}

This set of six parameters plus time form the classical Keplerian parameter set. Figure 6.9 shows the parameters.

In reality, the problem is more complex; for example the Earth's mass is not uniform and additional celestial bodies influence the orbit (mainly the moon and sun). To accommodate this, GPS, BeiDou, and Galileo introduced additional parameters:

- Δ_n: mean motion difference from computed value
- $\dot{\Omega}$: rate of right ascension
- $\dot{I}$: rate of inclination
- C_{rs}: amplitude of the sine correction term to the orbit radius
- C_{uc}: amplitude of the cosine harmonic correction term to the argument of latitude
- C_{us}: amplitude of the sine harmonic correction term to the argument of latitude
- C_{ic}: amplitude of the cosine harmonic correction term to the angle of inclination
- C_{is}: amplitude of the sine harmonic correction term to the angle of inclination
- C_{rc}: amplitude of the cosine harmonic correction term to the orbit radius

Together with a data set ID that changes typically every 2 hours, the transmitted ephemeris consists of 15 elements plus reference time. Modernized messages like CNAV and CNAV2 contain some more parameters to refine the accuracy even more (e.g., the change rate of the $\sqrt{A}$).

To calculate the satellite location, a set of equations is published for each of the four GNSS systems.

6.4.2.3 GLONASS Orbit Description

As mentioned earlier, GLONASS utilizes a different approach to specify the satellite's trajectory. The GLONASS ICD describes the ephemeris as "immediate information," while for the almanac, GLONASS also employs a Keplerian representation. The main information in the GLONASS ephemeris message consists of

- X (satellite position in ECEF's X-direction at time t_b)
- Y (satellite position in ECEF's Y-direction at time t_b)
- Z (satellite position in ECEF's Z-direction at time t_b)
- $\dot{X}$ (satellite velocity in ECEF's X-direction at time t_b)
- $\dot{Y}$ (satellite velocity in ECEF's Y-direction at time t_b)
- $\dot{Z}$ (satellite velocity in ECEF's Z-direction at time t_b)

- $\ddot{X}$ (satellite acceleration in ECEF's X-direction at time t_b)
- $\ddot{Y}$ (satellite acceleration in ECEF's Y-direction at time t_b)
- $\ddot{Z}$ (satellite acceleration in ECEF's Z-direction at time t_b)
- T_b (calculated from t_b, P1 and P2)

Depending on the required accuracy, the GLONASS ICD details a complex approach and a simplified form for the numeric integration that results in the satellite's position at a certain time.

The interested reader is referred to the GLONASS ICD [5], an exemplary implementation can be found as part of the GPS Toolkit [6] in the file GPSTk/core/lib/GNSSEph/GloEphemeris.cpp.

6.5 GNSS MULTIFREQUENCY

6.5.1 GNSS Legacy Signals and Modernized GNSS Signals

From the start of the GNSS age in the late 1970s until the early 2000s, there were only two civil GNSS signals available: GPS L1 C/A and GLONASS FDMA L1OF. Both signals were initially only of limited use for different reasons: before May 2, 2000, the GPS L1 C/A signal was deliberately degraded to a precision of approximately 100 m by introducing artificial jitter and/or imprecise navigation messages (selective availability). The GLONASS system on the other hand could not provide constant global coverage before the year 2011, when its modernization was finished.

The years from 2010 to 2020 saw the renewal and emergence of multiple GNSS systems: GLONASS global coverage reestablished in 2011, BeiDou regional operation started in 2012, Galileo initial services started in 2016, and Navic and QZSS started in 2018. Both BeiDou and Galileo are scheduled to reach global coverage by the end of 2020. The receiver industry recently extended its GPS receivers to multi-GNSS receivers accordingly. However, until 2018, virtually all mass-market receivers were using only the L1 frequencies: 1575.42 MHz (GPS, Galileo, QZSS), 1602 MHz (GLONASS), and 1560 MHz (BeiDou), which are very close to each other so that they could be covered by a single RF front-end.

All systems additionally broadcast civil signals on other bands, which remain to date unused by mass-market chipsets:

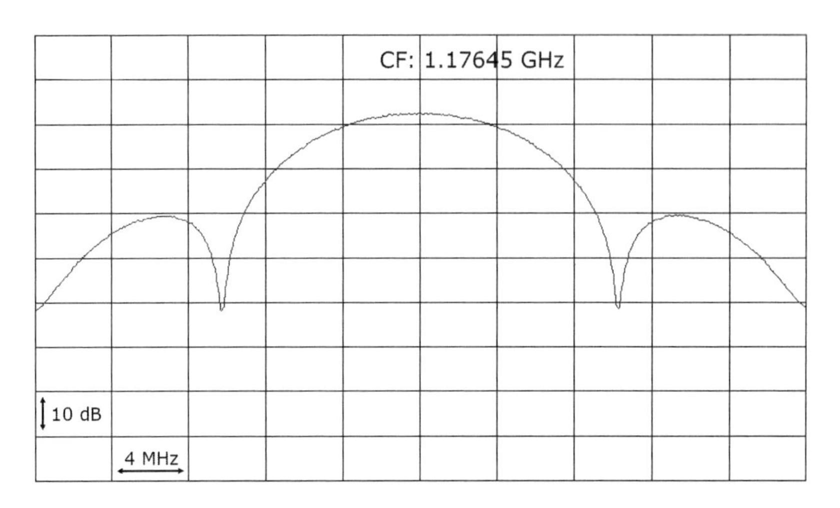

Figure 6.10 Spectrum of the GPS L5 signal.

1. GPS, QZSS, Galileo, BeiDou-3: L5/E5/B2I/B2a
2. GPS, QZSS, GLONASS: L2C/L2OF
3. Galileo, QZSS: E6/L6

These signals were originally intended for surveying and aeronautics, but are also beneficial for any kind of receiver. As discussed in Section 6.2.4, single-frequency receivers cannot remove the ionospheric error completely. Only a dual-frequency receiver can remove the ionospheric error without the need for ionospheric models. For this purpose, a clone of the L1 signals on a frequency sufficiently separated would already be working. This approach was used by GLONASS L2OF. The GPS/Galileo signals on L5/E5 and L2, however, are not simple copies of the L1 signals. The L5 chip rate for example is higher by a factor of 10, leading to a higher spreading gain, more robust antijamming capabilities, and a better multipath resolution due to the narrow correlation peak. Due to the significant advantages the new signals deliver to end users, multifrequency capable multi-GNSS receivers will appear in the mass market in the years after 2018. An overview of the various signal characteristics is provided in [7].

6.5.1.1 Modernized GPS: L2C and L5

In the case of GPS, the new signals also transmit a modernized navigation message, which is more robust to bit errors due to forward error correction (FEC) schemes and reduces the time-to-first-fix in standalone receivers. The new message format for the L2C and L5 signals is called CNAV, and is highly modular by using various message types that can be scheduled with a large degree of freedom. As of 2018, the following message types are transmitted by the newer GPS spacecraft:

- Message Type 10 (Ephemeris part 1)
- Message Type 11 (Ephemeris part 2)
- Message Type 12 (Reduced Almanac)
- Message Type 30 (Clock-corrections + Iono. Model + TGD-corrections)
- Message Type 33 (Clock-corrections + UTC-Model)

Furthermore, the following messages are specified in the ICD, but currently not transmitted:

- Message Type 31 (Clock-correction + Reduced Almanac)
- Message Type 32 (Clock-correction + Earth-Orientation-Parameters)
- Message Type 34 (Clock-correction + Differential corrections)
- Message Type 35 (Clock-correction + GPS-Galileo time difference)
- Message Type 36 (Clock-correction + Text)
- Message Type 37 (Clock-correction + Midi Almanac)
- Message Type 13 (Clock-correction + Differential corrections)
- Message Type 14 (Ephemeris + Differential corrections)
- Message Type 15 (Text)

While the midi Almanac is quite similar to the legacy almanac, the reduced almanac is a further compressed variant to save bandwidth.

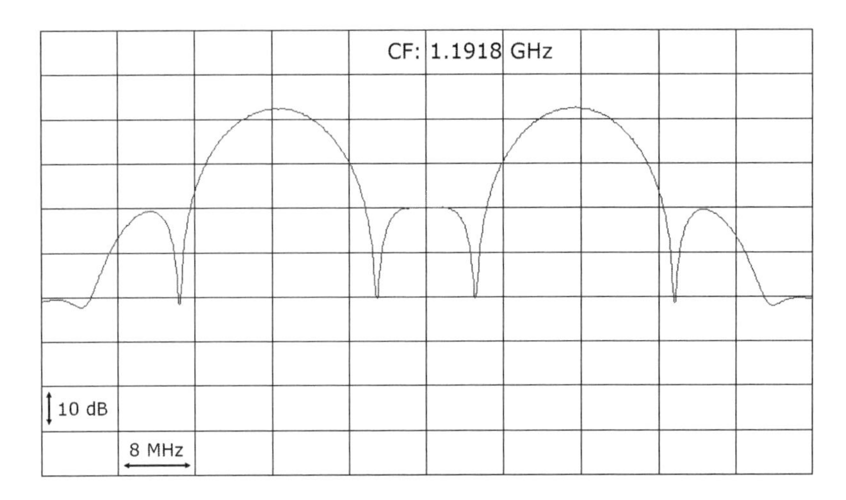

Figure 6.11 Spectrum of the Galileo E5a and E5b signal.

6.5.1.2 GPS III: L1C

GPS Block III satellites launched since 2018 will introduce yet another new signal: L1C (additionally to the legacy L1C/A on the same frequency). It will be similar to the Galileo E1 signal, as it uses multiplexed binary offset carrier (MBOC) modulation. The navigation message will differ from the L2C and L5 signals and is therefore called CNAV/2. It is specified in IS-GPS-800. It is not as modular as the CNAV, but rather optimized for a fast standalone time-to-first-fix.

6.5.1.3 Modernized A-GPS

The usage of the CNAV messages in the A-GNSS mode is not mandatory: if a slightly inferior resolution is acceptable, the L1 C/A legacy navigation messages can also be used for L2 and L5. However, some optional parameters are unique to the CNAV message (e.g., the group-delay differences between L1 and L5). Furthermore, for data-bit wipe-off, the transmission of CNAV messages is beneficial.

6.5.1.4 Galileo

Galileo supported E1 and E5 signals from the beginning. An additional service, not yet officially specified but capable of being enabled via a software update in the satellites, is the E6 signal. It is a much different signal than the ones broadcast on the other frequencies. It will provide PPP correction services at high data rates and offers authenticated ranging codes, which are spoofing-resistant. PPP will be explained in detail in Chapter 7 of this book.

6.5.1.5 QZSS

QZSS transmits signals that are exactly same as the GPS L1 C/A, L1C, L2C, L5 [8], and a signal similar to the Galileo E6 [9].

6.5.1.6 BeiDou-2 Multifrequency

The second phase of the BeiDou system (the first one to go global) already included three civil frequencies in the BeiDou-2 variant. The second frequency B2I is supposed to be replaced by the B2a signal in the near future. The B3I signal is a BPSK signal with a larger spreading factor.

6.5.1.7 BeiDou-3 Multifrequency

BeiDou-3, the third phase of the BeiDou system, will support the legacy BeiDou-2 signals B1I and B3I. Additionally, the BeiDou-3 will transmit compatible signals aligned to GPS and Galileo: the B2a signal is aligned to the GPS L5 and Galileo E5a and the B1C signal is aligned to GPS L1C and Galileo E1. This international cooperation will simplify the design of multi-GNSS L1+L5 receivers.

6.5.1.8 GLONASS Multifrequency

The L2OF signal is a simple clone of the L1OF FDMA signal. Several new CDMA-based signals are envisioned, which will simplify receiver design. The first CDMA signal is the L3-OC signal, which is broadcast from three satellites as of 2018. In the future, a L1-OC signal might be added, colocated at the international 1575.42 MHz frequency.

6.5.1.9 Implementation in 2018/2019 Mass-Market Chipsets

The first dual-frequency L1/E1+L5/E5 capable smartphone receiver became available in 2018. It is expected that L2C will also be added to mass-market receivers. Together with the modernized BeiDou-3 and GPS III signals, a mass-market receiver in the year 2020 will be able to take advantage of 14 global civil signals from four constellations in a frequency range from 1100 MHz to 1602 MHz: L1 C/A, L1C, L2C, L5, E1, E5a, E5b, E6, L1OF, L2OF, B1I, B1C, B2a, and B3I.

6.5.2 GNSS Signal Characteristics Overview

6.5.2.1 Legacy L1

Legacy L1 signals (GPS L1 C/A, GLONASS L1OF, B1I) of the first generation have the following features in common (see [5, 10, 11]):

- BPSK modulation scheme (the characteristic BPSK spectrum can be seen in Figure 6.8 for the B1I signal).
- Small spreading factor (GPS: 1.023 Mchips/s/50 symbols/s).
- No forward error correction on NAV message.

6.5.2.2 Modern L1

Modern L1 signals (E1, B1C, L1C) added the MBOC modulation scheme with the following advantages [12, 13, 15]:

- MBOC modulation scheme, backward-compatible with GPS L1 C/A, sharper correlation peak.
- Forward error correction on Nav message.
- Slightly higher bandwidth leading to better multipath mitigation.
- Dedicated pilot signal component (data-less), longer coherent correlation time possible.
- GPS L1C: interleaving.

6.5.2.3 L2C

The GPS/QZSS L2C signal again uses the simpler BPSK with the same chip rate as L1 C/A, but still offers additional benefits [10]:

- BPSK modulation scheme.
- Several hundred MHz apart from L1 for ionospheric correction.

- Forward error correction.
- Small spreading factor for GPS: 1.023 Mchips/s/50 symbols/s, process gain: 43 dB.
- Dedicated pilot signal component (data-less), longer coherent correlation time possible.
- Longer code: CM and CL, better code separation in high-dynamic range situations, narrowband interference suppression.
- More accurate orbit data in messages.

6.5.2.4 L5/E5/B2a

The L5/E5a/E5b/B2a signals' main benefit is the significantly higher spreading factor that can be seen in Figure 6.10) and Figure 6.11 [12, 14, 16]:

- BPSK modulation scheme.
- Several hundred MHz apart from L1 for ionospheric correction.
- Forward error correction on Nav message.
- Large spreading factor 10.23 MChips/s/100 symbols/s, process gain: 50 dB.
- Dedicated pilot signal component (data-less), longer coherent correlation time possible.
- Long codes/higher chiprate: 10230 chips per ms.
- Higher power than L1 C/A and L2C.

While some of the advantages of these new signals are only relevant in standalone operation (e.g., FEC or the new navigation message), there are also significant advantages for A-GNSS-capable devices to utilize these signals.

In open-sky conditions, the largest source of error today is the ionospheric delay. Using two frequencies, which are separated several hundred MHz from each other, it is possible to remove the ionospheric delay due to the dispersive nature of the delay. Basically, the delay is a function of the carrier frequency; hence, a receiver can combine the measurements on two frequencies to correct the delay.

In urban environments, the largest error source in the range measurement is caused by multipath reflections. Ideally, only LOS measurements should be taken into account, but in the case of a reflection, it is not always possible to separate the LOS and the reflection in the correlation. The correlation peaks of both components smear into each other. The new signals help multipath mitigation in two ways:

- L5 signals use a higher chip-rate, which spreads the signal wider in the frequency domain and leads to a narrower correlation peak in the time domain.
- L1C, E1 and B1C utilize BOC modulations, which spread the signal power on a wider bandwidth (as can be seen in Figure 6.7) and leads to a narrower correlation peak.

Furthermore, the L5 signal is broadcast with higher power and with a larger bandwidth, leading to better indoor penetration and more resistance to RF interference or jamming.

Another advantage is the pilot signal component. While the classic L1 signals were always modulated with data bits, the new signals either carry a data-less component in Q-phase (L5, E5) or use time multiplexing to broadcast the pilot. BPSK-modulated data bits lead to phase jumps of pi/2, which limit the coherent correlation time (unless they are anticipated and removed using data wipe-off).

Finally, the modernized signals broadcast longer codes (L2C CL: 767250 chips/CM: 10230 chips, L5 10230 chips, E1 4092*25=102300; compared to L1: 1-ms period at 1.023 Mchips/s). This property helps reducing the cross-correlation between signals from different satellites. Scenarios with high dynamic range will especially profit from this feature, since the lower cross-correlation allows detection of low-power signals even if a high-power satellite dominates the received signal. Nevertheless, the advantage of the longer code length comes at the expense of an increased number of correlators. State-of-the-art L1 C/A receivers are capable of correlating all possible code positions for each satellite at the same time. According to [2], this means 2*1023 complex correlators per satellite. To search for 10 visible satellites simultaneously during acquisition stage, 10*2*1023=20460 complex correlators are required. If a L5 or L2C signal should be acquired directly in the same amount of time, the number increases to 10*2*10230=204600. A straightforward implementation would lead to a massive increase in silicon area, which means additional cost and power consumption. However, some alternatives exist, which can reduce the silicon complexity significantly:

- Using L1 C/A code to determine the code-phase of L5 and L2C.
- A-GNSS with code-phase assistance and fine-time aiding.

Both options have limitations: the need for a mobile network connection in the latter and the restriction to acquire L1 first in the former. If L1 acquisition is required, it is the weakest link in the chain and the improved indoor penetration and RF interference resistance of L5 are lost. However, other advantages remain: improved multipath mitigation and removal of ionospheric effects. Since modern smartphones can use complementary techniques for indoor location (e.g., WLAN RSSI, as seen in Chapter 9), and jamming resistance might not be a paramount design goal, the L1 aiding strategy may be a viable solution for standalone GNSS or outside mobile network coverage. With A-GNSS available, a receiver can utilize all benefits of the L5 and L2C signals.

6.5.3 GNSS Frequency Bands

The frequencies used today for civil GNSS are almost exclusively in the L-band between 1164 MHZ and 1350 MHz and 1559-1610 MHz. Navic is currently the only system that nominally transmits an additional signal in the S-band (2483.5-2500 MHz) [17], but some BeiDou-3 satellites also carry an experimental signal. The L-band is further subdivided into L1, L2, L5 and L6. The numbers of the subbands have no relation to the actual frequency. If sorted from lowest to highest frequency, the order is L5, L2, L6, L1. While L5, L2, and L6 are adjacent to each other, there is a gap of more than 200 MHz between L6 and L1. Depending on the GNSS system, the nomenclature differs (e.g., Galileo specifications specify the L1 and L5 bands as E1 and E5, respectively). BeiDou's B2a signal is located exactly at the GPS L5 frequency. For the sake of clarity, band names follow the GPS nomenclature and the Galileo E6 band is referred to as L6 in the following section. The L3 and L4 bands are not used for navigation purposes as of today. Besides the L-band and the S-band, the ITU allocated a third frequency band for GNSS usage: C1. So far, no GNSS transmits civil ranging signals in the C1 band (5010-5030). The international coordination and protection of the GNSS bands is conducted by the ITU and codified in resolutions, which are typically agreed in World Radio Conferences (WRC). A list of GNSS-related resolutions can be found in an *Inside GNSS* article [18].

6.5.3.1 In-Band Interference Impacting GNSS

The L1 and the L5 bands are highly protected internationally as Aeronautical Navigation Service Bands (ARNS). The L1 band is even exclusively allocated to

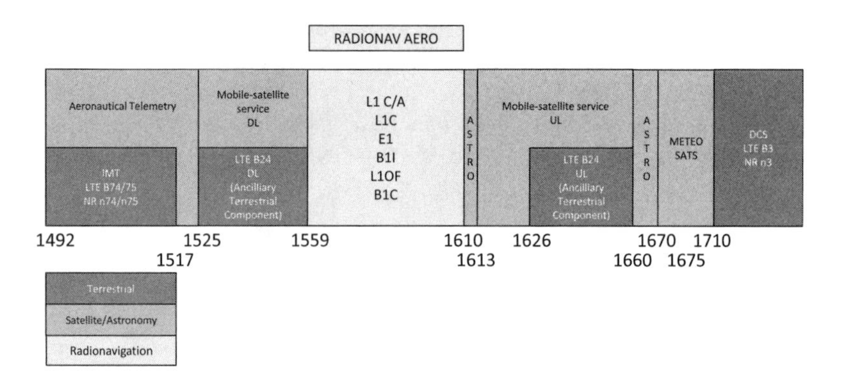

Figure 6.12 Spectrum allocations around L1.

GNSS, and consequently there is no legal interference within the band, as can be seen in Figure 6.12.

L5 is primarily allocated to ARNS, with distance measuring equipment (DME) systems (used by aircraft to measure the distance to ground beacons for navigation purposes) being the most prominent and powerful user of the band [19]. The principle behind DME is based on round-trip time measurements. Aircraft send short pulses, which are answered by DME ground stations after a well-defined time. The peak power of a typical DME ground transmitter is 1-2 kW. DME interrogators installed in airplanes do not transmit inside the L5 frequency, but a GNSS receiver near a DME ground station in a busy airspace may experience interference with a worst-case duty cycle of 4.9 % of the time per DME station. As of 2018, the FAA operates 800 DME stations [20] in the United States. In the San Francisco Bay area, it is possible to receive 8 DME ground stations at the same time, leading to significant interference [21,22]. DME is foreseen to be one major backup system for GNSS [23] in aeronautical navigation. Hence, it cannot be expected that this interference will decline in the future and GNSS receivers therefore need to

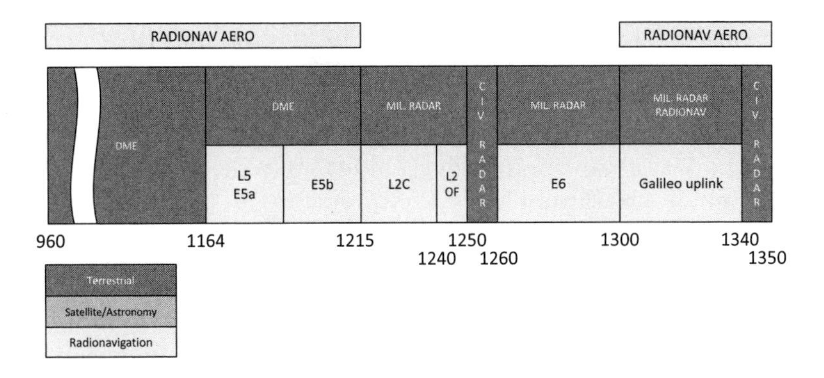

Figure 6.13 Spectrum allocations around L1, L2, L6.

be capable of handling the interference caused by DME systems by employing techniques like pulse blanking or notch filtering. While DMEs interfere with the L5 signals, the designation of the band to ARNS on a worldwide basis provides a high legal protection against other interference. However, there may be military radar stations transmitting directly above 1215 MHz, which might impact receivers, especially those using the E5b band.

L2 and L6 are less protected internationally and primarily allocated to civil and military radar surveillance services outside the United States. Civil ATC L-band radars usually operate in the 1250-1260 MHz range. They are located between GLONASS L2OF spectrum and the Galileo and BeiDou L6 band and well above the GPS L2 band. In the United States, radars of all kind avoid the GPS L2 center frequency to reduce the effects ([24, 25]), but this behavior is not guaranteed by ITU regulations world-wide.

The L6 band is allocated to military surveillance radars on a primary basis, and on a secondary basis to radio amateurs, which use it for services like digipeaters. Since the interference caused by military radar services and radio amateurs is not always known in advance and might change over time, interference mitigation in GNSS receivers may be more challenging in L2 and L6 than in the relatively benign L5 band. A exemplary description of the effects and mitigation strategies in the receiver can be found in [26, 27].

The S-band used by Navic is located between the WLAN 2.45-2.83 GHz band and the uplink of LTE band 7 starting at 2.5 GHz. The band itself is not used by other high-power systems in most parts of the world. However, in Japan, legacy 802.11b WLAN devices might interfere if they are configured to use channel 14.

The C-band envisioned for future signals is primarily assigned to aeronautical navigation. This band was originally reserved for the Microwave Landing System (MLS), which was intended to replace ILS systems. MLS never saw widespread adoption due to the advent of GPS-based instrument landing systems, which are significantly cheaper to operate for airports. The bands adjacent to 5010-5030 MHz are satellite services, and therefore the environment is relatively silent and suitable for GNSS signals.

6.5.3.2 Interference Caused by GNSS

Due to the very low power of GNSS signals, typically no interference is caused for other applications except radio astronomy and highly sensitive radar receivers. Nonetheless, some measures have been taken to further mitigate the interference. For instance, GLONASS no longer uses channel numbers higher than +6 to protect the radio astronomy band above 1610 MHz on L1OF and civil ATC in the bands above 1250 on L2OF.

6.5.3.3 Self-Interference and Spurious Emissions

A challenge for mobile phones is self-interference from the cellular modem. For example, LTE bands 13 and 14 have a second harmonic of their uplink signal (787.5 MHz) in the L1 band. Similarly, future potential LTE or NR bands with an uplink signal around 588 MHz, 614 MHz, and 639 MHz might cause self-interference in the L5, L2, and L6 bands (as of 2018, no bands in this frequency range have been defined by the cellular industry). Devices incorporating Navic and WLAN or LTE/NR band 7 might need additional precautions to avoid self-interference.

6.5.3.4 Adjacent Band Compatibility

The ever-increasing demand for mobile broadband data leads to high pressure to use all available radio spectrum resources, especially in the lower bands below 6

GHz. To balance protection of GNSS receivers with the desire for additional data bandwidth, several initiatives in the United States and Europe are ongoing. As of 2018, the LTE bands 3, 24, 74, and 75 are in the vicinity of the L1 band as can be seen in Figure 6.13. The U.S. Department of Transportation conducted the Adjacent Band Compatibility Study [28], which identified the interference tolerance masks (ITMs) of existing GPS receivers.

In Europe, ETSI was tasked with the development of a GNSS receiver standard (EN 303 413 [29]) as part of the European Commission's Radio Equipment Directive (RED).

6.5.3.5 Implications on Receiver Design

A mass-market RF front-end can cover all civil legacy signals within one band (e.g., L1 C/A, B1I, E1, L1OF). A gap of more than 200 MHz between L1 and the other bands, however, typically results in the need for a second RF chain in dual-frequency receivers. Furthermore, the antenna(s) need to be specially designed for two or more bands. The gap between L5/L2/L6 and L1 is used by various other applications, auch as radar, satellite communication links (Iridium, etc.), cellular networks, and radio astronomy. Using a second RF chain compared to a single broadband RF has the advantage that a strong blocking radar signal will not deteriorate the receiver's performance. On the other hand, the phase relation of the signals is lost with multiple RF front-ends, which might restrict some high-precision algorithms.

6.6 RELIABILITY AND REDUNDANCY

Since GNSSs are complex systems designed and operated by humans, there is always a risk of failures, even in the ground or space segment.

6.6.1 Receiver Autonomous Integrity Monitoring

To add an additional safety net, a concept known as receiver autonomous integrity monitoring (RAIM) [2][Chapter 8.4.3] can be used, given that the number of satellites is sufficiently high: if the equation system is overdetermined, the Gaussian least-squares method also provides the residual error. If one satellite provides incorrect data, it will degrade the residual error of the solution. The RAIM approach detects this and performs calculations based on subsets of the original measurements, sorting out one satellite after the other. If only a few satellites are affected, this method can safely detect and remove it from the solution and indicate the error. This approach can be extended to multi-GNSS receivers, so that even a systemwide error of one GNSS could be detected in a RAIM-capable multi-GNSS receiver.

6.6.2 SBAS

The need for higher accuracy and integrity, driven by civil aviation, lead to the development of various SBASs around the globe: WAAS (United States), EGNOS (Europe), GPS-aided GEO augmented navigation (GAGAN) (India), and multi-functional satellite augmentation system (MSAS) (Japan). These systems augment the major GNSSs by transmitting regional ionospheric corrections and integrity information over geostationary satellites. While the GNSSs require specialized satellites carrying atomic clocks, the SBASs can be transmitted via commercial TV satellites. Hence, the SBAS signal is typically not used for ranging. The main goal here is the payload. SBAS transmits the payload at a significantly higher bit rate. SBAS provides a grid of ionospheric corrections, which allows a much better ionospheric correction for single-frequency receivers. Together with the integrity information, it allows commercial airliners to use GNSS for instrument landing without the expensive Instrument Landing System (ILS). Although intended for aviation, the SBAS signal can also be utilized by many mass-market receivers to improve the accuracy. Discussion is ongoing whether L5 signals are added to SBASs. SBASs can be considered as a simple form of PPP, although the resolution of dedicated PPP services is much higher.

6.6.2.1 SBAS Integrity Monitoring

SBASs offer an additional error detection capability by monitoring the system's integrity at several ground stations. In the case of a failure, a warning is broadcast via dedicated geostationary satellites, which are independent of the GNSS. The Galileo system also plans to offer an in-built integrity information in the I/NAV message broadcast on E1 and E5b (Safety-of-life service).

6.6.2.2 SBAS Ionospheric Integrity Estimates

While it was already mentioned that SBASs transmit an ionospheric grid of VTEC values to improve the ranging error, the GEO satellites also transmit an uncertainty value linked to the VTEC values, to allow the receiver estimation of a worst-case bounding box around the vehicle (typically for aircraft in final approach).

6.6.3 GPS Week Rollover

Before the year 2000, there was great concern in the IT industry that some systems might not handle date rollover correctly, since many SW components used only two digits for the year. Analogously, there are rollover dates in some GNSS signals which might also lead to problems in receivers if not handled correctly. One such issue is the GPS week rollover, which comes from the fact that the GPS week number is only represented by 1024 values in the GPS L1 C/A navigation message. 1024 weeks equal approximately 19.7 years and since GPS time started in January 1980, the first rollover occurred in 1999 (with only a limited number of civil GPS devices in use). The second rollover happened on the night of April 6/7 2019 and lead to misleading UTC dates in some implementations.

6.6.4 UTC Leap Second

Another common source of error is the leap-second transition. Depending on the Earth rotation rate, which is not perfectly predictable, approximately every 2 years a leap second is added or removed to UTC. While GPS, Galileo, and BeiDou specs are clear on the procedure, some receivers might experience issues handling this transition appropriately. GLONASS is a special case, since it does not use a leap-second free time system, but uses UTC directly. The behavior of GLONASS during leap-second events was initially not clearly specified. Hence, receivers need to handle this event carefully.

6.7 GNSS, THE RAN, AND THE CORE NETWORK

Albeit less prominent than the role GNSS plays in mobile handsets, its usage in the network is of equal, if not higher importance. The main aspect of interest in that case is the time and frequency synchronization. During the evolution of five generations of cellular radio networks so far, the requirements for synchronization became increasingly important. Techniques like coordinated multipoint (CoMP), intercell interference coordination (ICIC), and the TDD variants of LTE and NR allow the efficient usage of spectrum and rely in many cases on an accurate timing of all cells in a radio network. Cellular network based positioning technologies like OTDOA also depend on a low synchronization error in the range of <50 ns. The core network relies on precise timing for multiple aspects (e.g., accounting and billing). The accuracy required by these techniques can be achieved by wire-based protocols like the IEEE.1588 PTP protocol in the case of modern Ethernet backhauls (or ATM for legacy networks). However, in most cases it is easier and more cost-effective to use a GNSS-disciplined oscillator at a base station. The reliance on GNSS-based timing makes many cellular networks vulnerable to GNSS-jamming and spoofing, which lead the U.S. Department of Homeland Security to publish a guide on how to reduce this risk: "Improving the Operation and Development of Global Positioning System (GPS) Equipment Used by Critical Infrastructure" [30]. If an atomic clock is used as the GNSS-disciplined oscillator, the base stations are able to bridge GNSS outages of several hours before going out of sync.

6.7.1 Dynamic Spectrum Use with the Aid of GNSS

Historically, spectrum for cellular networks was assigned on a primary basis to one operator in a certain region for several years. The cellular network was subsequently planned and built out in a relatively static manor. With the advent of single-RANs and software-defined radios, base stations could act more dynamically and share spectrum with other users. A prominent example is the Citizens Broadband Radio Service (CBRS) band in the United States, where three user classes share the band dynamically in space and time. The primary user of the band is the U.S. Navy, which employs shipborne radar systems. The secondary user class is a priority access license (PAL), which can be bought at an auction. The third class of users is called general authorized access (GAA). The radar systems are protected by a environmental sensing capability (ESC), which is a sensor network detecting radar signals from ships in coastal regions. If no radar signal is detected, the PAL holders may use the band for cellular networks. If no PAL holder utilizes the spectrum according to a dynamic database, a GAA user can use the spectrum free of charge at a certain location. Sophisticated licensing schemes like CBRS will increasingly require knowledge of the base station location and it is likely that GNSS systems will be used for positioning.

6.8 GNSS LIMITATIONS AND COMPLEMENTARY SYSTEMS

GNSS can be considered the backbone of PNT applications today. It provides an absolute position and time wherever sky view is available, free of charge. Its accuracy (especially using RTK or PPP) is unmatched by any other mass-market technology. The technical availability of all GNSS is very high and systemwide outages are extremely rare. Nevertheless, today's GNSS have limitations as well:

- Very low signal power at receiver.
  - Limited indoor penetration.
  - Easy to be jammed with low-power equipment.
- No integrity protection for most of the civil signals.
  - Spoofing possible (exception: Galileo E6, E1B, with Message Authentication).
- Vulnerable to space weather conditions.
  - Service could be impacted by a strong solar storm.
- Legislative obstacles.
  - Foreign GNSS might be barred from use within some countries.

The spoofing threat might be solved technically with the latest generation of signals (E1B Navigation Message Authentication, E6 with encrypted ranging code). The limitations that are a consequence of the low received power (limited indoor penetration, jamming) will persist with the current GNSS constellations as the transmit power is limited by the power provided by the satellite's solar cells.

6.8.1 STL - LEO Satellite Based Positioning

The private company Iridium, which operates a fleet of approximately 60 LEO satellites for global communication, also launched a commercial timing and location service called STL. Due to the fact that Iridium's LEO satellites are 25 times closer to the receiver than traditional GNSS satellites, the STL signals are 30 dB more powerful than today's civil GNSS L1 signals. The encryption of the STL signals can be considered as an additional benefit, as it makes the system spoofing resistant. Iridium is optimized for communications, and therefore only 1-2 satellites are typically visible to a receiver. Furthermore, the satellites do not carry atomic clocks. Thus, the trilateration approach is not feasible with STL. The system rather relies on the Doppler effect to derive the position (like the Transit system, a predecessor of GPS) and typically measures one satellite at several points in time. As low Earth orbit (LEO) satellites have a high velocity relative to the user, during a period of 10 minutes, the receiver will measure significantly differing Doppler shifts while the satellite passes over it and can calculate a position from that information. Naturally, a moving receiver also creates an additional offset to the measured Doppler, which impacts the accuracy. STL claims to achieve an accuracy of approximately 50 m for a moving vehicle and a time-to-fix of several minutes. This accuracy may be sufficient as an additional mean for integrity checking or in indoor cases.

6.8.2 Terrestrial Technologies and IMUs

There are also a number of terrestrial technologies that are employed to provide deep indoor coverage and increase jamming resistance. In the mobile phone market, WLAN reference signal strength indicator (RSSI) database matching and LTE/ECID+OTDOA are widely used to support deep indoor positioning and integrity (see Chapters 8 and 9). Several private companies deploy metropolitan beacon systems (pseudolite systems), which provide high-power, integrity-protected GNSS-like signals on dedicated frequencies (Chapter 9). The aviation industry can rely on their pre-GNSS radio navigation technologies like VHF omnidirectional radio range (VOR)/DME and IMUs to backup GNSS. The United States, United Kingdom and South Korea are currently testing the terrestrial eLORAN long-wave system as a complementary GNSS backup. The automotive industry uses map-matching, odometers, and steering-wheel inputs to bridge GNSS signal gaps in parking garages and tunnels.

6.9 CONCLUSION

This chapter has introduced GNSS and described its principle of operation and applications to cellular network positioning. It began with standalone GNSS systems, detailing the different parameters such as the ephemeris or the almanac and explaining the influence of the different perturbation factors like atmospheric effects. From there, this chapter moved on to assisted GNSS, clarifying the motivation for why to send assistance data for A-GNSS over the cellular network.

The second part of the chapter presented the different GNSS deployments around the world as well as the multiple frequency bands and signals of each. Next, the reliability and redundancy of the GNSS signals were studied, also explaining additional methods to improve the robustness of GNSS.

Finally, the limitations of GNSS were analyzed. Although it is a powerful positioning technology, GNSS alone cannot cover all the use cases needed in cellular network's location-based services. Thus, the next chapters of this book will explain further improvements (differential GNSS) and complimentary technologies (e.g., OTDOA, ECID, WiFi RSSI). All together, these technologies will form the LBS ecosystem in 5G and future networks to come.

References

[1] Spilker Jr, J.J., et al, *Global Positioning System - Theory and Applications*, Volume 1, American Institute of Aeronautics and Astronautics, Inc., August, 2012.

[2] van Diggelen, F. *A-GPS: Assisted GPS, GNSS, and SBAS*, Norwood, MA: Artech House, 2009.

[3] European Commission, "Ionospheric Correction Algorithm for Galileo Single Frequency Users," *European GNSS (Galileo) Open Service*, 2016.

[4] 3GPP TS 37.571-1, *UTRA and E-UTRA and EPC; UE Conformance Specification for UE Positioning; Part 1: Conformance Test Specification*, V16.3.0, January, 2020.

[5] Global Navigation Satellite System Glonass, "Navigational Radiosignal in Bands L1, L2," *Interface Control Document*, Edition 5.1, Moscow, Russia, 2008.

[6] https://github.com/SGL-UT/GPSTk/, accessed on December 2019.

[7] Eurocontrol, "Deliverable C5: Compatibility Criteria and Test Specification for GNSS," *FCI Technology Investigations: L Band Compatibility Criteria and Interference Scenarios Study*, August, 2009.

[8] Quasi-Zenith Satellite System, "Satellite Positioning, Navigation and Timing Service (IS-QZSS-PNT-003)," *Interface Specification*, November, 2018.

[9] Quasi-Zenith Satellite System, "Centimeter Level Augmentation Service (IS-QZSS-L6-001)," *Interface Specification*, November, 2018.

[10] Global Positioning Systems Directorate, Systems Engineering & Integration, "Interface Specification IS-GPS-200J," *NAVSTAR GPS Space Segment/Navigation User Segment Interfaces*, April, 2018.

[11] BeiDou Navigation Satellite System, "Signal in Space Interface Control Document, Open Service Signal," *China Satellite Navigation Office*, Version 2.0, December, 2013.

[12] European GNSS (Galileo) Open Service, *Signal in Space Interface Control Document (OS SIS ICD)*, Issue 1.3, December, 2016.

[13] Global Positioning Systems Directorate, Systems Engineering & Integration, "Interface Specification IS-GPS-800," *NAVSTAR GPS Space Segment/User Segment L1C Interfaces*, April, 2018.

[14] Global Positioning Systems Directorate, Systems Engineering & Integration, "Interface Specification IS-GPS-705," *NAVSTAR GPS Space Segment/User Segment L5 Interfaces*, April, 2018.

[15] BeiDou Navigation Satellite System, "Signal in Space Interface Control Document, Open Service Signal B1C," *China Satellite Navigation Office*, Version 1.0, December, 2017.

[16] BeiDou Navigation Satellite System, "Signal in Space Interface Control Document, Open Service Signal B2a," *China Satellite Navigation Office*, Version 1.0, December, 2017.

[17] India Regional Navigation Satellite System, "Signal in Space ICD for Standard Positioning Service," Version 1.1, August, 2017.

[18] http://insidegnss.com/rnss-and-the-itu-radio-regulations/, accessed on February, 2020.

[19] Gao, G.X., et al., "DME/TACAN Interference Mitigation for GNSS: Algorithms and Flight Test Results," *GPS Solutions* Vol. 17, pp. 561-573, 2013.

[20] https://www.faa.gov/air_traffic/flight_info/aeronav/criticaldme/, accessed on December 2019.

[21] European Organization for the Safety of Air Navigation, "GNSS Sole Service Feasibility Study," *Eurocontrol Experimental Center*, May, 2003.

[22] Gao, G.X., "DME/TACAN Interference and its Mitigation in L5/E5 Bands," *Proc. 20th Int. Tec. Meeting of the Inst. of Navigation, ION GNSS*, Fort Worth, TX, United States, Sep. 25-28, 2007, pp. 1191-1200.

[23] Lilley, R.W., "DME/DME for Alternate Position, Navigation, and Timing (APNT)," *Federal Aviation Administration*, white paper, 2012.

[24] FCC, DA-02-2786A2, *Measured Emissions Data for Use in Evaluating the Ultra-Wideband (UWB) Emissions Limits in the Frequency Bands used by the Global Positioning System (GPS)*, 2002.

[25] NTIA Compendium 1215.00-1240.00, *NTIA Band Introduction 1215-1240 MHz*, March, 2014.

[26] Heinrichs, G., Loehnert, E., and Wittmann E., "User RAIM Integrity and Interference Mitigation Test Results with Upgraded German Galileo Test Range GATE," *5th ESA Workshop on Satellite Navigation Technologies and European Workshop on GNSS Signals and Signal Processing (NAVITEC)*, Noordwijk, 2010, pp. 1-7.

[27] DLR, *Galileo Test- und Entwicklungsumgebungen in Deutschland*, January, 2011.

[28] United States Department of Transportation, *Global Positioning System (GPS) Adjacent Band Compatibility Assessment*, April, 2018.

[29] ETSI EN 303 413, *Satellite Earth Stations and Systems (SES), Global Navigation Satellite System (GNSS) Receivers, Radio Equipment Operating in the 1 164 MHz to 1 300 MHz and 1 559 MHz to 1 610 MHz Frequency Bands, Harmonized Standard Covering the Essential Requirements of Article 3.2 of Directive 2014/53/EU*, V1.1.1, 2017.

[30] U.S. Department of Homeland Security, *Improving the Operation and Development of Global Positioning System (GPS) Equipment Used by Critical Infrastructure*, Unclassified.

[31] Kaplan, E.D., and Hegarty, C.J. *Understanding GPS, Principles and Applications*, Second Edition, Norwood, MA: Artech House, 2006.

Chapter 7

High-Precision GNSS in 5G

7.1 INTRODUCTION

The previous chapter introduced the concept of GNSS and the mobile network enhancement, A-GNSS. Modern GNSS-based algorithms typically offer an accuracy in the meter range, which is more than sufficient to support regulatory positioning requirements as well as the initial requirements of commercial applications. However, as more demanding commercial use cases (e.g., D2D, V2X) started to arise (mainly as of Evolved UMTS Terrestrial Radio Access (E-UTRA) Release 15), the positioning accuracy requirements went from a few meters to the centimeter range. With the advent of 5G and the plethora of new applications, detailed in Chapter 5, highly accurate positioning became even a more relevant topic.

Getting to the centimeter accuracy range was not feasible with existing LTE positioning technologies or A-GNSS. Thus, 3GPP Release-15 specifications standardized the used of advanced GNSS-based technologies such as RTK and PPP [1]. These technologies have been carried over to NR as of Release 16.

Both RTK and PPP are evolutions of the differential GNSS (D-GNSS) methodology, which has been used for all kinds of navigation applications in order to overcome conventional GNSS limitations. This chapter will explain the working principle of D-GNSS and its enhancements, focusing mainly on RTK, PPP, and a combination of both methods.

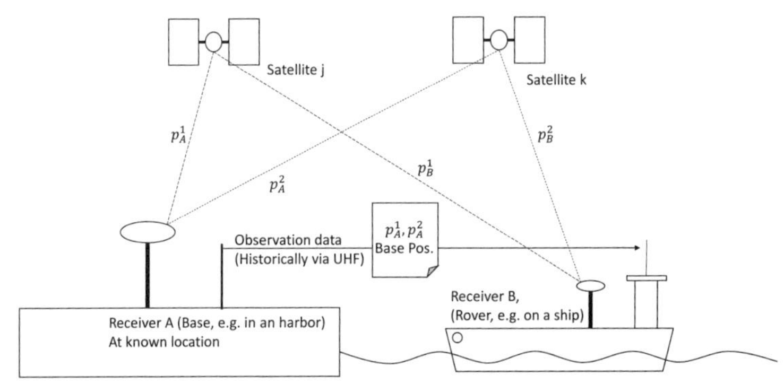

p_A^1 = Code Phase of satellite j observed at receiver A

$\Delta_1 = p_A^1 - p_B^1$: Single Difference for Sat 1→ Sat clock errors and bias cancelled, Tropospheric and Ionospheric error reduced

$\Delta_2 = p_A^2 - p_B^2$: Single Difference for Sat 2→ Sat clock errors and bias cancelled, Tropospheric and Ionospheric error reduced

$\Delta_1 - \Delta_2$: Double Difference: $p_A^1 - p_B^1 - (p_A^2 - p_B^2)$ → Sat and receiver clock errors and bias cancelled, Tropo. and Iono. error reduced

Figure 7.1 Principle of double differencing used in D-GPS and RTK.

7.2 THE PRINCIPLE OF DIFFERENTIAL-GPS

Until 2001, the civil GPS signal was artificially degraded to 100 m accuracy (95th percentile level) by the Selective Availability feature of GPS. Selective Availability typically introduced artificial inaccuracies by adding deliberate satellite clock jitter. For some safety-critical civilian maritime applications such as harbor entry, this accuracy was not sufficient. Hence a simple but powerful technique was developed to correct the signal: D-GPS.

7.2.1 Differential GPS

The fundamental idea is to observe the GPS signals from a station with known coordinates (base). These observations and the base location is provided to receivers (rovers) in the vicinity; for example via UHF broadcast or via the internet (see Figure 7.1). The receiver can compare the observations with its own and remove many error sources. The effect of the deliberate clock jitter could be largely removed with this approach, even for long distances between base and rover, as the clock error is not depending on the receiver location (see [2] Chapter 1).

If the rover and the base are sufficiently close to each other, not only can the effects of Selective Availability be removed, but other (unintentional) errors are also mitigated.

D-GPS takes advantage of temporal and spatial correlations of error sources. Errors that stem from the ionosphere or troposphere can be largely removed, as long as the signals to base and rover travel through areas of the atmosphere with similar properties. With increasing distance between base and rover (the baseline), these errors are less correlated and the performance of the approach degrades. For errors that only have a temporal correlation (e.g., satellite clock error) the performance does not decrease with the baseline length.

The typical error growth with baseline length is given in Table 7.1 (from [3]). The term ppm refers to parts-per-million (e.g., a value of 2 ppm equals to 2 mm per 1-km baseline length).

Table 7.1 gives an overview of the impact of baseline length on the error.

Table 7.1

Spatial Decorrelation by Error Type

Error Source	Relative Influence
Satellite orbit	0.1 ... 2 ppm
Satellite clock	0.0 ppm
Ionosphere	1 ... 50 ppm
Troposphere	0 ... 3 ppm

Similar to the spatial decorrelation with increasing baseline length, some error sources suffer from temporal decorrelation with increasing latency of the correction data link. This effect is, however, only relevant for significantly higher latency than those typically achieved by 4G or even 5G. Experiments from [4] (p. 13) show only a neglectable impact for a latency lower than 2 seconds.

7.2.2 Single Differencing and Double Differencing

The error canceling works by subtracting the observations from base and rover multiple times: The first difference of pseudoranges consists of observations of the same satellite by base and rover. Using only this difference, errors like satellite clock and ionospheric and atmospheric impact are already reduced. The difference in receiver clocks at the base and rover, however, remains as an error source. To mitigate this, two such single differences (for two different satellites) are typically differenced another time (see Figure 7.1). Using the delta of these deltas (the double difference), one can obtain the relative offset between the rover and the base. The base coordinates are transmitted along with the base observations to the rover, so that the absolute position of the rover can be determined from the relative offset.

As the original application was a maritime one, the correction data format was standardized ([5–7]) by the Radio Technical Commission for Maritime Services (RTCM). This set of standards became the de facto standard for D-GPS (and later RTK) applications.

At the time of writing, the original code-based D-GPS concept is only rarely used for several reasons:

- When Selective Availability was turned off in 2001, the standalone GPS precision was sufficient for many applications
- For more precise applications wide-area D-GPS systems came into existence, which achieved a similar performance as D-GPS with less effort on the rover side (see Section 7.4.1)
- For very demanding applications, code-phase based D-GPS evolved into carrier-phase based RTK (see Section 7.3.1)

7.3 RTK: OBSERVATION STATE REPRESENTATION

7.3.1 Carrier-Phase Measurements

Scientists and engineers in the fields of geodesy and construction were trying to utilize GPS for their high-precision applications to avoid time-consuming triangulation. The concept of D-GPS was subsequently extended to RTK, which uses the same concept of a base station providing corrections to a rover, but goes one step further by transmitting not only the pseudoranges, but also the carrier-phase observed by the base station, leading to the method's name, OSR. While the chips of the PRN-codes used in normal GPS have a length of 300 m, the GPS L1 C/A carrier frequency is orders of magnitudes shorter with 19 cm. For a receiver, which can track both with an accuracy of 1%, this results in 3 m for code-based GPS, but only 1.9 mm for the carrier-phase. There is a caveat, however: while the PRN code is designed to repeat only every 300 km and an unambiguous position solution can be easily obtained, the carrier-phase, which is a sine, repeats every 19 cm. The receiver needs to solve this ambiguity before the location can be precisely determined.

7.3.2 Integer Ambiguity Resolution

The equations in this chapter are from [8], which provides an detailed overview over the principles of double differencing for the interested reader. The equation for a carrier-phase observation $\Phi_A^j(T_A)$ at a receiver clock time T_A is given in 7.1 and 7.2. Superscript j means that an element is related to satellite j, while subscript A indicates that it is related to observer (base or rover) A.

$$L_A^j(T_A) = \lambda_0 \Phi_A^j(T_A) = \rho_A^j(t_A, t^j) + c\tau_A + c\tau^j + Z_A^j - I_A^j + B_A^j \quad (7.1)$$

$$B_A^j = \lambda_0(\varphi_{0A} - \varphi_0^j - N_A^j) \quad (7.2)$$

The term we are actually interested in is the pseudorange $\rho_A^j(t_A, t^j)$.

λ_0 is the nominal wavelength of a signal. t^j is the true transmission time, when the signal is sent from the satellite, and t_A is the true receive time. Clock errors occur at receiver τ_A and satellite τ^j. The atmospheric impacts are given by I_A^j (ionosphere) and Z_A^j (troposphere). B_A^j is the (unknown) phase bias, where N_A^j is an integer number of whole cycles, φ_{0A} is the unknown fractional phase offset of the receiver, and φ_0^j is the fractional phase offset of the satellite.

To solve for $\rho_A^j(t_A, t^j)$, all other terms need to be estimated or removed from the equation. Modeling the parameters is the approach followed by PPP and its variants (see Section 7.4.2), while RTK aims at removing the unknowns by using double differencing.

Building the single difference, we obtain Equation 7.3, where the satellite clock error τ^j canceled out (as it is same for both observations) and the ionospheric error ΔI_{AB} and the tropospheric error ΔZ_{AB} consists only of the differences experienced on the paths to base and rover.

$$\Delta L_{AB}^j = \Delta\rho_{AB}^j + c\Delta\tau_{AB} + \Delta Z_{AB}^j - \Delta I_{AB}^j + \Delta B_{AB}^j \quad (7.3)$$

Once we perform the double differencing according to equation 7.4, we see that the receiver clock error τ_{AB} is also fully removed. The ∇ operator is used here like in the GNSS literature instead of a second Δ to indicate that the second difference is between two different satellites (mnemonic: two upper edges means delta between satellites, two lower edges means delta between ground observers).

$$\nabla\Delta L_{AB}^{jk} = \nabla\Delta\rho_{AB}^{jk} + \nabla\Delta Z_{AB}^{jk} - \nabla\Delta I_{AB}^{jk} + \nabla\Delta B_{AB}^{jk} \quad (7.4)$$

While the property of removing the clock errors and reducing the ionospheric and tropospheric errors is already quite useful, double differencing has another important advantage: The carrier-phase error term $\nabla\Delta B_{AB}^{jk}$ is always an integer multiple of the nominal wavelength! This becomes obvious when we build the double difference for this term in Equation 7.5:

$$\begin{aligned}\nabla\Delta B^j_{AB} &= \Delta B^j_{AB} - \Delta B^k_{AB} = (B^j_A - B^j_B) - (B^k_A - B^k_B) = \\ &= \lambda_0((\varphi_{0A} - \varphi^j_0 - N^j_A) - (\varphi_{0B} - \varphi^j_0 - N^j_B)) \\ &- ((\varphi_{0A} - \varphi^k_0 - N^k_A) - (\varphi_{0B} - \varphi^k_0 - N^k_B)) = \\ &= -\lambda_0(N^j_A - N^j_B - N^k_A + N^k_B) = -\lambda_0(\nabla\Delta N^{jk}_{AB})\end{aligned} \tag{7.5}$$

As the integer ambiguities are non stochastic, a correct integer ambiguity resolution results in a solution with a very high precision [9].

Several algorithms exist for the integer ambiguity fixing. Teunissen (1993) described this problem as the integer-least-squares-problem. A simple approach to solve it is by first ignoring the integer constraint and solving the regular least-squares problem [9]. This solution is also termed the float solution. Then, the real-valued estimate of the integers is rounded. Based on the rounded values, the float solution is corrected and becomes a fixed solution. This approach is not ideal; modern receivers typically implement more sophisticated algorithms, such as the LAMBDA method [10], which go beyond the scope of this book.

For a better understanding, a visualization of the integer fixing is given in Figure 7.2: a receiver obtains many candidate locations for the precise location within the uncertainty radius of the float solution. The outer circle marks the uncertainty of the float position fix, while every intersection marks a potential candidate for an integer position fix. If only two satellites are observed for one moment, it is impossible to determine which one is the correct one. With an increasing observation time or a larger number of satellites, more and more candidates can be removed as they are not intersecting precisely. Alternatively, signals at different wavelengths can be observed to resolve the ambiguity. Once the integer ambiguity is resolved, the satellites need to be continuously observed. If the signals are lost, a new integer ambiguity search is required.

RTK was the first technique that achieved carrier-phase accuracy with integer ambiguity resolution. Its characteristic sign is the transmission of carrier and code observations from a base to a rover, and hence it is also called OSR. In its simplest form, RTK consists of a single base and a single rover. The working principle of an OSR correction service is as follows: the base receives the GNSS signal, measures the pseudoranges and the carrier-phase offsets, and sends to a centralized location server. The rover also calculates its approximate location (without using any correction data) and transmits it to the server, which in turn provides the

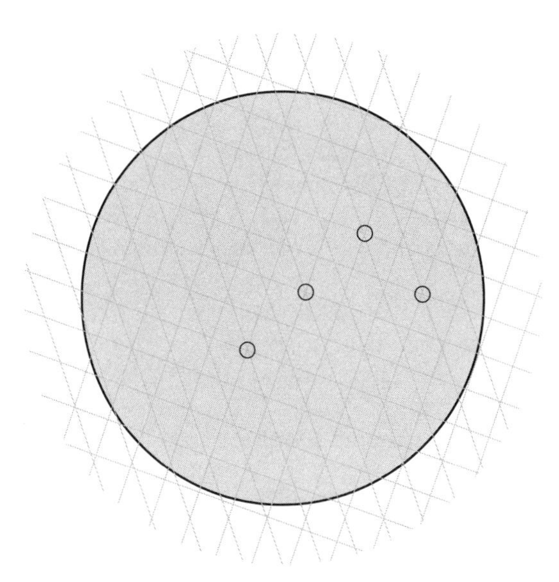

Figure 7.2 Ambiguity resolution: The grey area is the uncertainty of the float solution; only a few carrier wave fronts are intersecting in a precise point.

correction data of the nearest base station. Using this correction data, the rover is able to obtain much better positioning accuracy, up to the centimeter range, as seen in the previous section.

Even though they are a very powerful tool for increasing the GNSS positioning accuracy, OSR correction services present a few limitations, as they offer only local coverage and they need bidirectional communication.

7.3.3 Reduced Convergence Time with Multi-GNSS and Multifrequency

When RTK became available, it was used for surveying and construction, where a GPS L1-C/A only receiver reduced the time for a precise fix to tens of minutes, compared to a time-consuming and infrastructure-dependent triangulation. To reduce this time even further, GPS+GLONASS receivers were developed, leading to more visible satellites and consequently shorter convergence times. Using more frequencies was a major obstacle for civil users, as early generations of GPS and GLONASS satellites only broadcast one unencrypted signal each. Researchers found ways to utilize the encrypted military L2P signal without knowledge of the code (codeless and semi-codeless receivers). The price to pay was a loss of sensitivity, which restricted this method to benign sky-view scenarios. The last few years, however, dramatically improved the situation for RTK users: GLONASS introduced a second civil signal on L2OF, GPS added the L2C and the L5 signals and the newer GNSS systems Galileo and BeiDou each provide three civil signals. The codeless techniques are therefore becoming largely obsolete and are not further addressed in this book.

With the large amount of visible satellites, each broadcasting 2-3 civil signals, it is now possible to reduce the ambiguity search time to a few seconds. The dramatic reduction in convergence time together with inertial-sensor-aided receivers opens completely new applications for RTK: automotive, UAVs, and virtual reality applications can now benefit from accuracies below 10 cm.

7.3.4 Scaling up RTK

The appearance of new applications like autonomous driving added new requirements for RTK. When RTK was mainly used for land surveying, it was acceptable to have local coverage regions with not-so-dense RTK base stations. However, autonomous driving requires country- or even continent-wide coverage. Furthermore, to keep an acceptable convergence time, the typical radius for a RTK base coverage area is 10-20 km. Larger radii lead to less accuracy and slower convergence times. Many countries already have RTK bases, but usually with larger distances between the stations than required for instantaneous convergence.

Another challenge for scaling up RTK is the distribution of the correction data. In most cases today, an IP-based unicast protocol is used for this purpose (e.g., network transport of RTCM via Internet Protocol (NTRIP)). As already mentioned, this method requires bidirectional communication and increases the overall traffic with every additional user. As long as the number of RTK users is low (e.g., surveyors), the data can be easily provided via 3G or 4G cellular networks. However, with millions of cars or other vehicles potentially using RTK, the unicast distribution becomes extremely inefficient. Thus, the 3GPP studied ways of multicasting or broadcasting the correction data, as it will be seen in later sections of this chapter.

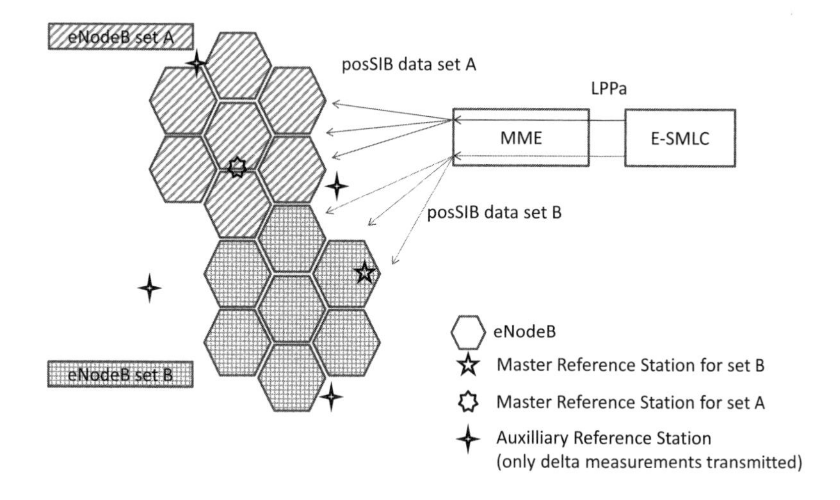

Figure 7.3 Concept of RTK with master-auxiliary stations in 4G/5G.

7.3.5 Network-RTK

The original concept of RTK involved only one base-rover pair, which is a suitable approach for static applications. However, a rover in an automotive platform will frequently move from one RTK base station coverage region to another, meaning it must be able to perform handovers between RTK bases. This could lead to the potential loss of a precise fix. By allowing a multiple bases per rover concept, interpolation can be applied and the effects of a handover can be mitigated.

The extensions to multibase RTK are summarized under the term network-RTK. Depending on which side (the network or the rover) does the interpolation, there are three major variants of network-RTK. All are detailed in the following section and have been added to the LTE and NR specifications by 3GPP during the Release-15 timeframe. The paper referenced in [11] gives a good overview of the three methods and is the basis for the following sections.

7.3.5.1 Master-Auxiliary Concept

In the master auxiliary concept (MAC), the observation of code and carrier phases of a master station and of multiple auxiliary stations are collected and transmitted to the rover as shown in Figure 7.3.

The rover receives all the data from the real stations and calculates the interpolated correction values. These interpolated correction values, together with the approximate location of the rover, are used to simulate a virtual observation coming from a virtual base station in the proximity of the rover. The virtual observations can then be used for a single-base RTK solution.

In the MAC model, most of the computational cost and the additional complexity is placed on the rover's side. Furthermore, the amount of data to be transferred is high, as the observations from each RTK station (the master and all the auxiliaries) is transmitted to the rover.

7.3.5.2 Virtual Reference Station

In the virtual reference station (VRS) method, the rover does not receive the observations from the real RTK base stations. Instead, it receives the observations directly from a virtual base station resulting from interpolating the data of the real ones. Most of the calculation steps of the MAC are done on the network side, and only the end result, the virtual observations, are transmitted to the receiver. For this step, it is normally necessary to transmit first the approximate user location from the rover to the network.

Figure 7.4 shows how RTK with VRS can be implemented in 4G and 5G with SIB broadcasting. The real reference stations as shown in the figure are transparent to the rover. Thus, from the rover point of view, the VRS mode is basically a single-base RTK in which the base is the virtual reference station.

The main advantages of this method are that it does not require big amounts of data to be transmitted and that it can be done without additional complexity on the receiver side.

7.3.5.3 FKP Mode

An intermediate solution between the MAC and the VRS mode is based on the transmission of area correction parameters, typically known as FKP, which is the abbreviation from the German name (Flächenkorrekturparameter).

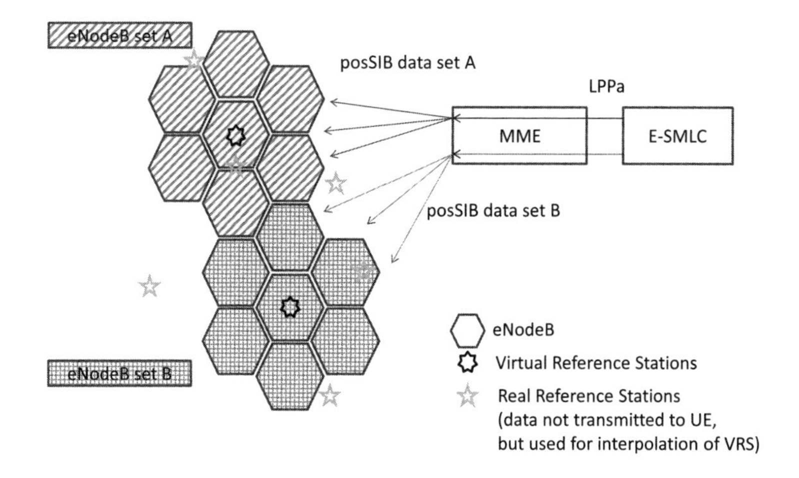

Figure 7.4 Concept of RTK with virtual reference station in 4G/5G.

In FKP, the network calculates auxiliary values used for the interpolation, the area correction parameters. These values are transmitted to the rover together with the observation data for the reference station. The rover calculates the virtual observation values by combining the FKP with the real observation. The 3GPP variant using SIB broadcasting is shown conceptually in Figure 7.5.

FKP transmits bigger amounts of data than VRS and requires a slightly more complex receiver on the rover side. As a counterpart, the network does not require a priori knowledge about the rover approximate location, so it can be done without uplink data.

7.3.5.4 Network-RTK Variants Summary

Table 7.2 gives an overview of the three variants previously described and their merits. Each presents advantages and disadvantages. Hence, the 3GPP has decided to standardize all and give flexibility to the implementation to decide which one to use depending on the application.

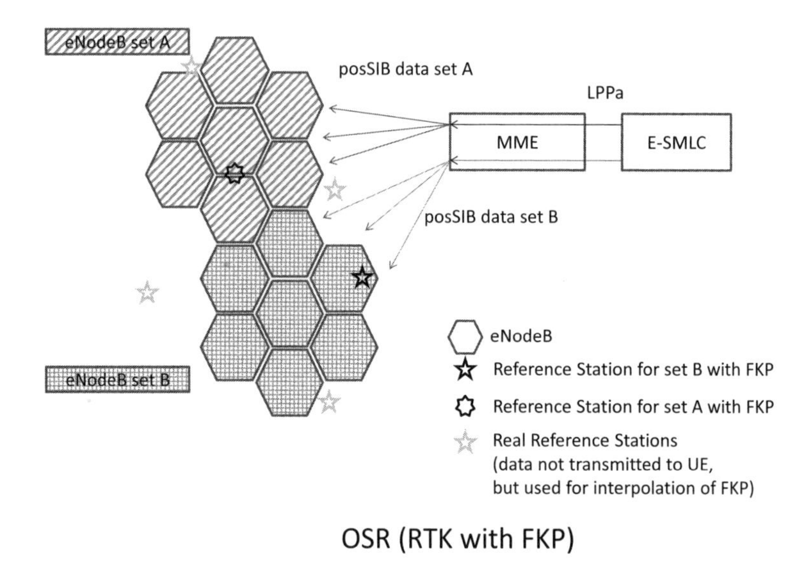

Figure 7.5 Concept of RTK with Flächenkorrekturparameter in 4G/5G.

7.4 PPP, PPP-AR AND PPP-RTK: SSR

7.4.1 Wide Area DGPS

While D-GPS, RTK, and its variants are all based on transmitting observations from a base to a rover (OSR) and using double differencing to eliminate errors, a second concept exists, which aims at modeling the effects that degrade the GNSS signal and does not require a base station. This technique is Wide Area Differential GPS (WADGPS)/Wide Area Differential GNSS (WADGNSS) (if only codes are used) or PPP (if codes and carriers are used). It models the errors described in Chapter 6 with a much higher fidelity than the models broadcast by the GNSS satellites in the navigation message. The SBAS standard used in aviation is the most prominent WADGPS service and includes clock corrections, ephemeris corrections, and an ionospheric grid, although with a relatively low resolution (maximum about 5 degrees of latitude and longitude, which are around 550 km at equator), as it

Table 7.2
Network-RTK Variants

	MAC	FKP	VRS
Data bandwidth	High	Medium	Low
Rover complexity	High	Medium	Low
Interpolation model	Any	FKP	Any
Uplink required	No	No	Yes
RTCM version	≥ 3.0	≥ 3.0	≥ 2.3
RTCM types	1017, 1039	1034, 1035	
LPP types	Observations Reference Station Info Auxiliary Station Data Residuals	Observations FKP-Gradients Residuals	Observations Residuals

is only intended for code-based fixes. The Japanese sub-meter level augmentation service (SLAS) service provided by QZSS is another WADGNSS service.

7.4.2 Precise Point Positioning

PPP (see [12]) is similar to WADGNSS as it is an absolute positioning technique using non differential observations. Unlike WADGNSS, it also utilizes the carrier phases and works with significantly higher accuracy. The term SSR is also often used to contrast PPP from the differential techniques, which use OSR. The original PPP concept was introduced by Zumberge et al. at NASA's Jet Propulsion Lab in 1997. It is a method for obtaining high accuracy measurements without the need of a local reference station. Its fundamental features are a highly precise orbit and clock corrections. While the normal broadcast GPS ephemeris has a typical accuracy of 100 cm and 5 ns for the clock data, the PPP real-time precise ephemeris are much more accurate. For example, the data provided by the International Geospatial Service (IGS) free of charge is accurate to 5 cm and 3 ns, respectively, for real-time applications (see [13]). This highly accurate data is produced independently of the normal GPS ground control segment by a complex algorithm (e.g., Jet Propulsion Lab's GIPSY OASIS [14]), which uses observations from monitoring stations around the world. This algorithm takes into account many subtle effects like ocean loading, solid earth tides, solar radiation on the satellite, and many more.

While the derivation of the precise orbit and clock data is highly complex, the resulting correction data for the receiver is quite simple: a precise ephemeris and a precise satellite clock correction. Both are valid worldwide and can be broadcast, which is an unique advantage of this high-precision solution.

While PPP brings many benefits, the original concept has several limitations: Since the original PPP approach does not model the ionosphere, it is necessary to use a dual-frequency receiver, which can remove most of the ionospheric error. Furthermore, only a float solution is possible. As there is no double differencing involved, the receiver clock error, the tropospheric parameters, and the (noninteger) carrier-phase ambiguities need to be estimated iteratively [15]. Therefore, it takes much longer to converge to a high accuracy. The convergence time is typically massively longer (e.g., 1800s in [16]) compared to 10-50s for RTK. The integer ambiguity fixes (as typically used in RTK) are not obtained with PPP (see [17]), as real-valued receiver and transmitter specific phase delays cannot be separated from the integer ambiguities in Equation 7.2 ([12] see: Limitations).

As of today, several PPP service providers exist, which distribute their data via geostationary communication satellites for a subscription fee. If an internet connection is available, PPP corrections can be obtained via NTRIP protocol from IGS: http://www.igs.org/rts/access.

7.4.3 PPP-AR

A recent extension to traditional PPP is precise point positioning ambiguity resolution (PPP-AR), where the satellites' fractional cycle bias are estimated. These values can be absorbed in the PPP clock model [15] or transmitted via a navigation message (e.g., in QZSS). The receiver can remove the fractional receiver bias by taking the mean of all corrected ambiguities (see [18], 2.2). In benign environments, an ambiguity resolution is possible. This reduces the convergence time compared to PPP, but PPP-AR is still slower than an RTK or PPP-RTK solution. However it preserves the advantage of the large coverage area. Japan plans to provide a PPP-AR service for the whole region of Asia-Oceania via the QZSS satellites.

7.4.4 PPP-RTK

Much research is ongoing in the GNSS community to combine the advantages of RTK and PPP. To close the performance gap of PPP-AR, additional error sources are modeled, which leads to the PPP-RTK method (see [19], Figure 7.6). It adds precise models for ionosphere and troposphere. For both, a grid model is typically used.

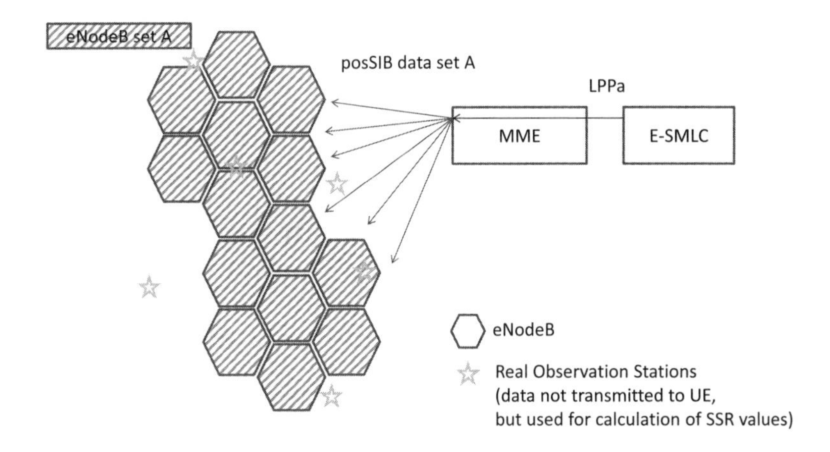

Figure 7.6 Concept of PPP-RTK (SSR) in 4G/5G.

Compared to traditional OSR-based RTK, SSR-based PPP-RTK can significantly reduce the data rate, as slow varying parameters may be transmitted in longer intervals and the number of data does not grow linearly with the number of observed satellite signals. As shown in Figure 7.7, SSR can schedule parameters with low dynamic less frequently, which reduces the data rate compared to OSR with many observations. In the right part of the figure, it can be seen that the clock corrections are sent every time, while the other corrections (tropospheric, ionospheric, code phases, and orbit) are sent only once every four transmissions.

In 2018, Japan started its QZSS PPP-RTK service, covering the Japanese islands, which is called centimeter-level augmentation service (CLAS). The European Union announced a plan to provide a similar service through the Galileo satellites for the area of the European Union (see also [20]). QZSS and Galileo both use the L6/E6 frequency to broadcast the SSR model parameters free of charge.

The bandwidth limitation of signals broadcast over L6 is noteworthy here: in the case of QZSS, the L6 data rate is 2000 bits/s. Japan's QZSS uses 212 grid points to cover the country with an area of around 380,000 km^2. The European

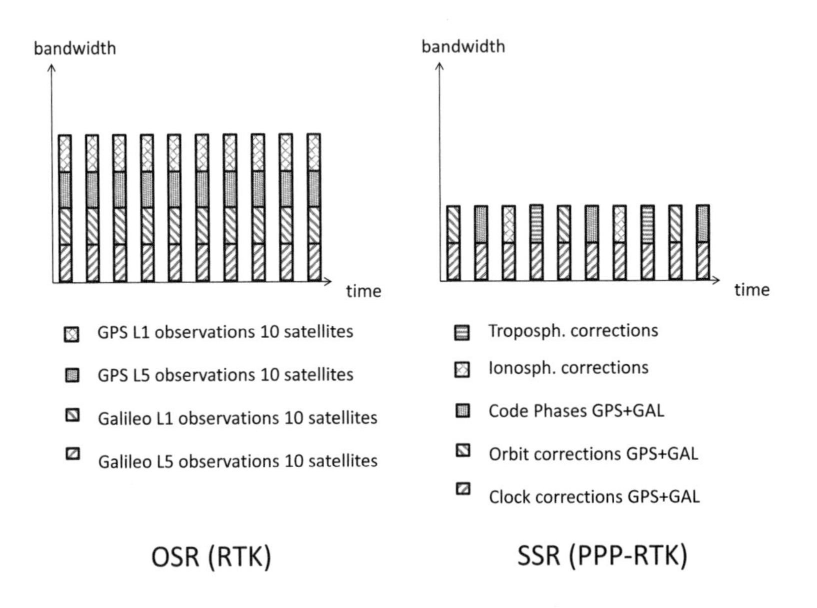

Figure 7.7 Comparison of the scheduling for OSR and SSR models.

Galileo E6 signal has a data rate of 500 bits/s. The area of the European Union is around 4,400,000 km^2, a factor of 10 larger than Japan. Transmitting PPP-RTK correction data via E6 with the same quality for a similarly dense grid network seems therefore quite challenging. Hence, the terrestrial 4G/5G networks can be a suitable (additional) channel for PPP-RTK correction data provision.

7.4.5 QZSS CLAS Message Elements: Compact SSR

The use of SSR in RTCM was discussed since 2011, but at the time of writing, none of the proposed extensions for SSR made it into the official RTCM specification (see [21]). The Japanese QZSS system architects decided to use a SSR representation, which is based on the proposed extensions by Geo++ and Mitsubishi Electric Corporation, the Compact SSR. This scheme is also discussed for the European L6 PPP-RTK service [22]. GEO++ proposed a different message format called SRRZ and Sapcorda published the similar SPARTN format in January 2020. Both Compact SSR and SRZZ are available as proprietary extensions to RTCM. 3GPP decided to base their PPP-RTK variant in Release 16 on the Compact SSR scheme, but encoded in ASN.1. At the time of writing, Compact-SSR seems to be the first officially published, universal SSR representation, hence this chapter gives an introduction to this scheme. Table 7.3 gives an overview of the message types of the Compact SSR.

Table 7.3
Compact SSR Message Elements

Message Type	RTCM Proprietary Type
Compact SSR Mask	MT4073,1
Compact SSR GNSS Orbit Correction	MT4073,2
Compact SSR GNSS Clock Correction	MT4073,3
Compact SSR GNSS Satellite Code Bias	MT4073,4
Compact SSR GNSS Satellite Phase Bias	MT4073,5
Compact SSR GNSS Satellite Code and Phase Bias	MT4073,6
Compact SSR GNSS URA	MT4073,7
Compact SSR STEC Correction	MT4073,8
Compact SSR Gridded Correction	MT4073,9
Compact SSR Service Information	MT4073,10
Compact SSR GNSS Combined Correction	MT4073,11

7.4.5.1 Ionospheric and Tropospheric Grid Models

For the PPP-RTK specific atmospheric grid correction, QZSS uses predefined grid points, which are explicitly specified in the QZSS CLAS ICD section 4.1.4. [23] (e.g., compact network ID 1: grid point 1 is at latitude 24.75 and longitude 125.37, grid point 2 is at latitude 24.83, longitude 125.17, and so forth). This method is not suitable for a global 3GPP approach, as it is dependent on the geography of each country. At the UNNOSA meeting 39, Mitsubishi Electric Corporation presented generic methods for grid specifications [24], called Type-1, Type-2, and Type-3, which may be used by 3GPP. In effect, Type-1 is a simple rectangular grid pattern, where each grid point is transmitted sequentially from west to east, north to south. This method is not efficient if large bodies of water are within the coverage area, as the PPP-RTK service is usually only covering terrestrial applications. Type-2 adds a mask on top of the pattern of Type-1, hence avoiding to transmit grid points, which are of no interest (in the sea). Type-3 allows nonuniform grid offsets.

The resolution of the grid networks differs depending on the required service. For WADGNSSs like WAAS and EGNOS, the grid points have a distance of

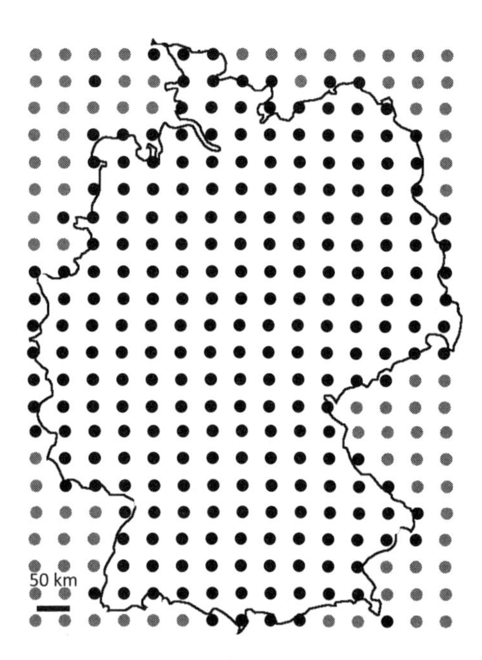

Figure 7.8 Example of a atmospheric grid that spans accross Germany, according to Type-2 with omitted grid points (grey).

around 500 km from each other [25]. On the other hand, for PPP-RTK a grid resolution that is considered to provide very good results is 50 km, according to the 3GPP discussion in [26]. An exemplary ionospheric grid that covers Germany is depicted in Figure 7.8. Countries like Japan, which feature a more complex geographic shape, may need multiple grid networks, each spanning a subsection of the country. Countries of the size of Japan or Germany require around 200 grid points (called grids in this context) to cover their whole territory. Larger countries obviously require many more grid points, which makes it challenging to broadcast the whole grid over bandwidth-limited channels. If the data rate is limited, such as when transmitting via a satellite's L6 signal, this might not even be possible. In the case of 4G/5G broadcast, the transmitted data can be geographically limited (e.g., base stations in California might only transmit the grid points for that state). In the most optimized case, each base station might only transmit one or a few grid points covering its region. Clearly, there is a trade-off in network management complexity and efficient bandwidth use. The LPP protocol at the time of writing is intended to give each operator the flexibility to choose this implementation detail.

7.4.5.2 VTEC and STEC: Ionospheric Correction Terms and Tropospheric Correction Terms

WADGNSS grid models only convey a simple correction term for the ionosphere: vertical total electron content (VTEC). Higher VTEC values mean higher signal delays. There is typically no tropospheric correction for WADGNSS.

QZSS's compact SSR by contrast conveys two specific tropospheric correction terms (wet and dry components) at each grid point. The tropospheric correction terms are assumed to be identical for all satellites in view. For the ionosphere, the compact SSR conveys the slant total electron content (STEC), which consists of up to four terms and depends on the latitude and longitude of the receiver and is satellite-dependent but not grid-point dependent. A residual term, which is both grid- and satellite-dependent, is transmitted as well to reduce the error even further.

7.5 RTK/PPP IN 5G

As part of the positioning enhancements in Release 15, RTK and PPP correction data was added to the 3GPP specifications. Even though the RTCM and NTRIP standards existed for a long time and are used via top-IP services, 3GPP favored to respecify the RTK and PPP data elements in the LPP specification instead of piggybacking RTCM payload. Consequently, significant additions were made to the LPP in Release 15 [27].

One of the main changes in LPP affects the way the assistance data can be transmitted. Before Release 15, LPP was a point-to-point unicast protocol. However, using LPP in unicast mode would have no advantage over RTCM in terms of efficiency and scalability. As the amount of data that needs to be transferred for RTK and PPP is quite significant, 3GPP decided to implement a broadcast mechanism with the introduction of positioning system information blocks (posSIBs) in the LTE RRC layer (see Figure 7.9).

A similar broadcast mechanism is intended for NR in Release 16 (see [28] with CR [30] and [29] sec 6.19.3). With these features, RTK and/or PPP over cellular networks have several significant advantages compared to RTCM and the L6 signals:

- Higher data capacity than L6

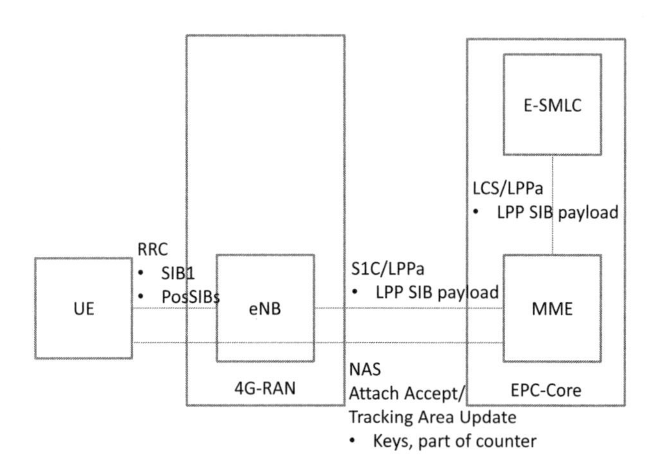

Figure 7.9 Extension to 4G architecture to support RTK/PPP broadcasting and key distribution.

- Higher power signal than L6
- Traffic independent of user number in contrast to NTRIP
- Small broadcast areas possible (RTK broadcast)

In many sources, it is stated that RTK requires an uplink channel, but with a cellular network's spatial reuse, even RTK corrections can be broadcast with different data in different cells. Alternatively, it is also possible to transmit the same information in a whole country (e.g., for PPP).

As of Release 15, 3GPP supports the following techniques:

- RTK
- Network-RTK with VRS
- Network-RTK with FKP
- Network-RTK with MAC

- PPP-AR with orbit, clock, and code phase corrections

LPP Release 15 does not contain any advanced ionospheric or tropospheric models, which means that PPP-RTK is not yet possible. The addition of these models is studied in R-16 under 3GPP Work Item "NR pos" (WID 830077, [31], section RAN2 centric objectives, and [32]). The work item proposes to use the very same compact SSR representation for PPP-RTK that is already utilized for QZSS CLAS and described in Section 7.4.5.

More detailed information about positioning SIBs and other LPP protocol mechanisms will be given in Chapters 13 to 16 of this book.

7.6 CONCLUSION

7.6.1 Comparison and Evaluation of the Technologies

A mass-market, high-accuracy GNSS technology that is useful for autonomous driving, UAVs, and augmented reality applications needs to meet several key performance indicators such as centimeter-level accuracy, convergence time within one minute or less, and countrywide coverage. Only two of the technologies introduced in this chapter meet all these requirements: network-RTK and PPP-RTK. As 5G leaves it up to the implementation which of these technologies to use, this section will compare these two technologies and their merits.

7.6.1.1 Network-RTK Summary

All N-RTK methods (VRS, FKP, MAC) provide very good convergence times and accuracies. Countrywide coverage is possible, but comes at a cost: either a bidirectional link is required to select a suitable VRS location, excluding broadcast methods, or a extremely high data rate is needed to convey observations and parameters from all stations of the satellite's footprint, rendering network-RTK (N-RTK) distribution via satellite impractical. Cellular networks have the distinct advantage that they can broadcast only the observations from stations nearby each cell, effectively reducing the bandwidth to manageable values (tens of kbps for all users of a cell).

7.6.1.2 RTK-PPP Summary

RTK-PPP is the most modern approach for high-precision GNSS. It promises all benefits of N-RTK at a significantly lower bandwidth (2 kbit per s for a mid-size country like Japan, <1 kbit per s if utilizing cellular broadcast). The low data rate allows broadcasting the correction data via a satellite link for medium-sized countries (QZSS L6). Furthermore, different service levels may be easily implemented with SSR by using different keys in the 5G broadcast for different message types (see the encryption of posSIB in Chapter 14). RTK-PPP correction data can be converted to N-RTK correction to allow legacy receiver designs to be reused. Whether the converge times of RTK-PPP are really on par with the excellent values of N-RTK needs to be seen, but initial results are promising.

7.6.2 Compatibility N-RTK and RTK-PPP Methods

Not all receivers might support all forms of (N-)RTK, PPP or RTK-PPP. A PPP-capable receiver will not work with RTK correction data and vice versa. A RTK-PPP correction stream may, however, be converted into a PPP correction stream as well as a RTK correction stream (i.e., conversion algorithms exist to convert from SSR to OSR). This means that existing RTK receivers may be used with little extra effort in a PPP-RTK system. The QZSS web page provides such an algorithm that converts the compact SSR navigation messages to OSR-based RTK.

A conversion from OSR to SSR is not directly possible. To obtain SSR data, many spatially distributed monitoring stations and sophisticated algorithms are required to obtain the model parameters for SSR. These algorithms are typically highly complex and to the author's knowledge, there is no open-source software available as of today for this process. Commercial products for this are offered by GEO++, Trimble, JPL, and others.

7.6.3 Summary: (N-)RTK, PPP, RTK-PPP, and 5G

This chapter introduced the concept of high-precision GNSS techniques, in particular carrier-phase measurements, ambiguity resolution, and the variants of correction data like OSR and SSR. It compared the most promising techniques for a large-scale, mass-market roll out (network-RTK and PPP-RTK) in terms of network complexity, required bandwidth, and convergence time. Centimeter-level accuracy provided by GNSS and 5G networks may be enablers for new verticals and business models, for example in the field of autonomous driving. The new 5G features may create new markets for network operators, device manufacturers, and industries.

References

[1] 3GPP RAN RP-171508, *UE Positioning Accuracy Enhancements for LTE*, June, 2017.

[2] Parkinson, B. et al. *Global Positioning System - From Theory to Practice*, Volume II, American Institute of Aeronautics and Astronautics, Inc., 1996.

[3] Wuebenna, G., "GNSS Network-RTK Today and in the Future Concepts and RTCM Standards," *International Symposium on GNSS, Space-based and Ground-based Augmentation Systems and Applications*, Berlin, Germany, Nov. 11-14, 2008.

[4] Song, J., "A New Approach of Latency Error Compensation Using Compact RTK for GPS/GLONASS signals," *57th Meeting of the Civil GPS Service Interface Committee*, Portland, Oregon, United States, Sep. 25-26, 2017.

[5] RTCM 10402.3 *RTCM Recommended Standards for Differential GNSS (Global Navigation Satellite Systems) Service*, V2.3, Amendment 1, May, 2010.

[6] RTCM 10403.3, *Differential GNSS (Global Navigation Satellite Systems) Services*, V3, October, 2016.

[7] RTCM 10410.1 *Standard for Networked Transport of RTCM via Internet Protocol (Ntrip)*, V2.0, Amendment 1, June, 2011.

[8] Blewitt, G., "Basics of the GPS Technique: Observation Equations," in *Geodetic Applications of GPS*, Swedish Land Survey, 1997.

[9] Teunissen, P.J.G., "Success Probability of Integer GPS Ambiguity Rounding and Bootstrapping," *Journal of Geodesy*, Vol. 72, pp. 606–612, October, 1998.

[10] Teunissen, P.J.G., "The Least-Squares Ambiguity Decorrelation Adjustment: a Method for Fast GPS Integer Ambiguity Estimation," *Journal of Geodesy*, Vol. 70, pp. 65-82, July, 1997.

[11] Wuebbena, G., Bagge, A., and Schmitz, A., "Network-Based Techniques for RTK Applications," *GPS Symposium, GPS JIN 2001, GPS Society, Japan Institute of Navigation*, Tokyo, Japan, Nov. 14-16, 2001.

[12] Zumberge, J.F., et al., "Precise Point Positioning for the Efficient and Robust Analysis of GPS Data from Large Networks," *Journal of Geophysical Research: Solid Earth*, Vol. 102, Issue B3, March, 1997.

[13] International Geospatial Service Products Overview, http://www.igs.org/products, accessed on February 2020.

[14] GIPSY-OASIS, *GPS Inferred Position System and Orbit Analysis Simulation Software*, NASA JPL, https://gipsy-oasis.jpl.nasa.gov/, accessed on February 2020.

[15] Li, H., et al., "Introduction of the Double-Differenced Ambiguity Resolution into Precise Point Positioning," *Remote Sensing*, Vol. 10, p. 1779, November, 2018.

[16] Wuebbena, G., Schmitz M., and Andreas Bagge, "GEO++, PPP-RTK: Precise Point Positioning Using State-Space Representation in RTK Networks," *ION GNSS*, Long Beach, California, United States, Sep. 13-16, 2005.

[17] Wanninger, L., "State-of-the-Art of cm-Accurate Real-Time GNSS Positioning," *Tutorial/EUREF Symposium 2017* Wroclaw, Poland, May 16, 2017.

[18] Zhang, X., Li, P., and Guo, F., "Ambiguity Resolution in Precise Point Positioning with Hourly Data for Global Single Receiver," *Advances in Space Research*, Vol. 51, pp. 153-161, January, 2013.

[19] Wübbena, J., et al, "SSR-Technologie für skalierbare GNSS Dienste Prinzipien, Anwendungen, Standardisierung," *Intergeo 2016*, Hamburg, Germany, Oct. 13, 2016.

[20] Hirokawa, R., "Recent Activity of International Standardization for High-Accuracy GNSS Correction Service," *Standards and Interoperability of Precise Point Positioning Services, Workshop on the Applications of Global Navigation Satellite Systems*, Suva, Fiji, June 27, 2019.

[21] Wuebbena, G., Schmitz, M., and Wuebbena, J. "SSR and RTCM–Current Status," *4th EUPOS Technical Meeting*, Bratislava, Slovakia, Nov. 21-22, 2017.

[22] Fernandez-Hernandez, I., "Galileo High Accuracy Service and its Importance for Mobility Applications," *INTERGEO*, Frankfurt, Germany, Oct. 16, 2018.

[23] Quasi-Zenith Satellite System, Interface Specification, *Centimeter Level Augmentation Service (IS-QZSS-L6-001)*, November, 2018.

[24] 3GPP RAN2 R2-1907147, *SSR Messages for: Carrier Phase Bias, URA, Atmosphere, Grid Definition*, April, 2019.

[25] Juan, J.M., et al, "Accurate Reference Ionospheric Model for Testing GNSS Ionospheric Correction in EGNOS and Galileo," *Proc. 7th ESA Workshop on Satellite Navigation Technologies: Era of Galileo IOV (NAVITEC 2014)*, ESA/ESTEC, Noordwijk, The Netherlands, Dec. 3-5, 2014.

[26] 3GPP RAN2 Email discussion on SSR Grids, https://list.etsi.org/scripts/wa.exe?A2=3GPP_TSG_RAN_WG2;473dd7ee.1909C, accessed on May 2020.

[27] 3GPP TS 36.355, *LTE Positioning Protocol (LPP)*, V15.5.0 , September, 2019.

[28] 3GPP TS 23.273, *5G System (5GS) Location Services (LCS); Stage 2*, V16.2.0, December 2019.

[29] 3GPP TR 23.731, *Study on Enhancement to the 5GC Location Services*, V16.0.0, December 2018.

[30] 3GPP SA S2-1908105, *Broadcast of Assistance Data for NR*, April, 2019.

[31] 3GPP RAN RP-191156, *NR Positioning Support (NR POS)*, June, 2019.

[32] 3GPP RAN2 R2-1901078, *TP for Additional GNSS Enhancements: Completion of SSR*, February, 2019.

Chapter 8

Terrestrial Positioning Technologies: Cellular Networks

8.1 INTRODUCTION

Long before satellite-based positioning systems were established, techniques using terrestrial beacons had already been installed and used for location determination. The LORAN system developed to locate naval vessels dates back to the early 1940s. In LORAN, low-frequency signals, typically around 100 KHz, were used as the basis for the position determination. The advantage of low frequencies is that they allow large geographical area coverage with a limited number of transmitters and offer good coverage over land and sea.

With the advent of mobile networks and the large-scale deployment of GSM and CDMA, similar terrestrial positioning techniques have been developed. These technologies leverage the existing base station deployment and offer a cost-effective way to reuse the infrastructure to locate a mobile device by triangulation, trilateration, or multilateration. The time synchronization of base stations, offered first by CDMA2000 networks and then adopted by LTE, increases the accuracy of the position fixes achieved.

Today, GNSS receivers are increasingly present in many consumer devices. Nevertheless, their availability is limited in dense urban and indoor scenarios. Network-based terrestrial positioning methods are widely used to complement GNSS-based positioning. They can provide a position fix in scenarios where a GNSS signal is not available, improve the accuracy of GNSS location fixes or provide robustness against attacks to the integrity and availability of the GNSS signal like jamming or spoofing.

This chapter will focus on positioning techniques based on cellular networks like GSM, UMTS, CDMA, or LTE. These are also referred to in 3GPP as RAT-dependent positioning technologies.

The positioning methods explained in this chapter include methods based on downlink signals as well as others based on the uplink. Downlink methods (except cell ID) require UE support to perform positioning measurements. These include mainly cell ID and E-OTD, used in GSM networks [1]; advanced-forward link trilateration (A-FLT) [2], defined for CDMA networks; and ECID and OTDOA, introduced for LTE. The uplink positioning methods used in all networks from GSM to LTE are uplink time of arrival (U-TOA) and U-TDOA. The location estimate is calculated entirely on the network side. Hence, they do not require support of any specific feature by the UE.

The above-listed terrestrial positioning systems are adversely affected by several error sources like noise, synchronization errors, and other sources that will also be studied in this chapter. Among all other error sources, in urban and indoor locations, multipath and non-line of sight (NLoS) propagation conditions have the most critical effects on the position calculation. Propagation delay calculations based on delay locked loops (DLLs), although sufficient for wireless communication purposes, severely affect position accuracy estimates. Multipath mitigation algorithms, which usually estimate the channel impulse response to calculate a LoS delay to the base station, can augment the terrestrial techniques. Typically, the path estimate with the smallest delay is considered to be the direct LoS path. These algorithms improve location estimates and perform better with wider signal bandwidths. Under multipath conditions, time difference techniques like E-OTD and OTDOA tend to slightly outperform other techniques because multipath tends to affect similarly multiple base stations and can be partially averaged out.

Due to the above-mentioned error sources and other properties inherent to cellular network positioning, like the measurement resolution, the position accuracy achieved with such techniques will be typically in the range of tens to hundreds of meters. A comparison of the different technologies considered throughout this chapter can be seen in Figure 8.1. The methods are classified by the network for which they have been designed and deployed and whether they are downlink or uplink: in oval shape with dashed border, GSM downlink positioning methods; in oval shape with solid border, CDMA methods; in oval shape with dotted border, LTE methods; and in rectangular shape, uplink positioning methods.

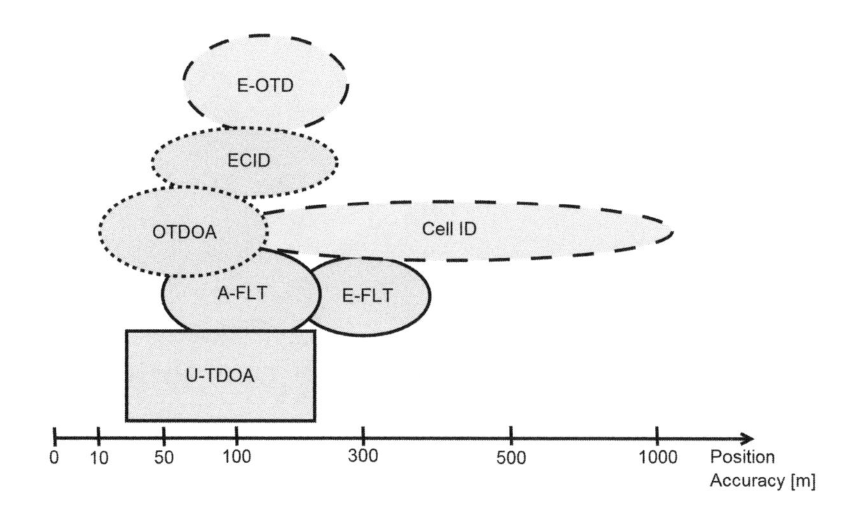

Figure 8.1 Comparison of positioning accuracy for different cellular network positioning techniques.

8.2 CELL ID

The location based on cell ID (CID) is one of the simplest positioning technologies for cellular networks. Every cell or base station transmits a unique ID. Using this ID, it can be determined that a UE is within the area of coverage of a certain base station, as shown in Figure 8.2. The UE is known to be located within the circular area. However, the accuracy of this method is low. The uncertainty in the position of the UE is equal to the size of the cell serving area or range. For GSM, the worst-case value is given by the maximum timing advance (i.e., the maximum time the uplink signal from a UE can travel before reaching the base station). This gives a serving area with a radius up to 35 km [3]. For LTE, the serving areas vary depending on the frequency band and on whether the environment is urban or rural, and they are classified as macrocells, microcells, picocells and femtocells. In rural environments with low frequency bands, macrocells can go up to a size of 100 km with acceptable performance, although the optimal cell size is between 5 and 10 km. In urban environments, macrocells are typically smaller, between hundreds of meters and a few kilometers. Microcells and picocells are normally used to increase the network capacity in areas with dense phone usage like train stations, shopping malls, and sport stadiums and their maximum size is around 2 km and 200 m, respectively. Femtocells are deployed typically indoors, at home or in office buildings, and its

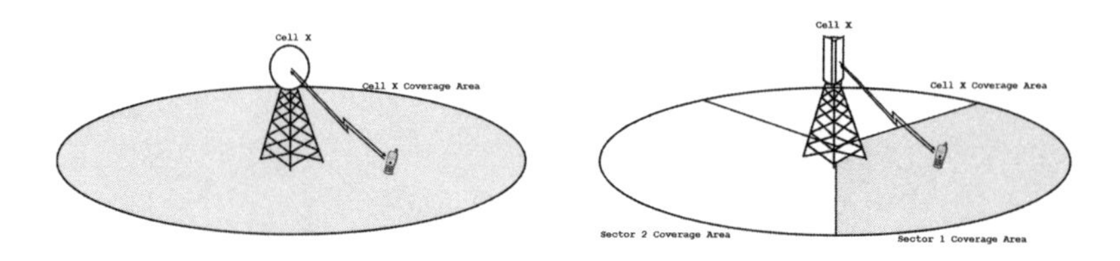

Figure 8.2 Cell ID positioning. The UE is located within the delimited area.

range is in the order of 10 m. In the case of sectorized networks, where different antenna arrays are used to cover different cell sectors, the cell ID positioning can be further refined, as shown in the right part of Figure 8.2.

This method can be used for GSM, WCDMA, CDMA, or LTE, although the definition of unique ID varies depending on the network. For GSM, WCDMA, or LTE, the unique cell ID is constructed with the MCC, MNC, location area code (LAC), and cell ID. For CDMA, the ID is formed by the system ID, the network ID, and the base station ID.

8.2.1 The Timing Advance

The precision of cell ID positioning can be improved if the base station takes into account the timing advance (TADV or TA) measurement. Since GSM is a TDMA network, it requires to know very accurately at what time stamp the signal from a particular UE will be received. The TADV measurement is used by the network to schedule the UE uplink signal to arrive at a particular time slot. The TADV value is the time that an uplink signal coming from the UE needs to reach the base station. Hence, the TADV is a ToA measurement and it can be useful to locate the UE within a circumference centered at the base station position.

This procedure is shown in Figure 8.3, where the UE can be located at any point of the circumference. However, the accuracy is limited by the resolution of the TADV measurement and the circumference of location is in fact a ring or annulus. This is represented by the uncertainty ring-shaped area in Figure 8.3. For GSM, the TADV measurement is encoded with 8 bits, taking values from 0 to 63 bit periods (~ 3.69 μs) [3]. The distance in meters equivalent to 1 TA is the bit period multiplied by the propagation speed of the radio frequency wave. In free space, $v_p = c \approx 3e8\frac{\text{m}}{\text{s}}$ so the radio signal can travel a round-trip distance $3e8\frac{\text{m}}{\text{s}} \cdot 3.69$ μs $\approx$ 1107m during one bit period, equivalent to distance d $\approx$ 553.5 m in a single direction (1107·0.5). Thus, the location uncertainty for a TADV measurement is $\text{d}/2$ or 276.75 m. In

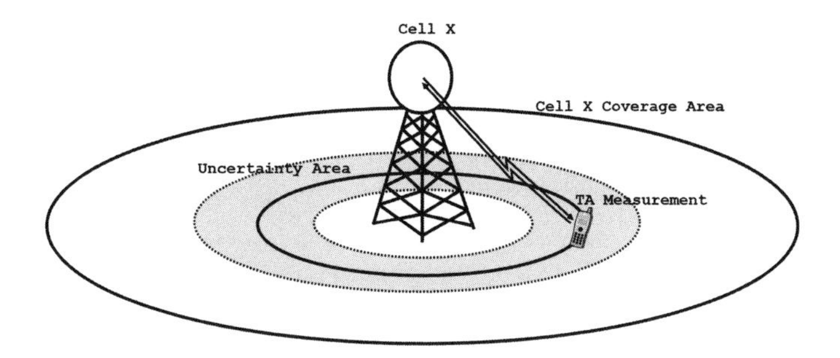

Figure 8.3 Cell ID positioning using TADV. The UE is located within the ring-shaped area based on the TOA measurement.

Figure 8.3, this value is the difference between the outer radius and the inner radius of the ring of location. The combination of CID with TADV can only be used in GSM and LTE networks, as CDMA2000 and WCDMA are based on CDMA and do not require TADV measurements. This method will be explained more in detail in Section 8.6.

8.3 ENHANCED OBSERVED TIME DIFFERENCE

E-OTD is a positioning method introduced for GSM and continued for WCDMA [4, 5]. In E-OTD, the UE measures the difference on the arrival time of signals received from different base stations. Each base station (BTS for GSM and Node B for WCDMA) periodically transmits a synchronization burst. The UE receives this burst and calculates the TDoA between bursts from different base stations, defining a localization hyperbola for each pair of base stations. The base stations are at the focal points of the hyperbola and the UE can be located at any point on the curve. Hence, E-OTD is a multilateration method, as seen in Section 2.3.4.

Recalling the multilateration equation from Section 2.3.4 and particularizing it for the E-OTD method, Equation 8.1 is obtained:

$$d_2 - d_1 = TD_{2,1} \cdot v_p, \tag{8.1}$$

where $d_i = \sqrt{(x - x_i)^2 + (y - y_i)^2 + (z - z_i)^2}$ is the distance between the UE and each of the base stations, v_p is the speed of propagation of the radio frequency waves and $TD_{2,1}$ is the time difference measurement for the base station pair (1,2).

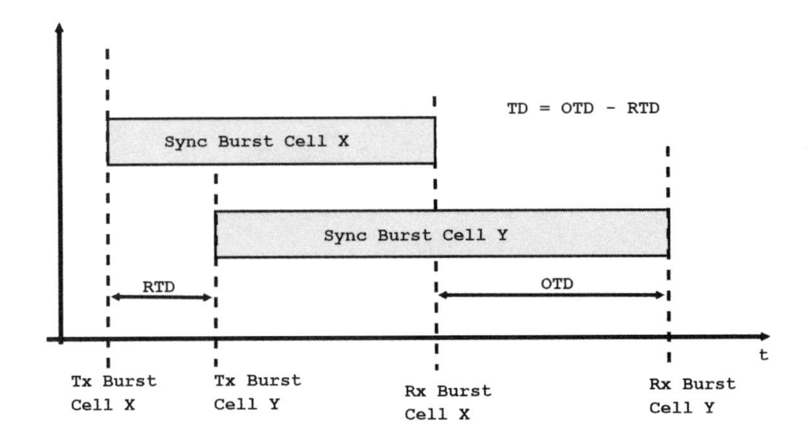

Figure 8.4 Relation between the TD, RTD, and OTD measurements.

For GSM and WCDMA, the base stations are not synchronized, which means that bursts from different base stations are not transmitted at the same time. Hence, the TD measurement in Equation 8.1 is not exactly the OTD measured by the UE; it is also influenced by the real-time difference (RTD), the difference in the transmission time between the bursts from two base stations [6–9]. This relationship can be seen in Figure 8.4: the physical time difference between the bursts from two base stations X and Y at the UE position is the time difference observed by the UE (OTD) minus the time difference between the transmission of each of the bursts (RTD). This value is referred to in the GSM standards also as geometrical time difference (GTD) [2].

$$GTD_{2,1} = OTD_{2,1} - RTD_{2,1} \tag{8.2}$$

The E-OTD method can operate in two different modes: the positioning can be calculated on the network side (UE assisted) or entirely on the UE side (UE based). For UE assisted, the network requests the UE to perform E-OTD measurements and report them back. The measurements are transferred to the GMLC, which estimates the UE position. For UE based, the network requests the UE to calculate and report its own position. In order to do that, the UE needs some information from the network, as the exact location of each base station required for the positioning algorithm, and the RTD value for each pair of base stations required to estimate the physical time difference seen in Equation 8.2. These values are measured at the location measurement unit (LMU) and transmitted as part of the assistance data [4].

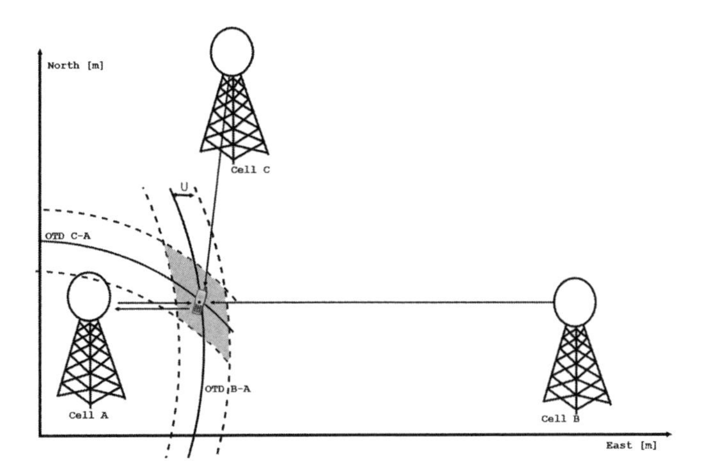

Figure 8.5 E-OTD positioning. The UE is within the area formed by the intersection of the localization hyperbolas.

In order to obtain a 2D location, at least two OTD measurements are required to calculate the two unknowns x and y. By convention, the time differences are always measured with respect to a reference base station. Accordingly, at least three base stations must be visible to the UE for E-OTD. This process is shown in Figure 8.5. Cell A is in this example the reference base station. An OTD measurement is performed between the two neighbors, Cell B, and Cell C and the reference. This results in two localization hyperbolas, which form an intersecting area where the UE is located.

Similar to cell ID + TA, the accuracy of E-OTD positioning is bounded by the resolution of the OTD measurement. For GSM, the resolution of OTD is the same as the resolution of the TA measurement, resulting in the same uncertainty U of 276.75 m for each of the hyperbolas [10]. This is represented in Figure 8.5, where the solid lines represent the actual OTD measurements and the dashed lines represent the uncertainty. This translates into an uncertainty on the determined position represented by the shaded area. Nonetheless, the studies in [7, 8] claim that E-OTD positioning can accomplish an accuracy typically between 60 and 150 m in practice.

8.4 MULTILATERATION TA AND MULTILATERATION OTD

In GSM, there are two other positioning methods similar to E-OTD and TA, called multilateration timing advance (MTA) and multilateration observed time difference (MOTD) [1]. These techniques do not require any LMU; the whole procedure is done by the base station and the UE.

For MTA, the base station acquires TA information from a set a of neighbor cells selected by the UE. Upon receiving the MTA Request the UE selects a subset of the nearby cells based on received signal strength and performs a packet access procedure to allow the serving base station to estimate a TA for each of the cells. Combining the measurements for three or more neighbor cells, the serving base station is able to calculate the UE position. Although the method is called "multilateration," it is in fact a trilateration technique, as it is based on time of arrival measurements.

The MOTD procedure is quite similar to E-OTD, the main difference being that instead of the RTD measurements performed by the LMUs, MOTD relies on TA measurements to the serving base station in combination with the OTD measurements of the neighbor cells. This approach combines TOA and TDOA measurements and it is often referred to as mixed multilateration [9].

8.5 ADVANCED FORWARD LINK TRILATERATION

A-FLT is a network-based positioning technology developed for CDMA networks. It was introduced in the CDMA IS-95 standard, in which the base stations are synchronized using GPS. As a consequence of this synchronization, the UE can measure accurately the phase of arrival of a CDMA pilot signal and calculate its ToA. This allows a more accurate measurement compared to GSM E-OTD and it does not need any additional RTD information from the network. The position is calculated by trilateration combining TOA measurements from three or more base stations.

Furthermore, the measurement resolution is better than for OTD measurements. The chip duration in CDMA networks is 0.813 μs, and the A-FLT reporting resolution is $\frac{1}{8}$ of the chip duration, resulting in an uncertainty much smaller than for GSM and WCDMA. Doing the same derivation as in Section 8.2.1, it can be calculated that the A-FLT reporting uncertainty is $\sim$ 30.5 m. Based on field trials, it has been shown that A-FLT positioning provides better results than E-OTD [11,12].

The A-FLT calculation requires software support in the UE to perform the phase measurements, which is not available for all handsets. Consequently, CDMA networks also allow a slightly different variant of FLT measurement called enhanced forward link trilateration (E-FLT), compatible with all CDMA UEs. The accuracy of E-FLT is substantially worse than for A-FLT. If the latter is able to reach a typical accuracy between 50-200 m, E-FLT can only reach 250-350 m [13].

8.6 ENHANCED CELL ID

ECID is a network-based positioning method for LTE defined as part of the LPP in TS 36.305 [14] as an enhancement of the timing advance procedure. The method calculates the distance between a mobile device and a base station by measuring the RTT of a signal [15]. The RTT is defined as the time between the moment a downlink signal is sent by the base station and the acknowledgment for the same signal is received. Being based on RTT measurements, ECID has the advantage that the clock errors from both receiver (UE) and transmitter (BS) cancel out. Hence, the ECID measurement is a *range* rather than a pseudorange as in GNSS.

ECID is based on the acquisition of Rx-Tx measurements. This measurement, defined in 3GPP specification TS 36.214 [16], is the difference between the transmission (Tx) time of a signal and the reception (Rx) of the acknowledgment. The process of calculating the RTT based on Rx-Tx measurements can be explained in three steps:

1. The eNB measures its own Rx-Tx time.

$$eNB\,Rx - Tx = T_{eNB,Rx} - T_{eNB,Tx}.$$

2. The eNB sends a TADV adjustment command to the UE to adjust UE's uplink timing.

3. The UE measures and reports its Rx-Tx time.

$$UE\,Rx - Tx = T_{UE,Rx} - T_{UE,Tx}.$$

In order to better understand the three steps above, it is worth remembering that the LTE standard is defined to achieve that the eNB reception and transmission times are simultaneous. Hence, in the ideal case, $eNB\,Rx{-}Tx = 0$. The eNB sends the UE timing advance commands to indicate how much in advance the UE needs

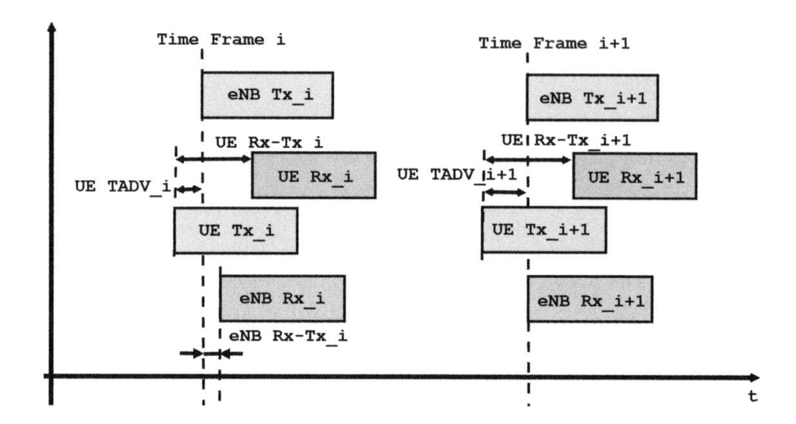

Figure 8.6 ECID RTT measurement process.

to send the uplink data, so that it arrives at the eNB at a certain time frame. This process is shown in Figure 8.6. In Time Frame i, the eNB measures the eNB Rx-Tx. This value should be the same as $\frac{UERx-Tx_i}{2} - T_{adv,i}$. As, $eNB\,Rx - Tx_i \neq 0$, the eNB sends a T_{adv} adjustment command to the UE to correct the UE UL timing to match $T_{adv} = \frac{UERx-Tx}{2}$. In Time Frame i+1, the UE uplink timing has been corrected and eNB Rx-Tx should be zero. Due to that, the UE can measure its UE Rx-Tx value accurately and report it back to the eNB.

ECID is a trilateration mechanism as seen in Section 2.3.3. Thus, an ECID range measurement can be represented as in Equation 8.3:

$$\rho_e = \frac{RTT}{2} \cdot v_p + \nu_{RTT} = \overline{BSUE}, \tag{8.3}$$

where v_p represents the propagation speed and ν_{RTT} is the error associated to the RTT measurement. The receiver and transmitter clock errors are canceled out by doing a round-trip measurement instead of a single direction time of flight. The ECID range represents the distance between the UE and the base station and defines a sphere of location.

$$\overline{BSUE} = \sqrt{(x_{bs} - x_{ue})^2 + (y_{bs} - y_{ue})^2 + (z_{bs} - z_{ue})^2}. \tag{8.4}$$

The UE Rx-Tx measurement is calculated in T_s. A T_s is the basic unit of time in LTE and it is defined as $T_s = \frac{1}{15000 \cdot 2048} s$, approximately 32.5 ns. The UE Rx-Tx measurement resolution is 2 T_s [17]. Thus, the uncertainty of the Rx-Tx

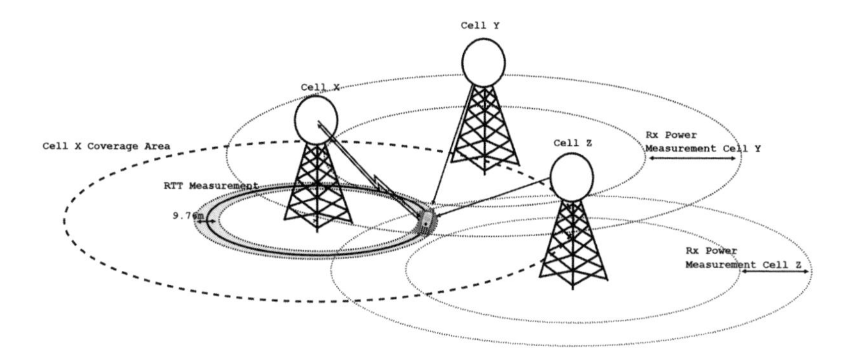

Figure 8.7 ECID Positioning with RTT and power measurements.

measurement is $U_{Rx-Tx} = 2 \cdot T_s \cdot v_p \approx 9.76$m. As it can be seen, the uncertainty is much smaller than with the methods used in previous RATs.

However, the RTT measurement is at the moment only defined between the UE and the serving base station. Hence, ECID based on ToA measurements cannot be used as the only source for position calculation. It has to be combined with other methods to get at least three measurements (three equations) to solve the position of the UE. One approach, if the UE is not moving, is that the network commands the UE to perform a handover (HO) to a neighbor cell. After the HO is completed, a new ECID session can be conducted to obtain a second RTT measurement. This process can be repeated with a third cell to be able to obtain a position. However, the whole process is time consuming and it requires the UE to be static; otherwise the first RTT measurement no longer remains valid.

Another approach is to combine the RTT measurement with distance calculated from power range measurements based on received power levels (reference signal received power (RSRP) and/or reference signal received quality (RSRQ)) from the surrounding neighbor cells, as depicted in Figure 8.7. The UE measures RTT to the serving cell and Rx power measurements from two neighbor cells. Measured RSRP and RSRQ values can be used to estimate the distance to the base station. A basic method to calculate the distance is to use the equation of free space path loss (FSPL), which is derived from the Friis transmission formula by assuming isotropic transmitter and receiver antennas. The calculation is shown in Equation 8.5, where P_{rx} is the received power, P_{tx} is the transmitted power, λ is the wavelength of the signal, and d is the distance between the transmitter and receiver antennas. Nonetheless, the FSPL is a simplification and cellular networks probably use a more advanced model of the propagation of the signal.

$$\frac{P_{tx}}{P_{rx}} = (\frac{4 \cdot \pi d}{\lambda})^2. \tag{8.5}$$

Using ECID, the UE position can be estimated to be within the dark shaded area in the picture. Nevertheless, the ranges obtained from power measurements are more inaccurate than the range from ToA measurements and including those equations into the system will lower the accuracy of the solution. The RSRP and RSRQ measurements have a reporting resolution of 1 dBm and 0.5 dB, respectively [17]. The amount of uncertainty in the distance calculation created by the RSRP and RSRQ resolution depends on other parameters like the frequency of operation, the actual distance between the base station and the UE, and the signal path: the path loss will be different if the signal propagates in free space, or if the signal goes through walls or other materials. In general, the uncertainty in the distance estimation using power ranging will remain much greater than the 10 m achieved with RTT.

A third approach is to combine ECID with other positioning technologies like GNSS or OTDOA to create a hybrid solution. This approach will be further discussed in the following sections.

8.7 OBSERVED TIME DIFFERENCE OF ARRIVAL

OTDOA is the evolution of the E-OTD for LTE networks. The main difference to its predecessor is the introduction in LTE of a specific reference signal for positioning, the positioning reference signal (PRS). The UE can still perform the TDOA measurement using any other downlink signal (e.g., primary synchronization signal (PSS), secondary synchronization signal (SSS), or cell reference signal (CRS)) if the received signal strength is enough for detection. However, these signals may not suffice to identify enough neighbor base stations to calculate a position. This is due to the properties of these signals, their transmission power, and other characteristics. The PRS signal has been specifically designed to improve the quality of the OTDOA measurements and increase the number of measurements available [15, 18].

The PRS in LTE is defined by three parameters:

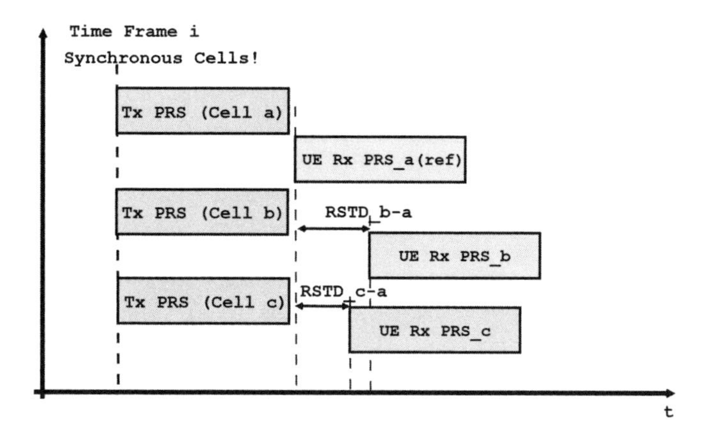

Figure 8.8 OTDOA RSTD measurement process.

- PRS configuration index: it defines the periodicity of the transmission of PRS T_{PRS}; that is, how often the PRS will be transmitted. It also indicates the subframe offset Δ_{PRS} of the PRS signal.
- PRS bandwidth N_{RB}^{PRS}: the frequency bandwidth used for the transmission of the PRS signal. It has to be smaller or equal to the BW of the cell and one of the valid LTE BW values: 1.4, 3, 5, 10, 15, or 20 MHz.
- Number of downlink subframes N_{PRS}: the number of consecutive subframes in which the PRS signal will be transmitted. It can be 1, 2, 4, or 6 subframes.

Given these three parameters, the mobile device knows when to expect a PRS signal. Each instant when a PRS signal can be received is called a PRS occasion. However, the eNB may not transmit the PRS at every single occasion. In order to spare some subframes, the 3GPP specifications have given the possibility to mute the PRS according to a predefined muting sequence. This muting sequence is a pattern of bits that repeats itself. At one particular PRS occasion, if the muting sequence has a value of 1, the PRS will be transmitted, while a value of 0 means the PRS will be muted. The muting sequence must as well be communicated to the mobile device with two fields: the length in bits of the sequence and the sequence itself.

The process to perform an OTDOA measurement is shown in Figure 8.8. The mobile device tunes to the PRS symbols of multiple base stations and selects one (typically the serving cell or the cell with better signal quality) as a reference cell. Then, it computes the time difference between each of the other eNBs and the reference cell and reports back these values. Each of these measurements is called reference signal time difference (RSTD) and is defined as the time of arrival of the PRS of the neighbor cell minus the time of arrival of the PRS of the reference cell. It is important to notice that for optimal OTDOA measurements, the base stations in the network should be perfectly synchronized. The RSTD measurement is reported in T_s ($T_s = \frac{1}{15000 \cdot 2048}$ s), analogously to the UE Rx-Tx measurement for ECID. The RSTD measurement for a neighbor base station n is defined in Equation 8.6:

$$\begin{aligned} RSTD_n[T_s] &= (PRS_n[s] - PRS_{ref}[s]) \cdot 15000 \cdot 2048 \\ &= T_{PRS,sf\ Rx\ n} - T_{PRS,sf\ Rx\ ref}. \end{aligned} \tag{8.6}$$

$T_{PRS,sf\ Rx\ n}$ is the time when the mobile receives the start of the PRS subframe from cell n [15,16]. The default resolution of the RSTD measurement is 1 T_s, which gives a measurement uncertainty $U_{RSTD} \approx$ 4.88m. As of E-UTRA release 13, 3GPP has introduced an optional higher-resolution RSTD measurement of 0.5 T_s. Using the high-resolution RSTD would produce a measurement uncertainty of $U_{RSTD} \approx$ 2.44 m [17].

Analogously to E-OTD, OTDOA is a multilateration method as seen in Section 2.3.4. Hence, each OTDOA measurement can be used to form hyperbolic equations as defined in Equation 8.7, where t_{Rx,PRS_n} is the reception time for the PRS signal for the neighbor n, τ_{Rx} is the receiver clock error, and $\tau_{Tx,n}$ is the clock error of base station n.

$$\begin{aligned} RSTD_n[s] = (&t_{Rx,PRS_n}[s] - \tau_{Rx} - \tau_{Tx,n} \\ &- (t_{Rx,PRS_{ref}}[s] - \tau_{Rx} - \tau_{Tx,ref})). \end{aligned} \tag{8.7}$$

The time error of the receiver of the mobile device in Equation 8.7 cancels out and the final equation can be rewritten as in Equation 8.8, where $RSTD_{n,true}$ represents the real RSTD value without timing errors.

$$RSTD_n[s] = RSTD_{n,true} - \tau_{Tx,n} + \tau_{Tx,ref}. \tag{8.8}$$

Recalling the basic equation of the hyperboloid from Chapter 2, $d_2 - d_1 = K$, K for OTDOA can be defined as in Equation 8.9, and the hyperboloid equation results in Equation 8.10.

$$K = (RSTD_{n,true} - \tau_{Tx,n} + \tau_{Tx,ref}) \cdot v_p + e_{RSTD}. \tag{8.9}$$

$$\begin{aligned} d_n - d_{ref} &= \sqrt{(x - x_n)^2 + (y - y_n)^2 + (z - z_n)^2} \\ &- \sqrt{(x - x_{ref})^2 + (y - y_{ref})^2 + (z - z_{ref})^2} \\ &= (RSTD_{n,true} - \tau_{Tx,n} + \tau_{Tx,ref}) \cdot c + e_{RSTD}. \end{aligned} \tag{8.10}$$

Regrouping Equation 8.10 to have a notation similar to A-GNSS, an OTDOA range can be defined by Equation 8.11, where ν_{RSTD} represents all the errors associated to the RSTD measurement including clock errors from the base stations and is modeled as an i.i.d. Gaussian random process.

$$\begin{aligned} \rho_{otdoa} &= RSTD_{n,true} \cdot c - \tau_{Tx,n} \cdot c + \tau_{Tx,ref} \cdot c + e_{RSTD} \\ &= RSTD_{n,true} \cdot c + \nu_{RSTD} = \sqrt{(x - x_n)^2 + (y - y_n)^2 + (z - z_n)^2} \\ &- \sqrt{(x - x_{ref})^2 + (y - y_{ref})^2 + (z - z_{ref})^2} \end{aligned} \tag{8.11}$$

The first thing to notice is that the RSTD measurement is not analogous to a pseudorange as was the case for GNSS measurements. As there is no unknown time variable to calculate, it is a range. However, as already pointed out, the clock errors for the base stations play a significant role, and in order to obtain good position estimates using OTDOA, the base station network needs to be very accurately synchronized [15]. This and other sources of error will be analyzed in detail in the following sections.

As OTDOA is based on ranges instead of pseudoranges, a 3D position can be calculated with three equations[1] (i.e., three RSTD measurements, which means four base stations) and a 2D position can be calculated with just two equations. The UE reports the RSTD measurements back to the serving cell, which forwards these values to the location server of the network. The location server calculates the position by multilateration as detailed in Figure 8.9. The UE measures RSTD from two pairs of base stations. The location server is able to locate the UE within the shaded area. During the position calculation, the coordinates of the eNBs used for the measurements are required. Up to now, the 3GPP specifications do not allow for any mechanism to send the base stations' coordinates to the network users. Hence, there is no possibility of UE-based OTDOA. All position calculations must be done by the network.

1 However, in practice OTDOA and other network-based positioning systems do not offer good altitude results, for reasons related to the geometry of the base station network (this will be analyzed further in Section 8.11).

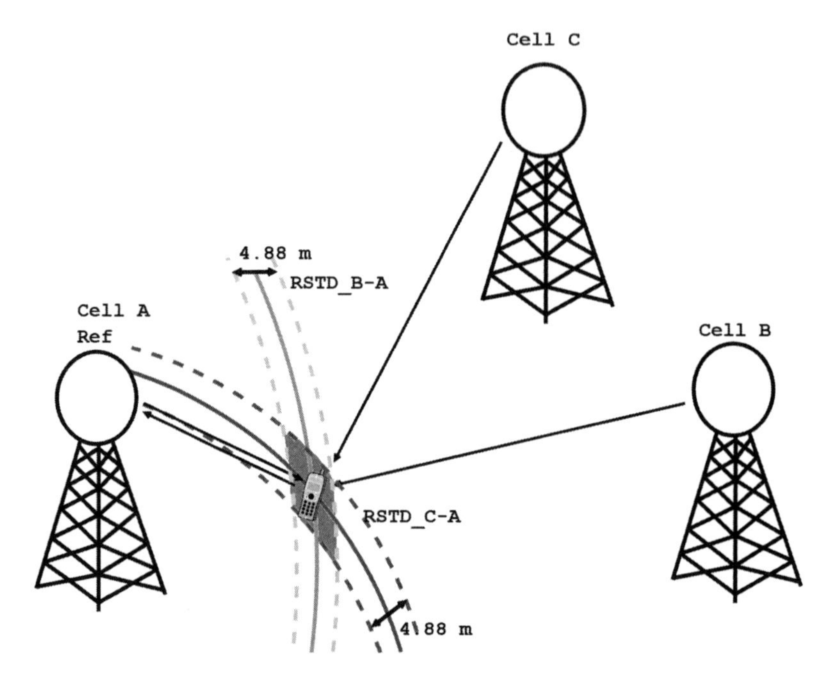

Figure 8.9 OTDOA positioning with RSTD measurements.

8.8 UPLINK TIME OF ARRIVAL

U-TOA is an uplink positioning method introduced for GSM [27]. The U-TOA method is based on the measurement of the TOA of the uplink signal received by a base station. The measurement is done in the access channel, as the UE transmits typically at a higher power for the initial access burst, and it does not require any dedicated U-TOA signal. The U-TOA measurement process is shown in Figure 8.10. In the figure, the TOA measurement for the same UE access burst to three different base stations is shown. The position is estimated in the LMU with measurements from three or more base stations using a trilateration method.

In order to induce the UE to send the access bursts required for U-TOA positioning, the network triggers an HO procedure [27]. After receiving an HO command, the UE will start to send the access burst to the target base station. As part of the positioning procedure, the target base station will not respond to the access bursts and the UE will continue sending bursts until the HO timer expires.

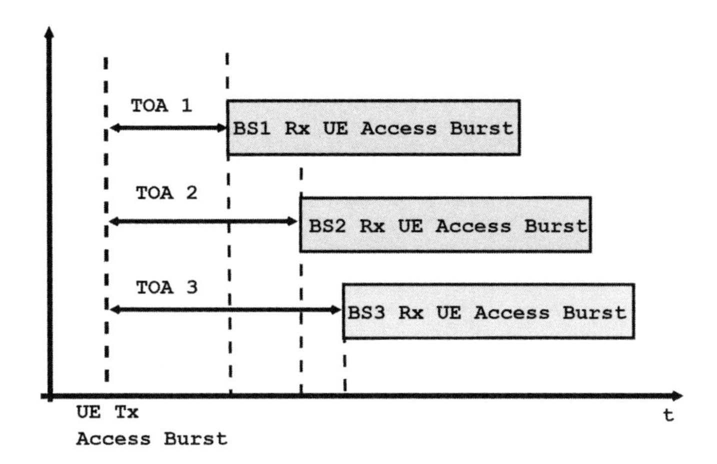

Figure 8.10 U-TOA measurement process.

8.9 UPLINK TIME DIFFERENCE OF ARRIVAL

U-TDOA is a positioning method used in cellular networks since GSM [28] as an upgrade to the U-TOA method described in the previous section. However, instead of time-of-arrival measurements, U-TDOA is based on time-difference-of-arrival measurements, as depicted in Figure 8.11. Another difference with respect to U-TOA is that the signals received during normal operation are sufficient for the U-TDOA positioning, and an artificial HO is not required. Hence, U-TDOA is more efficient and does not require additional signaling resources.

The working principle of U-TDOA is similar to E-OTD and OTDOA, with the fundamental difference that the TD measurements are performed on uplink signals. The measurement is done by the base station and the position is calculated in the LMU. This presents several advantages with respect to the downlink methods:

- All the complexity is on the network side: U-TDOA is compatible with every UE.
- RF resources efficiency: it does not require any specific signals, it relies on normal uplink data transfer during communication.
- It benefits from the frequency hopping in GSM to mitigate multipath effects.

A study of the accuracy of U-TDOA was developed by 3GPP in [28]. This study has been able to determine the lower bound of the root mean square (RMS)

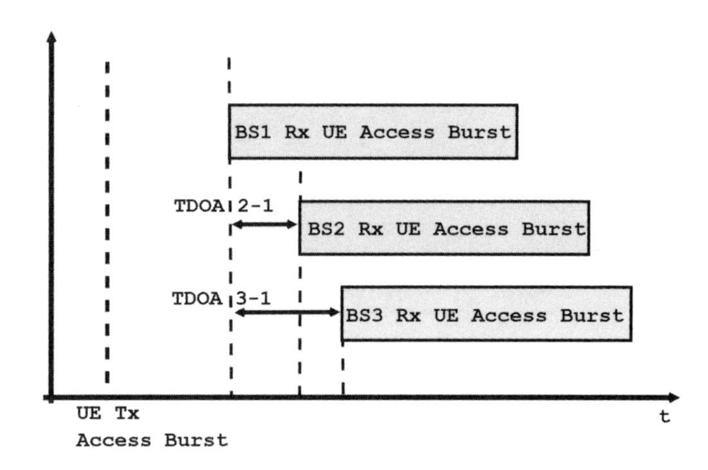

Figure 8.11 U-TDOA measurement process.

error in a U-TDOA measurement in Equation 8.12, called the Cramer-Rao bound. This RMS does not take into account the multipath effects.

$$U - TDOA_{rms} = \frac{\sqrt{12}}{2 \cdot \pi \cdot BW\sqrt{2 \cdot BW \cdot T \cdot SNR_y}}. \tag{8.12}$$

In Equation 8.12, BW is the bandwidth of the signal, T is the coherent integration period and SNR_y is the signal-to-noise ratio of the remote signal. Given that, the location error can be approximated in Equation 8.13, where P is the number of diversity antennas; N the number of base stations, with a minimum requirement of $N \geqslant 3$, and GDOP_c is the geometric dilution of precision relative to the GDOP at the center of a circular N-base station configuration [28].

$$Location_{rms} = U - TDOA_{rms} \cdot \sqrt{P}\sqrt{N} \cdot GDOP_c. \tag{8.13}$$

Based on this analysis and a field trial, the 3GPP has estimated that U-TDOA is between 75% and 140% more accurate than E-OTD [28]. Nevertheless, U-TDOA presents some challenges that have hindered its worldwide deployment. First, it requires very accurate network synchronization. The required installation is costly in comparison to other downlink technologies or leveraging existing technologies like A-GNSS. This problem can be partially overcome by the fact that not every base station in the network needs to support U-TDOA; it is sufficient if the technology is deployed to a subset of the network. Second, U-TDOA's accuracy relies on the network density and decays in rural areas.

U-TDOA is used in GSM, WCDMA, and LTE networks. In LTE, the individual measurements are defined as uplink relative time of arrival (UL-RTOA) [29]. Each UL-RTOA measurement can be used to form hyperbolic ranges similar to OTDOA in Equation 8.10. The only differences are the UL-RTOA measurement instead of the RSTD measurement and the fact that the base station clock errors are now receiver clock errors instead of transmitter clock errors, as seen in Equation 8.14:

$$\begin{aligned} d_n - d_{ref} &= \sqrt{(x - x_n)^2 + (y - y_n)^2 + (z - z_n)^2} \\ &- \sqrt{(x - x_{ref})^2 + (y - y_{ref})^2 + (z - z_{ref})^2} \\ &= (UL - RTOA_{n,true} - \tau_{Rx,n} + \tau_{Rx,ref}) \cdot c + e_{RSTD}. \end{aligned} \tag{8.14}$$

The U-TDOA measurement is defined in Equation 8.15:

$$\begin{aligned} \rho_{utdoa} \equiv &\sqrt{(x - x_n)^2 + (y - y_n)^2 + (z - z_n)^2} \\ &- \sqrt{(x - x_{ref})^2 + (y - y_{ref})^2 + (z - z_{ref})^2} \\ &- UL - RTOA_{n,true} \cdot c - \nu_{UL-RTOA} = 0. \end{aligned} \tag{8.15}$$

The resolution for UL-RTOA measurements is 2 T_s, as defined in LTE Positioning Protocol A (LPPa). This resolution gives a measurement uncertainty of $\approx$ 9.76 m, same as for ECID and bigger than for OTDOA.

8.10 HYBRID POSITIONING

The term hybrid positioning in this section refers to the combination of cellular positioning technologies, in particular ECID, OTDOA and U-TDOA, with other technologies like A-GNSS in order to estimate a position. This method can be used to obtain location estimates when very few GNSS satellites are available or also when the DOP of the satellite constellation is poor. Combining A-GNSS and ECID is quite straightforward, as both use trilateration methods based on localization spheres. OTDOA and U-TDOA are multilateration methods based on hyperbolic localization in 3D taking the shape of hyperboloids. Nonetheless, OTDOA and U-TDOA measurements can be normalized as well and all four methods together can be used in an extension of the positioning algorithm seen in Chapter 6 [15].

The combination of cellular and satellite positioning measurements presents several advantages in comparison to the standalone methods. It allows the position calculation with a minimum of four heterogeneous measurements (e.g, two satellite pseudo-ranges and two OTDOA RSTD measurements). Using the standalone methods, a UE position cannot be calculated with the set of measurements proposed in this example. Furthermore, it adds robustness to the positioning algorithm and improves the dilution of precision, as will be seen later in this section.

8.10.1 Hybrid Positioning Algorithm

The positioning algorithm described here is an extension of the A-GNSS WLS positioning solution defined in 3GPP TS 36.171 [19] for A-GNSS pseudoranges. A linearization process is applied to U-TDOA, OTDOA, and ECID ranges to be able to combine all the equations together in a single equation system and compute a hybrid positioning solution [15,20]. This procedure involves the representation of the range measurements as a Taylor series using Equation 8.16, where $f^{(n)}(x = a)$ denotes the nth derivative of the function f evaluated at a point a. For a multivariate equation, this term represents the partial derivatives with respect to each of the variables.

$$f(x) \approx \sum_{n=0}^{\infty} \frac{f^{(n)}(x = a)}{n!} \cdot (x - a)^n. \tag{8.16}$$

For congruence, the GNSS pseudorange ρ from Chapter 6 will be denoted here as ρ_g. The ECID and OTDOA ranges will be denoted as ρ_e and ρ_t, respectively. Finally, the term ρ will be used to refer to a generic (pseudo-)range that could take the form of any of the three technologies. For the purpose of this analysis, U-TDOA is completely equivalent to OTDOA and a separate analysis will not be performed. The U-TDOA range will also be denoted as ρ_t.

8.10.1.1 Linearization of the Ranges

For OTDOA, the first step is to linearize Equation 8.11 using the Taylor series representation in Equation 8.16 around a point $P_0 = \{x_0, y_0, z_0\}$. This process is detailed in Equation 8.17, where the notation has been simplified for clarity: $\rho_t(x_0, y_0, z_0) \equiv \rho_{t,0}$, $P = P_0 \equiv P_0$, and $(i - i_0) = \Delta i$.

$$\rho_t(x,y,z) \approx \rho_t(x_0,y_0,z_0) + \nu + (x-x_0)\left.\frac{\partial \rho_t}{\partial x}\right|_{P=P_0} + (y-y_0)\left.\frac{\partial \rho_t}{\partial y}\right|_{P=P_0} + \\ + (z-z_0)\left.\frac{\partial \rho_t}{\partial z}\right|_{P=P_0} = \rho_{t,0} + \nu + \Delta x\left.\frac{\partial \rho_t}{\partial x}\right|_{P_0} + \Delta y\left.\frac{\partial \rho_t}{\partial y}\right|_{P_0} + \Delta z\left.\frac{\partial \rho_t}{\partial z}\right|_{P_0} \tag{8.17}$$

The partial derivatives $\left.\frac{\partial \rho_t}{\partial i}\right|_{P=P_0}$, where i can be x, y, or z, are calculated in Equation 8.18, Equation 8.19, and Equation 8.20, where n represents the neighbor base station n and ref represents the reference base station.

$$\left.\frac{\partial \rho_t}{\partial x}\right|_{P=P_0} = \frac{1}{2}\cdot\frac{2\cdot(x_{ref}-x_0)}{d_{ref}|_{P=P_0}} - \frac{1}{2}\cdot\frac{2\cdot(x_n-x_0)}{d_n|_{P=P_0}} = \\ = \frac{(x_{ref}-x_0)}{d_{ref}(P_0)} - \frac{(x_n-x_0)}{d_n(P_0)} \tag{8.18}$$

$$\left.\frac{\partial \rho_t}{\partial y}\right|_{P=P_0} = \frac{(y_{ref}-y_0)}{d_{ref}(P_0)} - \frac{(y_n-y_0)}{d_n(P_0)} \tag{8.19}$$

$$\left.\frac{\partial \rho_t}{\partial z}\right|_{P=P_0} = \frac{(z_{ref}-z_0)}{d_{ref}(P_0)} - \frac{(z_n-z_0)}{d_n(P_0)} \tag{8.20}$$

The same procedure is applied to the ECID range from Equation 8.3, obtaining Equation 8.21. The same simplifications in the notation as for OTDOA have been directly included.

$$\rho_e(x,y,z) \approx \rho_{e,0} + \nu + \Delta x\left.\frac{\partial \rho_e}{\partial x}\right|_{P_0} + \Delta y\left.\frac{\partial \rho_e}{\partial y}\right|_{P_0} + \Delta z\left.\frac{\partial \rho_e}{\partial z}\right|_{P_0} \tag{8.21}$$

The partial derivatives are calculated in Equation 8.22, Equation 8.23, and Equation 8.24.

$$\left.\frac{\partial \rho_e}{\partial x}\right|_{P=P_0} = \frac{(x_0 - x_{ref})}{\rho_e(P_0)} \tag{8.22}$$

$$\left.\frac{\partial \rho_e}{\partial y}\right|_{P=P_0} = \frac{(y_0 - y_{ref})}{\rho_e(P_0)} \tag{8.23}$$

$$\left.\frac{\partial \rho_e}{\partial z}\right|_{P=P_0} = \frac{(z_0 - z_{ref})}{\rho_e(P_0)} \tag{8.24}$$

In order to be compatible with the A-GNSS algorithm seen in Chapter 6, a fourth variable, the receiver clock error τ, is required. It has already been explained that U-TDOA, OTDOA, and ECID are not affected by any UE clock errors, so $\tau_t = \tau_e = 0$. Thus, the partial derivative $\left.\frac{\partial \rho_t}{\partial \tau}\right|_{P=P_0} = \left.\frac{\partial \rho_e}{\partial \tau}\right|_{P=P_0} = 0$.

8.10.2 Hybrid Equation System

The U-TDOA, OTDOA, ECID, and A-GNSS (pseudo-)ranges are combined together in an equation system following Equation 8.25, based on the A-GNSS equation system in Chapter 6 [15]:

$$\Delta\hat{\rho} = A_h \cdot \Delta\hat{X} + \hat{\nu}. \tag{8.25}$$

The system in Equation 8.25 can be further derived as in Equation 8.26, where $\rho_i^{(n)}$ denotes the (pseudo-)range for a measurement n at iteration i and $\nu^{(n)}$ represents the errors associated with measurement n. ν is considered as Gaussian noise independent and identically distributed (*iid*) of mean $\langle\nu\rangle = 0$. The measurement n can be either a U-TDOA, an OTDOA, an A-GNSS, or an ECID measurement.

$$\begin{bmatrix} \rho_i^{(1)} - \rho_{i-1}^{(1)} \\ \rho_i^{(2)} - \rho_{i-1}^{(2)} \\ \vdots \\ \rho_i^{(n)} - \rho_{i-1}^{(n)} \end{bmatrix} = \begin{bmatrix} \left.\frac{\partial \rho^{(1)}}{\partial x}\right|_{P=P_0} & \cdots & \left.\frac{\partial \rho^{(1)}}{\partial \tau}\right|_{P=P_0} \\ \left.\frac{\partial \rho^{(2)}}{\partial x}\right|_{P=P_0} & \cdots & \left.\frac{\partial \rho^{(2)}}{\partial \tau}\right|_{P=P_0} \\ \vdots & \ddots & \vdots \\ \left.\frac{\partial \rho^{(n)}}{\partial x}\right|_{P=P_0} & \cdots & \left.\frac{\partial \rho^{(n)}}{\partial \tau}\right|_{P=P_0} \end{bmatrix} \cdot \begin{bmatrix} \Delta x \\ \Delta y \\ \Delta z \\ \Delta \tau \end{bmatrix} + \begin{bmatrix} \nu^{(1)} \\ \nu^{(2)} \\ \vdots \\ \nu^{(n)} \end{bmatrix} \tag{8.26}$$

The matrix A_h in Equation 8.25 is called the design matrix [15, 21] and is shown in Equation 8.27.

$$A_h = \begin{bmatrix} \left.\frac{\partial\rho^{(1)}}{\partial x}\right|_{P=P_0} & \left.\frac{\partial\rho^{(1)}}{\partial y}\right|_{P=P_0} & \left.\frac{\partial\rho^{(1)}}{\partial z}\right|_{P=P_0} & \left.\frac{\partial\rho^{(1)}}{\partial \tau}\right|_{P=P_0} \\ \left.\frac{\partial\rho^{(2)}}{\partial x}\right|_{P=P_0} & \left.\frac{\partial\rho^{(2)}}{\partial y}\right|_{P=P_0} & \left.\frac{\partial\rho^{(2)}}{\partial z}\right|_{P=P_0} & \left.\frac{\partial\rho^{(2)}}{\partial \tau}\right|_{P=P_0} \\ \vdots & \vdots & \vdots & \vdots \\ \left.\frac{\partial\rho^{(n)}}{\partial x}\right|_{P=P_0} & \left.\frac{\partial\rho^{(n)}}{\partial y}\right|_{P=P_0} & \left.\frac{\partial\rho^{(n)}}{\partial z}\right|_{P=P_0} & \left.\frac{\partial\rho^{(n)}}{\partial \tau}\right|_{P=P_0} \end{bmatrix} \tag{8.27}$$

The coefficients of the design matrix can take the form of the partial derivatives in Equation 8.18, Equation 8.19, and Equation 8.20 for OTDOA / U-TDOA; the partial derivatives in Equation 8.22, Equation 8.23, and Equation 8.24 for ECID; or the A-GNSS partial derivatives. These are reproduced in Equation 8.28, Equation 8.29, Equation 8.30, and Equation 8.31 for reference, where s_j is the satellite j and $\rho_g^{(j)}$ the pseudorange measurement associated to that satellite.

$$\left.\frac{\partial\rho_g^{(j)}}{\partial x}\right|_{P=P_0} = \frac{(x_0 - x_{s_j})}{\rho_g^{(j)}(P_0)} \tag{8.28}$$

$$\left.\frac{\partial\rho_g^{(j)}}{\partial y}\right|_{P=P_0} = \frac{(y_0 - y_{s_j})}{\rho_g^{(j)}(P_0)} \tag{8.29}$$

$$\left.\frac{\partial\rho_g^{(j)}}{\partial z}\right|_{P=P_0} = \frac{(z_0 - z_{s_j})}{\rho_g^{(j)}(P_0)} \tag{8.30}$$

$$\left.\frac{\partial\rho_g^{(j)}}{\partial \tau}\right|_{P=P_0} = c \tag{8.31}$$

The hybrid system in Equation 8.26 can be solved iteratively starting from an initial guess of the location of the mobile device $P_0 = \{x_0, y_0, z_0\}$. Each iteration i, a new estimate for the position of the mobile device $P_i = \{x_i, y_i, z_i\}$, will be calculated and used as input for the next iteration $i + 1$. The algorithm will continue iterating until $|\Delta P| = |P_i - P_{i-1}| < convergence\_limit$. The solution is typically reached by LSE [15,21]. The error vector $\hat{\nu}$ from Equation 8.25 is isolated in Equation 8.32:

$$\hat{\nu} = \Delta\hat{\rho} - A_h \cdot \Delta\hat{X}. \tag{8.32}$$

In order to find the LSE solution to Equation 8.32, the function in Equation 8.33 should be minimized:

$$J(X) = \sum_{i=1}^{n} \nu_i^2 = \nu'\nu = (\rho - A_h \cdot X)' \cdot (\rho - A_h \cdot X). \tag{8.33}$$

Following the derivation in [21,22], the same as done for A-GNSS, the solution to minimize Equation 8.33 is shown in Equation 8.34:

$$\Delta\hat{X} = (A_h' A_h)^{-1} \cdot A_h' \cdot \Delta\hat{\rho}. \tag{8.34}$$

8.10.2.1 Covariance and Cofactor

Analogous to Equation 8.34, errors in the measurements $\hat{\nu}$ will affect the final position estimation linearly according to Equation 8.35:

$$\hat{\nu}_X = (A_h' A_h)^{-1} \cdot A_h' \cdot \hat{\nu}. \tag{8.35}$$

The *covariance matrix* of the measurement errors $\hat{\nu}$ is defined as $C = E(\nu\nu')$, with coefficients $C_{ij} = E(\nu_i\nu_j)$. For $i = j$, $E(\nu_i\nu_i) = E(\nu_i^2) = \sigma_i^2$, which is known as variance. However, $\forall\ i \neq j$, the value of $E(\nu_i\nu_j)$ will depend on the type of the two measurements that are being combined, i and j. If the two measurements are from different types, or they are both ECID or GNSS measurements, and applying the definition of the error values as iid, the two measurements are uncorrelated and $E(\nu_i\nu_j) = 0$. However, if both measurements are OTDOA, they are not completely uncorrelated, as the reference base station used for both measurements is the same (see Equation 8.6). An error in the calculation for PRS_{ref} will equally affect both measurements. Thus, if ν_{ref} is the error associated to the PRS measurement for the reference cell, $E(\nu_i\nu_j) = E(\nu_{ref}\nu_{ref}) = \sigma_{ref}^2\ \forall\ i \neq j$, where i and j are OTDOA. Similar reasoning can be applied for U-TDOA. Hence, the off-diagonal terms in the covariance matrix for the hybrid positioning algorithm will be denoted as c_{ij}, and $c_{ij} = \sigma_{ref}^2$ if both measurements i and j are OTDOA or both are U-TDOA, and $c_{ij} = 0$ in any other case. The covariance matrix C is written as in Equation 8.36:

$$C = \begin{bmatrix} \sigma_1^2 & c_{12} & \cdots & c_{1n} \\ c_{21} & \sigma_2^2 & & \vdots \\ \vdots & & \ddots & c_{(n-1)n} \\ c_{n1} & \cdots & c_{n(n-1)} & \sigma_n^2 \end{bmatrix} \tag{8.36}$$

The covariance matrix of the calculated position error can be calculated as $C_X = E(\hat{\nu}_X \hat{\nu}'_X)$. Replacing Equation 8.35 into this formula, the solution for C_X depicted in Equation 8.37 is found [15,21]:

$$C_X = C \cdot (A'_h A_h)^{-1}. \tag{8.37}$$

In Equation 8.37, $(A'_h A_h)^{-1}$ is called the *cofactor matrix*, Q. C is strictly dependent on the measurement errors, which are influenced by many sources, and hence, very difficult to predict. However, the cofactor matrix Q defined in Equation 8.38 is purely geometrical, and it depends only on the coordinates of the base stations or satellites and the receiver, and the clock bias:

$$Q = (A'_h A_h)^{-1} = \begin{bmatrix} \sigma_x^2 & \sigma_{xy} & \sigma_{xz} & \sigma_{xT} \\ \sigma_{xy} & \sigma_y^2 & \sigma_{yz} & \sigma_{yT} \\ \sigma_{xz} & \sigma_{yz} & \sigma_z^2 & \sigma_{zT} \\ \sigma_{xT} & \sigma_{yT} & \sigma_{zT} & \sigma_T^2 \end{bmatrix}. \tag{8.38}$$

In Equation 8.38, it has been applied that $\sigma_{ij} = \sigma_{ji}$. This is the reason why A_h is called the design matrix, due to its influence in the cofactor matrix and the direct impact on the error of the position calculation. A_h can be used to design the base station and satellite constellation to reduce the error due to bad geometry. It is important to notice the inversion of the matrix resulting from $(A'_h A_h)$. If this product is not resulting in a matrix of range (or column space) equal to the number of unknown variables, four, the matrix is not invertible and the system has no solution. That implies, for instance, that satellites or base stations that are in the same direction to the receiver will not constitute two independent equations or that if all transmitters are in the same plane, a 3D position cannot be computed.

A possibility if the matrix Q cannot be calculated, which means that a position cannot be found, is to drop the altitude coordinate z and calculate a 2D position. This may be a valid solution in certain scenarios if the location of the UE can be assumed to be on the surface of the Earth, for instance [15].

8.10.3 Hybrid Dilution of Precision

The concept of DOP has been introduced in Chapter 2 and further enhanced in Chapter 6 for A-GNSS. The same concept can be defined for the hybrid positioning method. The DOP is calculated from the trace of the cofactor matrix Q, as shown in Equation 8.39.

$$DOP \equiv GDOP \equiv \sqrt{\sigma_x^2 + \sigma_y^2 + \sigma_z^2 + \sigma_T^2}. \tag{8.39}$$

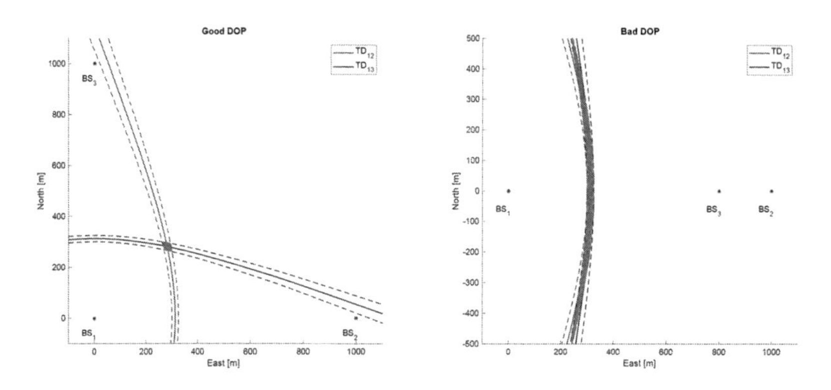

Figure 8.12 Dilution of precision for OTDOA.

Analogous to the definition of GDOP (geometrical DOP), PDOP, HDOP, VDOP, and TDOP can be defined to represent position, horizontal, vertical and time DOP, respectively:

$$PDOP \equiv \sqrt{\sigma_x^2 + \sigma_y^2 + \sigma_z^2} \tag{8.40}$$

$$HDOP \equiv \sqrt{\sigma_x^2 + \sigma_y^2} \tag{8.41}$$

$$VDOP \equiv \sigma_z \tag{8.42}$$

$$TDOP \equiv \sigma_\tau \tag{8.43}$$

A low value of DOP is considered to be a *good geometry* and it allows an accurate position calculation. A high value of DOP gives a *bad geometry* and it magnifies errors during the position calculation procedure. The effect of good and bad DOP in general was introduced in Chapter 2. In this section, the effect of good and bad DOP for terrestrial networks will be studied in more detail.

The effect of DOP for OTDOA positioning can be recognized in Figure 8.12, which represents a system with three base stations, BS1, BS2, and BS3, and one mobile receiver whose position is unknown [15].

On the left side of Figure 8.12, the receiver measures the TDoA of signals coming from BS2 and BS3 with respect to BS1, the reference base station. These measurements are represented with a solid line for both transmitters, BS2 and BS3. The dashed lines represent the uncertainty in the measurement. Hence, the location of the mobile device can be narrowed down to the relatively small shaded area in the figure. On the right, the mobile device is in the same position, but now the visible

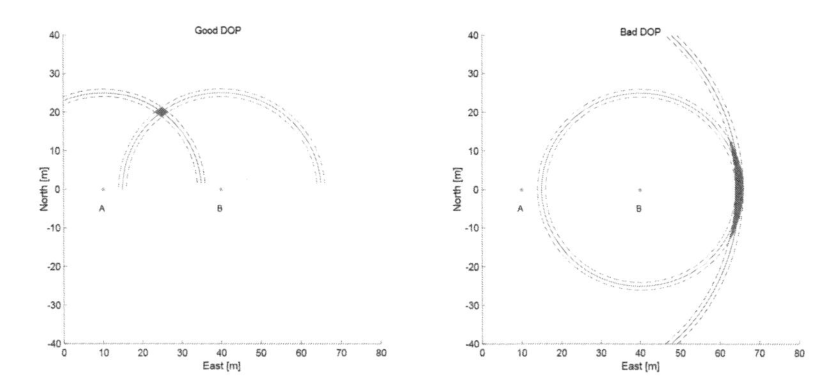

Figure 8.13 Dilution of precision for ECID.

base station constellation has changed. BS3 is no longer north of the reference cell but east in the same direction as BS2. It can be seen that in this second scenario the shaded area is comparatively much bigger. The same measurement uncertainty has a much bigger impact on the second scenario than on the first [15].

OTDOA and U-TDOA are hyperbolic methods and ECID is a spherical method, so the DOP of the base station constellation used for OTDOA or U-TDOA should a priori be different than the DOP of the same base station constellation used for ECID. For ECID, the measurements will form spheres (circles in 2D) instead of hyperboloids (hyperbolas in 2D), as depicted in Figure 8.13. However, apart from this detail, there are no further differences between the effect of DOP for OTDOA and ECID, and it has been proven that the values are identical $DOP_{ECID} = DOP_{OTDOA}$ [23]. This is a very important fact, because it implies that if a base station constellation is optimal for OTDOA it will also be optimal for ECID and vice versa, easing the cellular network planning for positioning.

8.10.3.1 Terrestrial HDOP in Typical Network Deployments

The HDOP of a cellular network deployment is tightly coupled with its topology, in particular with the geometrical distribution of the base stations, the distance between them, and the number of base stations that the UE will see at a certain time instant.

In Figure 8.14, the HDOP for several typical network deployments is shown. It includes deployments of four to seven visible base stations in different topologies: square, star, hexagon, and honeycomb. The location of the base stations is

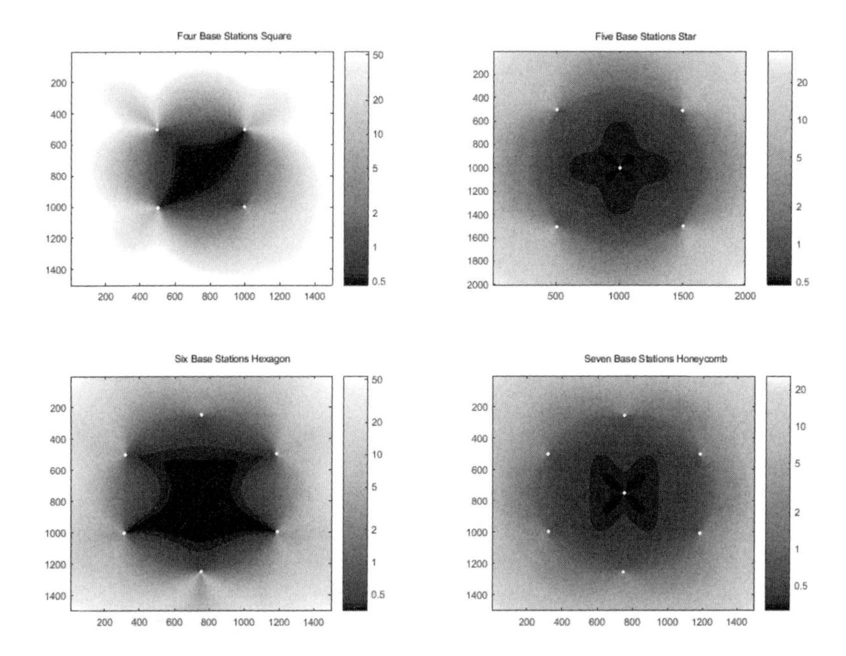

Figure 8.14 HDOP in typical network deployments.

represented by white dots. It can be noticed that the HDOP is good inside the area contained within the base stations and it worsens as soon as the UE moves outside that area. Another remark is that symmetric constellations, with the reference base station on the center (star, honeycomb) have slightly better HDOP than constellations where the reference base station is on one of the corners.

8.10.4 Weighted Least Squares Algorithm

At the beginning of this section, it was mentioned that the hybrid positioning algorithm is based on the WLS algorithm for A-GNSS. In order to reach that stage, a last step needs to be applied to the LSE solution in Equation 8.34. This last step consists of the addition of a weight matrix W, defined in Equation 8.44:

$$W_T = \begin{bmatrix} w_1 & 0 & \cdots & 0 \\ 0 & w_2 & & 0 \\ \vdots & & \ddots & \vdots \\ 0 & 0 & \cdots & w_n \end{bmatrix} \tag{8.44}$$

In Equation 8.44, w_i is the weight associated to measurement i. Applying the same procedure defined by the 3GPP for A-GNSS [19], the solution of the WLS algorithm is given by Equation 8.45:

$$\Delta\hat{X} = (A'_T \cdot W \cdot A_T)^{-1} \cdot A'_T \cdot W \cdot \Delta\hat{\rho}_T. \tag{8.45}$$

The weight w_i associated with each measurement should be proportional to the accuracy of the measurement. For instance, for A-GNSS the weight is the inverse of the RMS of the measurement for each satellite.

8.10.5 RAIM Enhancements

In outdoor positioning scenarios, it is often the case that the number of measurements from all four technologies exceeds the number of unknown variables used to calculate the position estimate by a hybrid positioning algorithm. The WLS algorithm attempts to find the best solution by minimizing the *residuals*. However, not all measurements will have the same accuracy. Trying to fit non accurate measurements will probably reduce the quality of the solution. This problem is partially mitigated by using the weighting matrix directly proportional to the reported measurement quality. However, it has also been seen that the same error does not affect different measurements equally. The geometry of the transmitter (satellite or base station) relative to the mobile device influences how critical a certain error is for that particular transmitter and how the inclusion of that transmitter in the solution calculation could impact the result. For instance, a 1-m measurement error for one transmitter might cause a 2-m meters error in the position calculation, while for a different transmitter it might cause just half a meter position error, as seen in Figure 8.12.

This asymmetric sensitivity to measurement errors will also affect the positioning algorithm, making it more sensitive to errors in some measurements than in others. Hence, it makes the algorithm suitable for a RAIM implementation, which helps detecting faulty measurements and improves the overall stability of the positioning.

RAIM methods have been commonly used by GNSS applications since the 1990s, especially for commercial aviation, military, and SoL applications [24–26].

There are several different algorithms, which can be grouped into two main families: the measurement rejection approach (MRA) and the error characterization approach (ECA). As an example, a variant of MRA called maximum solution separation (MSS) [24] will be briefly described here.

The MRA's core feature is the fault detection and exclusion (FDE) technique, which is based on detecting the faulty measurements and excluding them from the positioning calculation in order to improve the solution. The MSS method consists of comparing the separation between the position estimate using all available measurements and the position estimate on a series of subsets of the measurements, generated by a certain *subset filter*.

As example of MSS algorithm could be the following:

1. Calculate a position estimate using the full set of measurements.
2. IF the position estimate converges below a certain threshold, THEN return the calculated position and finish the algorithm.
3. ELSE IF the position estimate does not converge or the convergence is not better than the threshold THEN perform RAIM.
4. Compute a position estimate for each subset of measurements. The subsets are created by using all but one of the measurements. For n measurements, there will be n subsets.
5. Compare the different estimates obtained in step 4. Return the one that has converged with the best convergence.
6. IF none of the subset estimates have converged, the position calculation has failed.

The algorithm has converged at iteration k if the difference between the position calculated at that iteration and the position calculated at iteration $k - 1$ is below a certain threshold value. The convergence fails if the algorithm has not converged after a limited number of iterations or if the difference between the position calculated at iteration k and the position calculated at iteration $k-1$ exceeds a certain limit (the calculation diverges).

A possibility when the MSS algorithm has not given a position estimate is to continue one step forward and implement a multiple hypothesis solution separation (MHSS). The MHSS allows to include multiple faulty measurements. After creating the subsets excluding one measurement, it will continue creating all the subsets excluding two measurements, then three, and so forth, until it reaches a

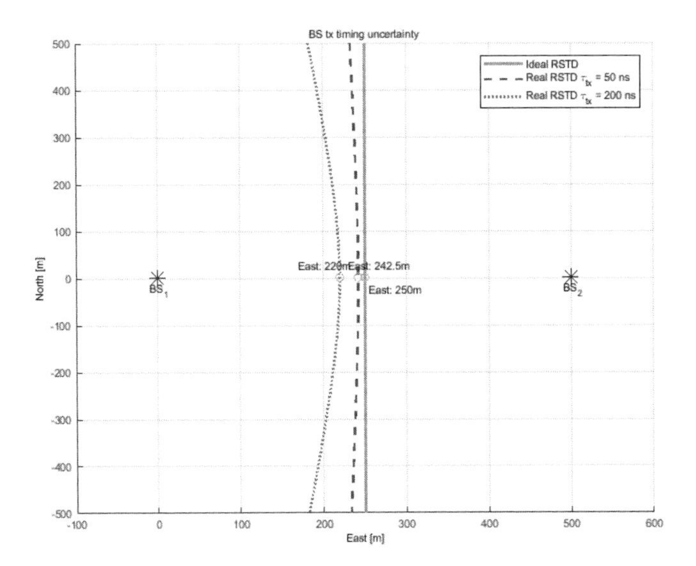

Figure 8.15 Analysis of the base station synchronization uncertainty for TDOA systems.

solution or there are not enough measurements left in the subsets to try a position calculation. However, the number of subsets to calculate greatly increases with the number of measurements to exclude. In a system with N total measurements and m measurements to exclude, the total number of combinations is $\binom{N}{m} = \frac{N!}{m!(N-m)!}$.

8.11 SOURCES OF ERROR IN CELLULAR NETWORK POSITIONING

8.11.1 Network Synchronization

TDOA methods (E-OTD, OTDOA, U-TDOA) rely on accurate synchronization between base stations. In particular, for OTDOA, the required synchronization is in the order of nanoseconds. UE manufacturers request a state-of-the-art synchronization uncertainty with a standard deviation of σ_T smaller than 50 ns [18]. However, the current uncertainty values in the network deployment can typically go up to 200 ns. This section analyzes the effect of base station synchronization errors in TDOA positioning algorithms using OTDOA as baseline.

Given two base stations BS_1 and BS_2 and a mobile device placed at an unknown position such that the time of flight of the signals coming from both base stations is the same (which means ideal RSTD is 0). Assuming that BS_1 has a timing error τ_{tx} and BS_2 is ideally synchronized, Equation 8.9 can be rewritten, neglecting the other error sources associated with the RSTD measurement, in Equation 8.46:

$$K = (RSTD_{n,true} + \tau_{Tx,ref}) \cdot v_p. \tag{8.46}$$

Figure 8.15 represents the hyperbolas calculated from the ideal RSTD measurement with $\tau_{tx} = 0$ and the real RSTD measurements contaminated with the timing error $\tau_{tx} = 50$ ns and $\tau_{tx} = 200$ ns. Defining the error induced in the measurement by the base station timing error as $e_{m,\tau}$ equal to the distance in meters between the ideal RSTD and the real RSTD, it can be seen that this error is not constant: it varies depending on at which point of the hyperbola the mobile device is placed. The error has a minimum if the mobile device is placed in the point where the distance between the hyperbolas is the smallest possible, in this example $North$: 0 m.

Calculating the minimum distance between the hyperbolas for Figure 8.15, $Min(e_{m,\tau}) = 7.5$m for $\tau_{tx} = 50$ ns and $Min(e_{m,\tau}) = 30$m for $\tau_{tx} = 200$ ns. This calculation is independent of the distance between base stations and of the ideal RSTD value. A lower bound for the measurement error induced by the base station timing uncertainty has been found. Hence, the measurement uncertainty created by the base station uncertainty also has a lower bound, defined in Equation 8.47:

$$\sigma_{m,\tau} \geq \frac{1}{2} \cdot \sigma_{tau,tx} \cdot v_p. \tag{8.47}$$

This lower bound applies to the measurement involving the base station with a synchronization error. Therefore, it affects the quality of one measurement. However, due to the WLS algorithm, the rest of the measurements might compensate partially or totally the error and the final position may result in being more accurate than individual measurements.

The next scenario to be analyzed is when all base stations are affected by synchronization errors. If BS_1 and BS_2 have both a positive timing error τ_{tx}, it can be seen from Equation 8.9 that they will cancel each other out and the total timing error will be reduced, as detailed in Equation 8.48:

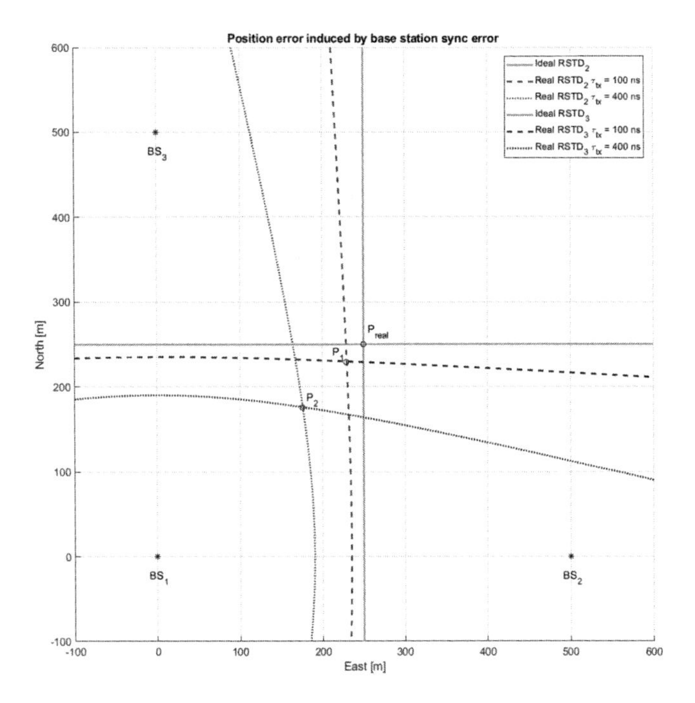

Figure 8.16 Position error induced by base station synchronization error for TDOA systems.

$$\begin{aligned} K &= (RSTD_{n,true} - \tau_{Tx,BS_2} + \tau_{Tx,BS_1}) \cdot v_p + e_{RSTD} \\ &= (RSTD_{n,true} - (\tau_{Tx,BS_2} - \tau_{Tx,BS_1})) \cdot v_p + e_{RSTD}. \end{aligned} \tag{8.48}$$

However, if the timing errors have the opposite sign, they will add up and the total timing error will increase.

Considering a base station constellation with three eNBs, BS_1, BS_2, and BS_3, in which the reference cell for the OTDOA measurement is BS_1, the worst possible scenario is the one where each individual OTDOA measurement is affected by the biggest possible timing error. That occurs when the timing error of BS_1 has the opposite sign to the timing errors from BS_2 and BS_3.

Considering $\tau_{tx,1} = 50$ ns and $\tau_{tx,1} = 200$ ns, as in the previous example, and $\tau_{tx,2} = \tau_{tx,3} = -\tau_{tx,1}$, and denoting the total timing error as τ_{tx}, the position error induced by the base station synchronization error is represented in Figure 8.16.

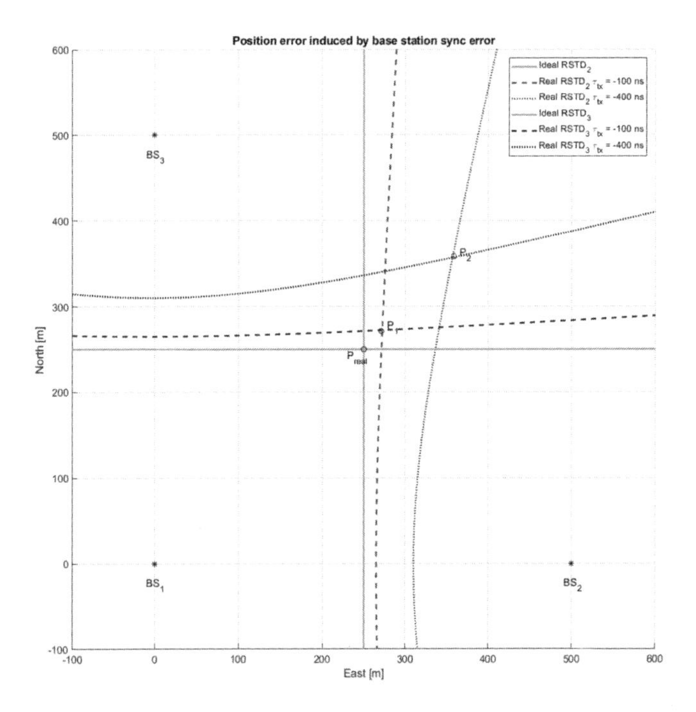

Figure 8.17 Position error induced by base station synchronization error for OTDOA systems II.

The mobile device is placed in P_{real}, while the positions calculated with $\tau_{tx,1} = 50$ and $\tau_{tx,1} = 200$ ns are P_1 and P_2, respectively. The position error for $P_1 = 29.82$m and for $P_2 = 105.32$m.

If the synchronization uncertainty of all the base stations in the network is the same, and considering the lower bound for the measurement uncertainty found in Equation 8.47, the individual measurement uncertainty if both base stations involved in the measurement have synchronization uncertainty is given by Equation 8.49:

$$\sigma_{m,\tau} \geq \sigma_{tau,tx} \cdot v_p. \tag{8.49}$$

In Figure 8.16, as the base stations are placed forming a right-angled triangle, the position error can be approached by the hypotenuse of the triangle whose other two sides are the measurement errors. Hence, Equation 8.50 is true for the right-angled base station constellation:

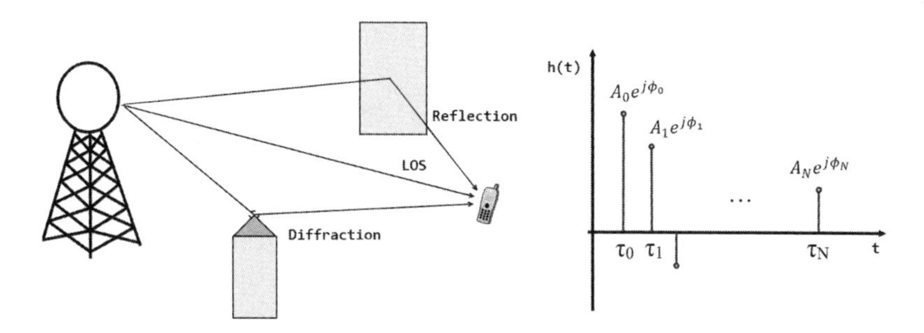

Figure 8.18 Multipath propagation effect.

$$\sigma_{P,\tau} \approx \sqrt(2) \cdot \sigma_{m,\tau} \geq \sqrt(2) \cdot \sigma_{tau,tx} \cdot v_p. \tag{8.50}$$

If the base stations are not forming a right-angled triangle, Equation 8.50 can be generalized applying the law of cosines as in Equation 8.51, where α is the angle formed by the vectors $\overline{BS_1BS_3}$ and $\overline{BS_1BS_2}$:

$$\sigma_{P,\tau} \geq \sqrt(2 - 2\cos\alpha) \cdot \sigma_{tau,tx} \cdot v_p. \tag{8.51}$$

The derivation above was considering that $\tau_{tx,1} > 0$ ns. If $\tau_{tx,1} < 0$ ns, keeping all other relations, the results are even worse, as can be seen in Figure 8.17. In this case, the position error for P_1 remains as 29.82 m but for P_2 it increases until 223.78 m.

A similar analysis could be applied as well for other hyperbolic methods like U-TDOA.

8.11.2 Multipath Propagation

Multipath is the effect by which a certain transmitted signal reaches the receiver antenna from more than one direction. This is caused mainly by reflections, diffractions, and scattering suffered by the signal, as depicted in Figure 8.18. In the left part of the figure, the signal transmitted by the base station arrives at the UE from three different paths: the LoS path, the reflection from a nearby building, and a diffraction. In the right part of the figure, the multipath effect is modeled as a series of impulses N arriving at discrete instants of time τ_k with a certain amplitude A_k and phase ϕ_k. The channel impulse response $h(t)$ is represented in Equation 8.52, where δ represents the Dirac delta function:

$$h(t) = \sum_{k=0}^{N} A_k \cdot e^{j \cdot \phi_k} \cdot \delta(t - \tau_k). \tag{8.52}$$

In the normal case, the LOS path should be received at a stronger signal level than the NLOS components, i.e. $A_0 \cdot e^{j \cdot \phi_0} > A_k \cdot e^{j \cdot \phi_k}, \forall k \in 1, N$. However, if the LOS path suffers scattering or other effects (e.g., destructive interference) that degrade the signal, it is possible that the LOS path received is lower than some of the NLOS components. This situation could potentially lead to errors in the RSTD measurement: if the receiver selects the strongest signal as the LOS path, the estimated time of arrival will not be correct.

Multipath is a phenomenon that will affect almost all real-life scenarios and it can be very damaging to the position calculation. The 3GPP has defined some reference multipath scenarios based on real-life scenarios in TS 36.111 Annex B [29], reproduced in Table 8.1. Taking as an example the extended pedestrian model A (EPA), it can be seen that the delay of the first multipath component is 30 ns. If the UE was wrongly selecting this component as the LOS path, the error in the range measurement would be of about 90 m. This error increases if the UE selects any of the other components. For different models like extended typical urban (ETU) model, it can be seen that the strongest path is actually not the LOS path, but the components at 200- and 230-ns tap delay. In this scenario, the positioning error induced by multipath would be much greater.

Table 8.1

Multipath Models in TS 36.111

Extended Pedestrian A	
Tap Delay [ns]	*Rel. Power [dB]*
0	0.0
30	-1.0
70	-2.0
90	-3.0
110	-8.0
190	-17.2
...	...

Extended Typical Urban	
Tap Delay [ns]	*Rel. Power [dB]*
0	-1.0
50	-1.0
120	-1.0
200	0.0
230	0.0
500	0.2
...	...

From: [29].

A multipath effect can be mitigated partially if the number of available measurements is high, as the positioning algorithm might be able to detect the faulty measurements and discard them by using RAIM (see Section 8.10.5) or a similar

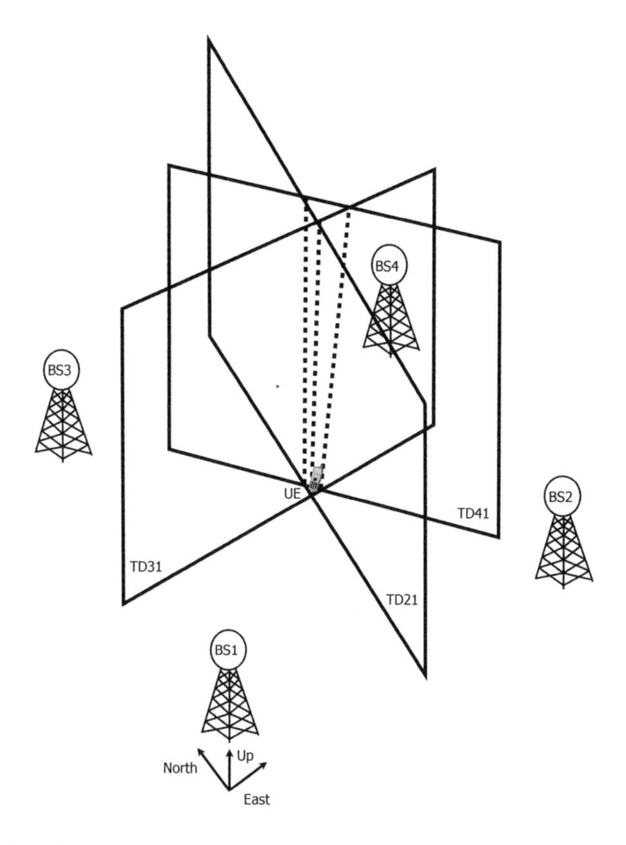

Figure 8.19 OTDOA 3D measurements in a high GDOP scenario.

algorithm. Thus, combining multiple measurement types in the hybrid algorithm described in Section 8.10.1 is an efficient way of reducing the error induced by multipath.

8.11.3 Geometry of the Base Station Network

In a common real-life scenario all base stations will likely be in similar altitudes. This scenario is not suitable to calculate a 3D position, as the base station constellation does not have enough altitude diversity to allow the algorithm to estimate the altitude of the mobile device correctly. In order to prove that affirmation, the analysis can be done mathematically or geometrically. Mathematically, if we go to

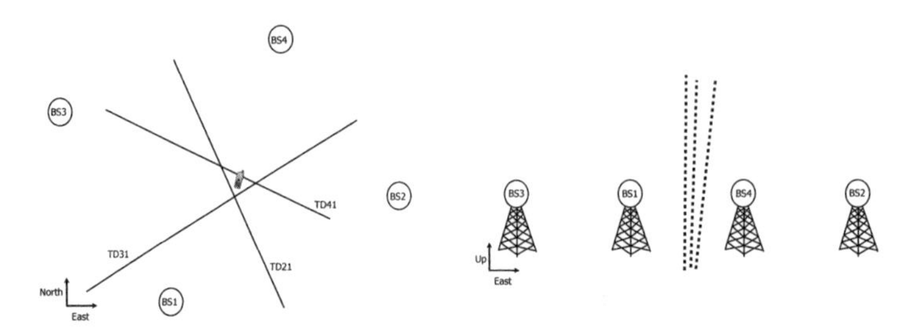

Figure 8.20 Horizontal and vertical sections of the OTDOA measurements in high GDOP scenario.

an extreme case where all base stations are at the same altitude, it can be seen that the design matrix A_h from Equation 8.27 is singular, and the equation system in Equation 8.26 cannot be solved.

Geometrically, the demonstration is much more intuitive. The geometry formed by such a constellation will present a very high value of the GDOP, tending to infinity if all base stations are at exactly the same altitude. In order to further explain this problem, the following scenario will be defined [15]: four base stations placed at the corners of a square of 1,000 m side and the mobile device placed at coordinates [480; 480]. All base stations and the mobile device are at the same altitude, 0 meters. The GDOP calculation for such a scenario diverges to infinity, while HDOP has a value of 0.47. This scenario and the hyperboloids resulting from the OTDOA RSTD measurements are depicted in Figure 8.19. As can be seen, the hyperboloids have degenerated to almost three vertical planes and cut in a relatively small triangular area in the horizontal plane, but they do not converge in the vertical axis. This is better illustrated in the two planar cuts in Figure 8.20.

Figure 8.20 depicts the horizontal (left) and vertical (right) cuts of the 3D image from Figure 8.19. As can be seen, the horizontal position can be calculated precisely from the measurements. However, in the vertical section, the three measurements do not converge to any point. The three dashed lines in the image represent the intersections between each pair of measurements. The lines are almost parallel and do not intersect.

Hence, 3D positioning using only U-TDOA, OTDOA, and ECID is only possible if the base stations and the mobile device are placed at different altitudes. For this reason, in many real-life scenarios U-TDOA, OTDOA, and ECID are seen rather as 2D positioning methods. The altitude information must be obtained from other technologies (e.g., using atmospheric pressure measurements, as will be

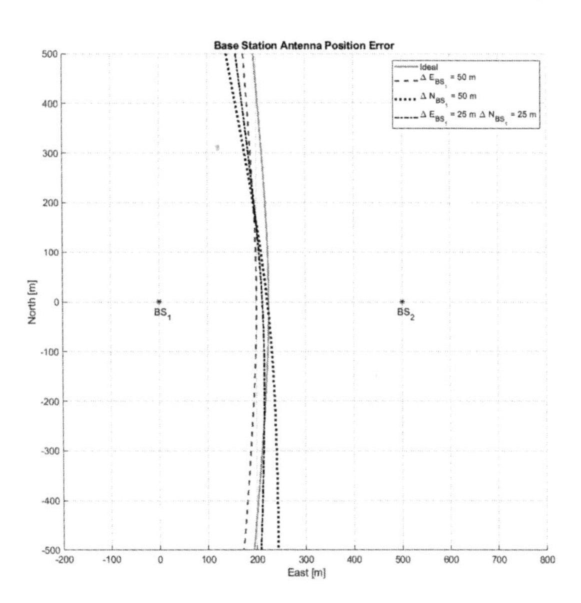

Figure 8.21 Measurement error induced by an error in the transmitter antenna coordinates.

explained in Chapter 11 of this book) or using a default value (e.g., using the Earth's surface as a boundary condition).

The improvement of the GDOP is another of the advantages of the hybrid solution combining A-GNSS and cellular network positioning. The GNSS satellites will add altitude diversity to the total constellation and allow to compute a 3D position where it would be otherwise impossible.

8.11.4 Location Database Error

The location database in the location server contains information about the different base stations in the network that is required by the positioning algorithm. This information can be, for example, the PRS configuration of each base station, the RTD between a pair of base stations, or the base station coordinates. An error in the database will produce errors in the positioning [15, 18].

In particular, a typical source of error in the database is the coordinates of the base station. The positioning algorithm needs the accurate position (normally in WGS-84) of the transmitter antenna. The reference point used to measure the base station coordinates may not always be the base station antenna, and this will directly translate into an error in the interpretation of the measurement by the location server, impacting the precision of the calculated position.

Figure 8.21 shows the influence of antenna coordinate errors in the interpretation of the position server of a measurement.

As can be appreciated, the effect is highly dependent on the direction of the coordinate errors. If the error in the antenna position is toward the other base station involved in the measurement (the dark line in the figure), the error is almost constant along the hyperbola with a standard deviation of approximately 25 m ($\Delta E/2$). However, if the coordinate error is toward a different direction, the measurement error presents a high variance for each point of the hyperbola. Hence, the effect of lack of precision of the transmitter antenna coordinates on the positioning algorithm will vary for each individual situation and cannot be predicted a priori.

8.12 CONCLUSION

This chapter analyzed a heterogeneous group of different terrestrial location methods defined as part of the cellular networks, also called RAT-dependent positioning technologies in 3GPP. These technologies are based on the measurement of cellular radio signals in order to calculate the position of the UE and they can be divided in two main groups: the TOA (e.g., ECID) and TDOA (e.g., OTDOA) methods. The most basic positioning technology for asynchronous cellular networks is the TOA of a particular downlink or uplink signal. More complex techniques involving round-trip time measurements or TDOA measurements offer higher accuracy but require either synchronous networks and/or complicated measurement procedures.

A second classification can be defined between downlink positioning methods (e.g., OTDOA) and uplink positioning methods (e.g., U-TDOA). Downlink positioning technologies require support in the UE and cannot be done with legacy UEs, which were developed before the technology was standardized. Uplink positioning technologies rely solely on the network side. They can be used with all UEs available in the field, but they require a certain network infrastructure. Thus, it can be concluded that downlink methods are preferable to keep the network cost and complexity reduced and uplink methods are preferable to keep the user equipment cost and complexity reduced.

The accuracy offered by cellular network positioning technologies is usually in the order of tens to hundreds of meters, depending on the particular technology used. Considering the most advanced methods defined for LTE (ECID, OTDOA, UTDOA), the accuracy error can generally be maintained under 100 m, unless the positioning signals are affected by critical disturbances like severe multipath propagation. These effects can be mitigated by using a hybrid solution approach and combining measurements from multiple sources, including satellite navigation.

References

[1] 3GPP TS 43.059, *Functional Stage 2 Description of Location Services (LCS) in GERAN*, V14.2.0., 2017.

[2] [LCS-010019], *Overview of 2G LCS Technologies and Standards*, Motorola Inc., 3GPP TSG SA2 LCS Workshop, London, January 11-12, 2001.

[3] 3GPP TS 45.010, *GSM / EDGE Radio Subsystem Synchronization*, V14.3.0., 2018.

[4] 3GPP TS 44.031, *Radio Resource LCS Protocol*, V14.3.0., 2017.

[5] 3GPP TS 25.111, *Location Measurement Unit Performance Specification; UE Positioning in UTRAN*, V14.0.0, 2017.

[6] Zhao, Y., "Standardization of Mobile Phone Positioning for 3G System," *IEEE Communication Magazine*, 2012, Vol. 40, pp. 108-116.

[7] Wylie-Green, M.P., and Wang, P., "GSM Mobile Positioning Simulator," *IEEE Emerging Technologies Symposium: Broadband, Wireless Internet Access*, 2000.

[8] Halonen, T., Romero, J., and Melero, J., *GSM, GPRS and EDGE Performance: Evolution Towards 3G/UMTS*, Second Edition, John Wiley & Sons, 2003.

[9] Glisic, S.G., *Advanced Wireless Communications and Internet: Future Evolving Technologies*, Third Edition, John Wiley & Sons, 2011.

[10] Gibson, J.D., *The Communications Handbook*, Second Edition, CRC Press, 2002.

[11] Caffery, J.J., and Stüber, G.L., "Overview of Radiolocation in CDMA Cellular Systems," *IEEE Communications Magazine*, 1998, Vol. 4, pp. 38-45.

[12] Nissani, D.N., and Shperling, I., "Cellular CDMA (IS-95) location, A-FLT Proof-of-Concept Interim Results," *The 21st IEEE Convention of Electrical and Electronic Engineers in Israel*, 2000.

[13] Brimicombe, A., and Li, C., *Location-Based Services and Geo-Information Engineering*, John Wiley & Sons, 2009.

[14] 3GPP TS 36.214, *Stage 2 Functional Specification of UE Positioning in E-UTRAN*, V14.3.0, 2017.

[15] Cardalda García, A., *Hybrid Localization Algorithm for LTE Combining Satellite and Terrestrial Measurements*, University of Oviedo, 2015.

[16] 3GPP TS 36.214, *E-UTRA Physical Layer Measurements*, V15.0.1, 2018.

[17] 3GPP TS 36.133, *E-UTRA Requirements for Support of Radio Resource Management*, V15.1.0, 2018.

[18] Fischer, S., *Observed Time Difference of Arrival (OTDOA) Positioning in 3GPP LTE*, Qualcomm Technologies Inc, 2014.

[19] 3GPP TS 36.171, *E-UTRA Requirements for Support of A-GNSS*, V14.0.0, 2017.

[20] Torrieri, D.J., "Statistical Theory of Passive Location Systems," *IEEE Transactions on Aerospace and Electronic Systems*, Vol. AES-20, No. 2, 1984.

[21] Blewitt, J., "Basics of the GPS Technique," *Geodetic applications of GPS*, 1997.

[22] Foy, W.H., "Position Location Solutions by Taylor Series Estimation," *IEEE Transactions on Aerospace and Electronic Systems*, 1976.

[23] Shin, D.H., and Sung, T.K., "Comparison of Error Characteristics between TOA and TDOA Positioning," *IEEE Transactions on Aerospace and Electronic Systems*, Vol. 38, No. 1, pp. 307–311, 2002.

[24] Brown, R.G., and McBurney, P.W., "Self-Container GPS Integrity Check Using Maximum Solution Separation," *Journal of the Institute of Navigation*, Vol. 35, No. 1, 1988.

[25] Brown, R.G., "Solution of the Two-Failure GPS RAIM Problem under Worst-Case Bias Conditions: Parity Space Approach," *Journal of the Institute of Navigation*, Vol. 44, No. 4, 1998.

[26] Brown, R.G., and Chin, G., "GPS RAIM: Calculation of the Threshold and Protection Radius Using Chi-Square Methods: a Geometric Approach," *Global Positioning System: The Institute of Navigation*, Vol. 5, 1997.

[27] 3GPP TS 03.71, *LCS Functional Description Stage 2*, V8.9.0, 2004.

[28] 3GPP TR 45.811, *Feasibility Study on Uplink TDOA in GSM and GPRS*, V6.0.0, 2002.

[29] 3GPP TS 36.111, *LMU Performance Specification; Network Based Positioning Systems in E-UTRAN*, V14.0.0, 2017.

Chapter 9

Terrestrial Positioning Technologies: Noncellular Networks

9.1 INTRODUCTION

The combination of satellite and cellular-based positioning technologies allow mobile location services to be able to calculate very accurately the position of a mobile device outdoors. However, indoor location and other scenarios such as dense urban environments remain a challenge for the aforementioned technologies. GNSS penetration rate indoors is poor, due to the very low received signal powers. In the best case, a mobile device may be able to detect a few GNSS satellites if the device is located near a window. Even in this case, the geometrical distribution of the satellites (likely coming from similar directions) would not allow accurate positioning. Similarly, the GNSS signal in dense urban scenarios is affected by multipath and a low number of visible satellites and does not guarantee an accurate location fix.

Multipath propagation is also a problem for the network-based positioning technologies like OTDOA and ECID. The accuracy of such technologies drastically reduces in such environments. Furthermore, as seen in the previous chapter, network-based positioning technologies are not suitable for altitude determination, primarily due to the geometrical distribution of the cellular base stations. Altitude determination is one of the most critical aspects of indoor positioning, as a few meters of error in the altitude can easily cause incorrect floor selection.

Precise outdoor positioning was initially sufficient to meet the requirements of commercial and emergency applications. However, pushed mainly by the FCC in the United States, indoor positioning requirements soon arose, proving the need for new technologies. This need was addressed by two standardization organizations, the 3GPP and the OMA, who developed standards for positioning techniques based on non-cellular networks. Similar to A-GNSS, these technologies are independent from the cellular network (e.g., LTE or 5G NR), but the network protocol has been extended to exchange the necessary positioning information. This includes Wi-Fi, short-range networks such as Bluetooth, and other types of networks, such as the terrestrial beacon system (TBS). These three technologies will be the main focus of this chapter.

Nonetheless, altitude determination required further location techniques such as atmospheric pressure measurements, which cannot be considered a network-based technology. Therefore, it has not been included in this chapter, but it can be found later in Chapter 12, with other sensor-based techniques.

9.2 NONCELLULAR NETWORK-BASED POSITIONING

Noncellular network-based positioning uses techniques very similar to the ones seen in Chapter 8 for mobile network-based positioning. The main difference is that the measured signals are not from a cellular technology, such as WCDMA or LTE. Instead, the mobile device can measure a wide variety of different networks, from the IEEE's Wi-Fi signal to short-range networks such as Bluetooth or even proprietary networks such as the TBS.

An example of a scenario with noncellular network positioning is shown in Figure 9.1. The mobile device is connected to the cellular network and measuring noncellular nodes (in the example, Wi-Fi access points and Bluetooth beacons) in the vicinity. These measurements are reported back to the network as part of the positioning session.

As these technologies are not under the control of 3GPP, they are treated mostly as a black box. The 3GPP LPP (or the competing OMA standard, LPPe) provides a mechanism for the necessary information exchange between the mobile phone and the cellular network, but it does not define any measurement or positioning algorithms. It is up to either the mobile phone or the location server of the network to use the available measurements to calculate a location fix by any means it deems suitable. Furthermore, it is likely that the cellular network will not have much a priori information about the noncellular networks. That means the cellular network will not be able to help the mobile device with any kind of assistance data (as it does for OTDOA or even A-GNSS), as for instance an indication of which Wi-Fi access points the mobile device should be seeing nearby. Another drawback

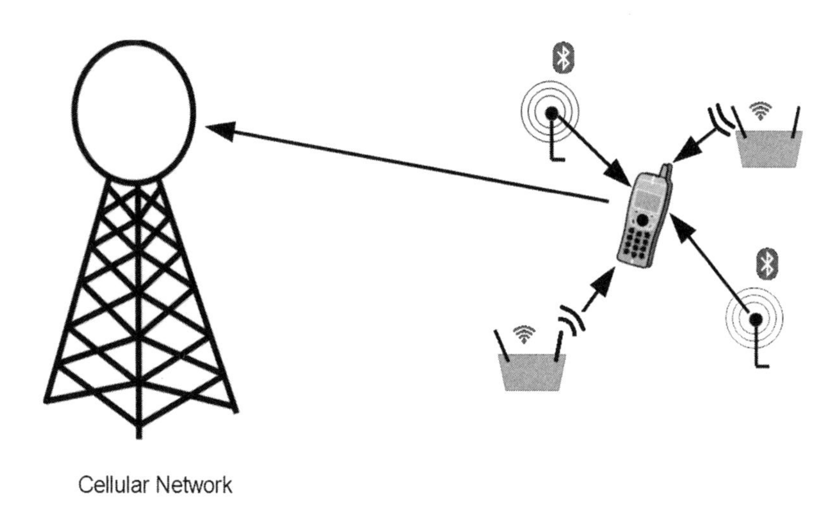

Figure 9.1 Typical noncellular network positioning scenario.

is that the network will most likely not have any information about the geographical location of the noncellular network nodes needed for most positioning algorithms. Thus, the network location server needs to rely on public databases, such as the NEAD in the United States, or third-party databases available online in order to calculate the position of the mobile device.

For all these reasons, it is difficult to estimate how accurate these technologies really are. An analysis and comparison to the RAT-dependent and A-GNSS technologies, based on publicly available studies, is collected in Chapter 11 of this book. Nonetheless, this chapter will give an overview of the different noncellular technologies that have been standardized, including the most widely used positioning algorithms and their average performance.

9.3 WI-FI

Currently, probably every person reading this book is more than familiar with Wi-Fi networks, as they are widely used to provide internet access for mobile phones, laptops, tablets, and smart home devices. Wi-Fi technology, which celebrated its 20-year anniversary in 2019 [1], has penetrated most of the households replacing or complementing cabled ethernet connections. It is difficult to determine how many

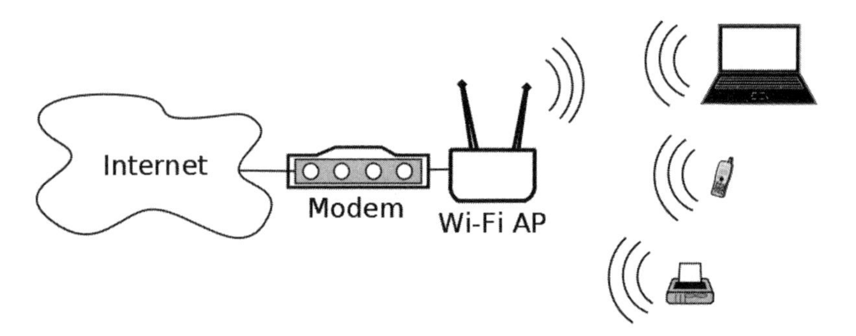

Figure 9.2 Topology of a typical Wi-Fi network.

houses are connected with Wi-Fi instead of (or in addition to) wired connections, but different studies [2, 3] give a range between 75% and 85% of U.S. households, while the number is even larger within the European Union.

Wi-Fi is based on the IEEE standard family 802.11 [4], which defines a technology for information exchange in WLAN. The topology of a typical Wi-Fi network is shown in Figure 9.2. The central point of the network is the Wi-Fi AP, which is typically connected to the internet through a modem. The rest of the devices, such as mobile phones, laptops, or smart home devices, are connected to the AP via the air interface. Devices connected to the same WLAN can transfer data between them or also access the internet through the router or access point.

However, Wi-Fi can be used for more than data transfer; it can also be used for location algorithms. Being an implementation of WLAN, a typical Wi-Fi network has a relatively small range. Depending on the version of the 802.11 standard supported, the range of an AP varies from about 50 m using the lowest Wi-Fi frequency band (2.4 GHz) to about a third of that distance in the 5-GHz frequency band with higher bandwidths and data rates. Thus, a device connected to a certain Wi-Fi network will be relatively close to the AP of the network. This principle is the key for most of the Wi-Fi based positioning algorithm, as well as for other short-range networks such as Bluetooth.

9.3.1 Wi-Fi Fundamentals

Before diving into Wi-Fi based positioning techniques, it is worthwhile to explain briefly some of the fundamentals of a Wi-Fi network. Wi-Fi networks are deployed in the industrial, scientific, and medical (ISM) frequency bands, an unlicensed part of the spectrum, which can be freely accessed. The ISM allocations vary from country to country, but the most widely used bands are at 2.4 GHz and 5 GHz. There have been studies on the usage of other frequency bands, such as the 3.6 GHz band in the United States, or the 60-GHz band, intended mainly for automotive applications, but these have not been widely used to date.

The 802.11 standard has multiple amendments, identified by letters of the alphabet (e.g., 802.11g or 802.11n) depending on the frequency band, bandwidth and transmission rates for which they have been defined. For example, a Wi-Fi AP that supports only 802.11g will transmit using the 2.4-GHz band, while the 802.11n supports both 2.4 and 5 GHz and higher data rates.

As there are no wired connections, there is a need for a mechanism that allows devices to detect an existing Wi-Fi network. This is typically called network scanning. The 802.11 standard defines two types of scanning: passive scan and active scan. In passive scan mode, the AP of the network periodically sends a signal, called a beacon, giving relevant information about the network, such as:

- BSSID, a 48-bit label following the MAC address conventions, which is used to uniquely identify the AP.
- Service set identifier (SSID), an identifier for the network that is typically written in natural language. It serves as the network name that can be seen, for instance, when searching for Wi-Fi networks with a mobile phone.
- Capability information, such as what transmission rates and frequencies the network supports.
- Beacon interval, a parameter describing how often the beacons are sent.

Any device that wants to detect a Wi-Fi network just needs to listen for the beacon frame. There, it can find all the information needed to identify the network and also to try to establish a connection to the network, if required.

In the active scan mode, the device that is searching for Wi-Fi networks does not wait for beacon frames. Instead, it can broadcast probe requests in each of the frequency channels. Upon receiving a probe request, the Wi-Fi access point sends a dedicated probe response to the requesting device. The contents of the probe response are very similar to the beacon frame.

Both methods have advantages and disadvantages in terms of battery consumption, network load, and efficiency in identifying all available Wi-Fi networks. However, for positioning applications, it does not matter which one of the two methods are used for the scanning. All that matters is that the mobile device is able to

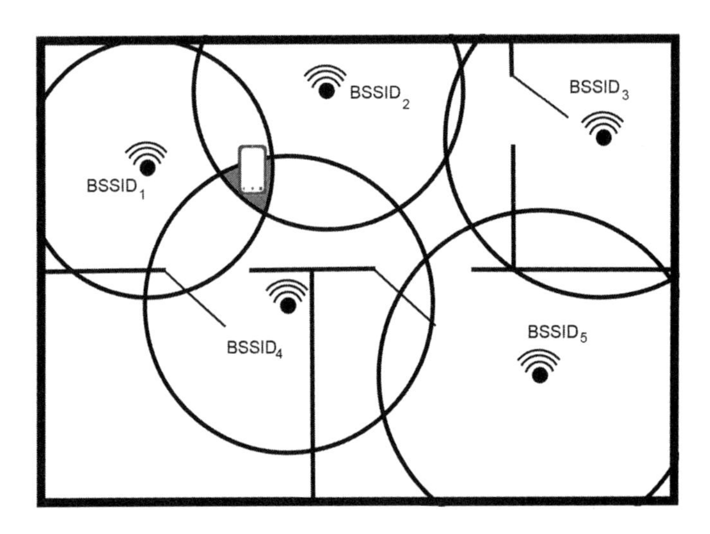

Figure 9.3 Example of Wi-Fi positioning using BSSID in an indoor scenario.

gather the BSSID of the AP and measure parameters such as the received signal strength.

9.3.2 BSSID-Based Positioning

The BSSID is enough for the most rudimentary form of Wi-Fi-based positioning. Similar to the Cell ID method described in Chapter 8, with the BSSID it is possible to know that a mobile device is within the coverage area of a particular Wi-Fi AP. There are, however, two fundamental differences here:

- The Wi-Fi AP coverage area is orders of magnitude smaller than a LTE cell coverage area. A mobile device being able to see a particular BSSID will be within tens of meters of the corresponding Wi-Fi AP, instead of within few kilometers in the LTE case.
- A mobile device could be able to see in the best case a hand full of LTE cells in the surroundings. This scales up substantially for Wi-Fi, where a mobile device can see tens of Wi-Fi APs simultaneously in some locations.

Taking into account both advantages mentioned above, BSSID-based Wi-Fi positioning is capable of giving much more accurate results than the Cell ID method, by collecting a list of all visible access points combined with some database matching algorithms at the location server. This is illustrated in Figure 9.3 for an indoor scenario. The mobile device in this example is able to see BSSID1, BSSID2, and BSSID4. Thus, a location database knowing the association between BSSID, coverage area and physical location of the AP, can easily conclude that the mobile device is within the shaded area in the image.

The downside, as already mentioned before, is that the cellular network typically does not have any information about the location and characteristics of the Wi-Fi APs. Thus, it needs to rely on a third-party database in order to obtain this crucial information. Such Wi-Fi databases already exist and are available online. Most of the smartphone operating systems today have access to their own Wi-Fi location databases. However, these are typically proprietary solutions that have been built based on crowdsourcing algorithms. Furthermore, most of the Wi-Fi APs belong to private users and are not necessarily static. For instance, a user could move between cities and carry over his or her Wi-Fi router and internet contract. In this case, the BSSID remains unchanged, but the physical location of the AP has changed. This creates an error in the database, and it can take some time (even weeks) until the crowdsourcing algorithm is able to update the Wi-Fi AP physical location.

For such a critical application as emergency services, the cellular network operator is not inclined to rely on proprietary solutions whose availability and accuracy is not guaranteed. Hence, the Cellular Telecommunications and Internet Association (CTIA) built the NEAD [5], a database containing information about Wi-Fi access points and Bluetooth beacons, meant solely for 911 calls.

9.3.3 BSSID+RSSI-Based Positioning

A further extension to the previous method is that the mobile device does not only report the BSSID of the visible APs, but also includes the corresponding RSSI. The RSSI is an estimated power measure of how much power the mobile device is receiving from a particular AP. The absolute accuracy of RSSI may not be particularly high. However, the relative accuracy (i.e., the difference between RSSI values from different APs) is much better, and RSSI can be used to determine which Wi-Fi APs are closer than others. This allows more complex positioning algorithms and improves the accuracy of the location fix.

An example of the added improvement of RSSI-based methods is shown in Figure 9.4. The scenario is the same as in Figure 9.3, but in this case the mobile device is only able to see BSSID1 and BSSID4. The device reports also RSSI1

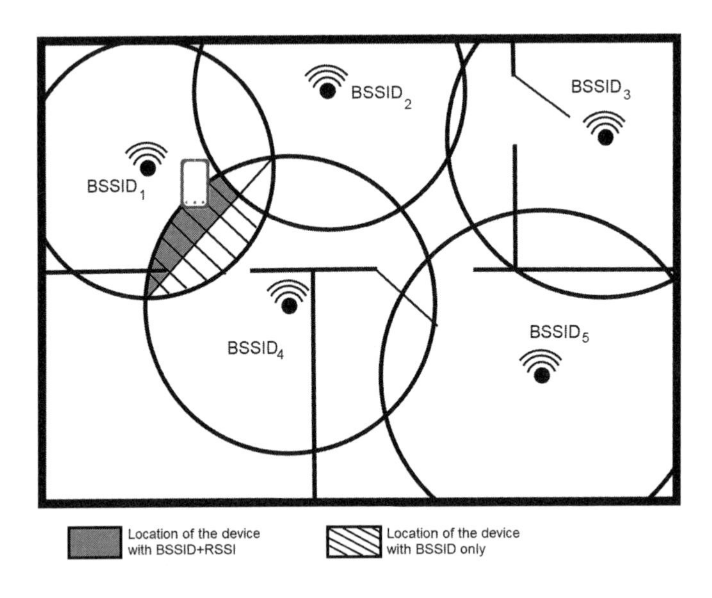

Figure 9.4 Comparison of Wi-Fi positioning using BSSID and adding RSSI.

and RSSI4, with RSSI1 being higher. A simple database matching algorithm could decide the device is closer to BSSID1, and the location area will be reduced from the lined area to the shaded area in the image.

However, as already discussed in Section 2.3.6 of this book, power measurements such as RSSI are only reliable if the transmitted power in the direction of the receiver is known and the propagation losses follow a known propagation model. This only occurs in direct line of sight conditions in the free space. The propagation path of the Wi-Fi signal is critical and the power measurement will be drastically affected by multipath conditions or if the signal crosses through walls, plants, and other obstacles. Using again the same example as before, where the mobile device reported BSSID1 with higher RSSI than BSSID4, the location server determined the device must be closer to BSSID1 and hence converged to the shaded area in Figure 9.4. However, the same effect (BSSID1 is received with higher power than BSSID4) can be true if the device is close to BSSID4, but the signal is received through the walls of the room, as depicted in Figure 9.5. In this case, the location fix estimated by the server is wrong and the device is actually outside of the shaded area.

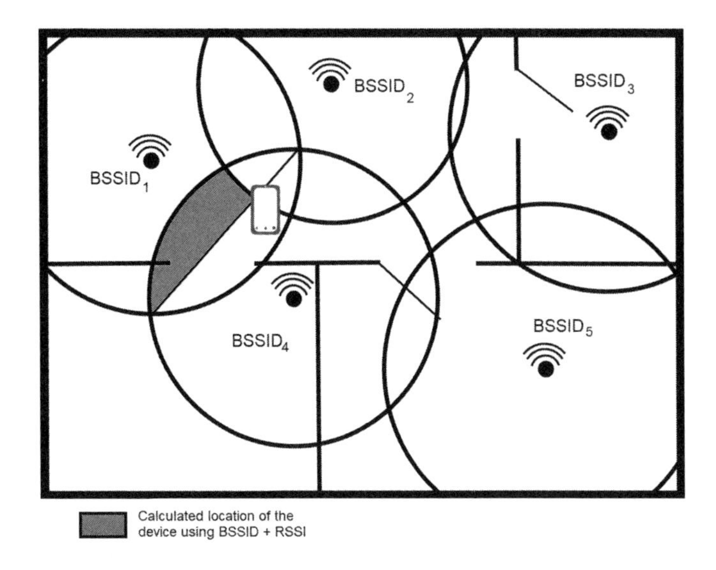

Figure 9.5 Example of a RSSI-based positioning error case due to propagation effects.

9.3.4 Wi-Fi RSS Fingerprinting

A possible solution to the error case shown in Figure 9.5 is the usage of the Wi-Fi RSS fingerprinting method [6, 7]. For Wi-Fi fingerprinting, the location server database needs to maintain a grid of reference points and the precise RSSI measurements of the Wi-Fi APs at each of the points. This is known as the fingerprinting map or the RSSI heat map. When the location server receives the RSSI observations from the mobile device, the location fix is calculated by comparing the RSSI observations to the fingerprinting map and selecting the closest match.

An example of Wi-Fi fingerprinting is shown in Figure 9.6. The location server has a grid of reference points represented by the stars. For each of the points, the location server has previously obtained RSSI measurements for all the visible Wi-Fi APs and are stored in the database. The mobile device sees BSSID1, BSSID2, and BSSID4 in order of decreasing signal strength. This is represented by the width of the signal propagation lines in the image. Thus, the location server is able to match the received RSSI reports with one of the points in the database, and the location of the device is estimated to be the circled area in the image.

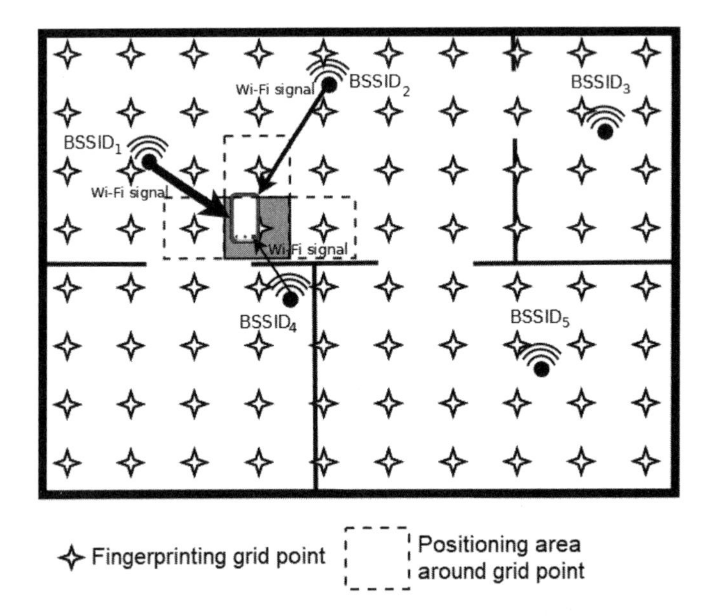

Figure 9.6 Example of Wi-Fi fingerprinting positioning.

The accuracy of Wi-Fi fingerprinting depends on the density of the grid (the more points used, the better accuracy can be obtained but also the more resources are needed to maintain the database) and on the quality of the stored RSSI information. After the initial set of data has been collected, this data can be continuously improved by crowd-sourcing. Thus, Wi-Fi fingerprinting can be a very good alternative for positioning, especially in specific indoor scenarios such as shopping malls or museums, where the availability of Wi-Fi AP is high, and the high number of visitors provide numerous measurements for the crowd-sourcing algorithm.

9.3.5 Wi-Fi RTT

A different approach not based on power measurements is Wi-Fi RTT. RTT is based on measuring the time a Wi-Fi signal needs to travel from the Wi-Fi AP to the mobile device and back. The concept is very similar to the ECID RTT measurement described in Section 8.5 of this book. Wi-Fi RTT is based on the FTM defined by IEEE 802.11mc group and incorporated into [4]. The process of how to perform

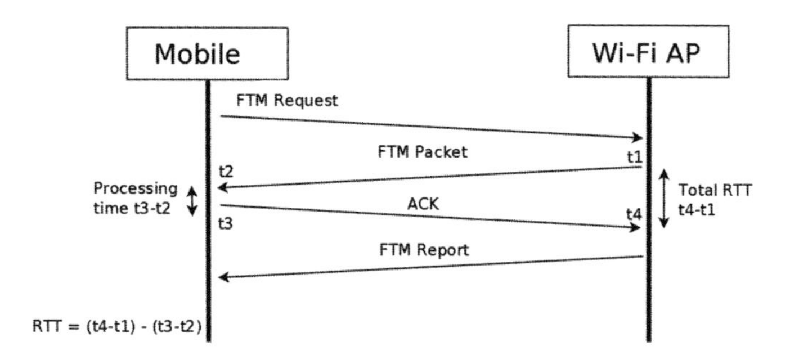

Figure 9.7 Procedure for Wi-Fi RTT based on FTM.

FTM is shown in Figure 9.7. The mobile device starts by sending a FTM request to the Wi-Fi AP. If the procedure is supported by the Wi-Fi AP, it will send a FTM packet at a time instant t1, which the mobile device should receive at time instant t2. The mobile phone processes the FTM packet and sends an acknowledgment at time instant t3, which is received by the Wi-Fi AP at t4. The AP reports the values of t1 and t4, which the phone uses to calculate the RTT, subtracting its own processing time (t3-t2).

The concept of Wi-Fi RTT measurements has been around for a while. However, its relevance for mobile phone positioning started mainly in 2018, when support for FTM was added to the Android API [8]. Nonetheless, Wi-Fi requires not only mobile phone support but also support on the Wi-Fi AP side. Legacy APs are not able to perform FTM procedures. However, if RTT-capable Wi-Fi APs find widespread use, this technology has the potential to become the reference for 5G indoor positioning solutions.

9.4 BLUETOOTH

Bluetooth (BT) is a wireless communication standard created in the 1990s and included in the IEEE 802.15.1 [9]. The IEEE is no longer updating this standard and the latest version was withdrawn in 2018. Bluetooth is currently maintained by the Bluetooth Special Interest Group (SIG) [10].

Bluetooth targets the wireless personal area network (WPAN) industry and has a typical range of a few meters but can reach up to 100 m. The Bluetooth signal is transmitted over the 2.4-GHz ISM band, coexisting with Wi-Fi. Classic Bluetooth was intended for data exchange over short distances between fixed and mobile devices. However, modern Bluetooth supports multiple different applications, mobile positioning being one of them.

9.4.1 Bluetooth Low Energy

Looking at the characteristics of classic Bluetooth, it is not apparent how this can be useful for positioning. Bluetooth features low ranges from typically a few meters, high data rates, and relatively high latency. For most of these parameters, a positioning application has the opposite requirements: positioning benefits from high ranges and low latency, it does not require high data rates and, as for any mobile phone application, power consumption is critical.

However, the version 4.0 release of the Bluetooth specification [11] introduced the concept of Bluetooth Smart, later renamed to BTLE, also known as BLE. BTLE was a significant revolution to the classic Bluetooth (both co-exist today), targeted at a completely different set of new applications, such as health and fitness, wearables, and beacons. A comparison between Classic Bluetooth and BTLE is given in Table 9.1.

Table 9.1

Comparison of BTLE and Classic Bluetooth

	Classic Bluetooth	BTLE
Range	Up to 100 m (typically lower)	$>$ 100 m
Max data rate	3 Mbps	2 Mbps
Latency	100 ms	6 ms
Power consumption	1	0.01 to 0.5

BTLE enables positioning applications by using BTLE beacons. A BTLE beacon is a piece of hardware that periodically broadcasts small amounts of information. Instead of using the whole range of the Bluetooth 2.4-GHz frequency band, a BTLE beacon typically sends in one of the three advertising channels, channel numbers 37 (2402 MHz), 38 (2426 MHz), and 39 (2480 MHz), shown in Figure 9.8. These three channels are ideal for beacons because they are quite far apart from each other, meaning that if one is occupied the others can still be free, and they do not overlap with the most commonly used Wi-Fi channels, numbers 1, 6, and 11.

The BTLE standard has given flexibility to the message format that a BTLE beacon can transmit. Thus, instead of a predefined format, the beacons use different

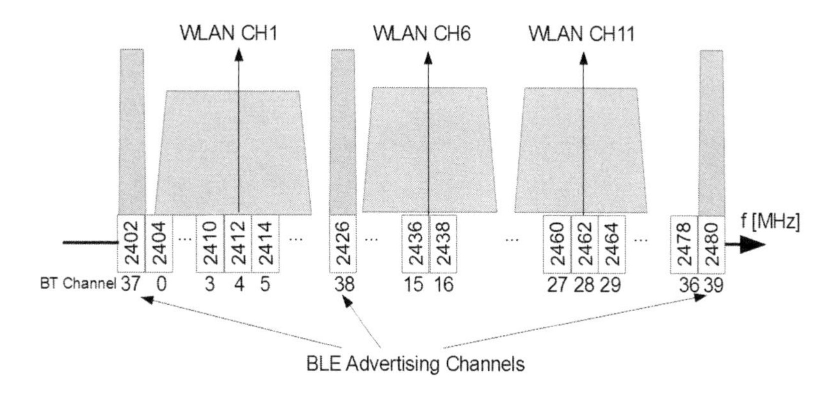

Figure 9.8 BTLE advertising channels in the 2.4-GHz ISM band.

manufacturer-specific formats. The most popular are iBeacon (Apple), AltBeacon (Radius Networks, Open source), and Eddystone (Google).

9.4.2 Beacon-Based Positioning

BTLE beacon-based positioning works on the same principle as Wi-Fi BSSID- and RSSI-based positioning. A BTLE beacon frame has the format shown in Figure 9.9. The header contains a 1-byte preamble and the access address. Then comes a PDU of variable length, which for a BTLE beacon is set to an advertising PDU, formed by a PDU header, the advertising address, and up to 31 bytes of payload. After the PDU comes the cyclic redundancy check (CRC) for error detection. The payload is flexible and has not been defined by the Bluetooth SIG. Each of the beacons mentioned above defined their own payload.

As an example, in Figure 9.9, the beacon formats for the iBeacon and the AltBeacon are shown. The advertising flags indicate important parameters about the beacon, such as whether it is connectable or not. Typically for positioning beacons are broadcast only and do not accept uplink connections. The iBeacon transmits a universally unique identifier (UUID), together with other fields (typically called major and minor) that allow to create a hierarchy between the different BTLE beacons as well as to identify each of them unambiguously. These three fields together play the role of the BSSID for the Wi-Fi AP. The AltBeacon does not distinguish between the three fields separately and just provides a Beacon ID that

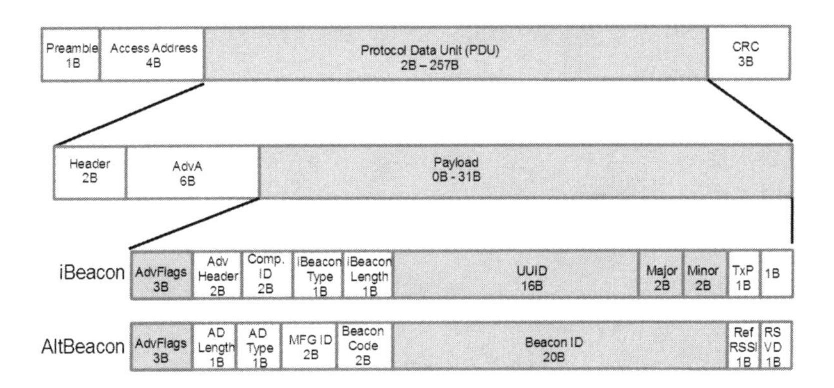

Figure 9.9 BLE beacon formats.

can be used with the same purpose. The important point is that independently on the beacon format, each BTLE beacon can be uniquely identified.

Analogous to Wi-Fi APs, the beacon identifier, together with other parameters such as the beacon geographical coordinates, can be stored in a database accessible to the location server. Thus, the same location methods seen in Section 9.3.2 to 9.3.4 are applicable to BTLE. Nonetheless, it is worth to notice that Wi-Fi APs are widely deployed and available in most households because they are used for other services such as connecting to the internet. On the other hand, BTLE beacons are much less spread and in most cases they would need to be specifically installed for positioning purposes.

9.4.3 Bluetooth Direction Finding

The next major advance for Bluetooth positioning came with V5.1 [12] and the Bluetooth Direction Finding feature. Bluetooth Direction Finding relies on antenna arrays to perform AoA or AoD measurements as seen in Section 2.3.5 of this book. Depending on which of the methods is used for the location, the complexity of the system resides on the receiver side (for AoA) or on the transmitter side (for AoD).

Angle-based positioning is also one of the new positioning features in 5G NR and will be explained in more detail in Chapter 10 of this book.

9.5 TERRESTRIAL BEACON SYSTEM

TBS is another of the noncellular network technologies that has been included to the LPP by 3GPP in Release 13 [13]. The aim of TBS is to deploy a GPS-like network on the surface of the Earth [14] using beacons instead of satellites. While a GNSS signal typically arrives to the Earth's surface around 30 dB below the noise floor, TBS-enabled devices can receive signal powers similar to LTE or other cellular technologies. This facilitates the reception and decoding of the signal, and also provides better coverage in both indoor and dense urban scenarios where GNSS is often not sufficient.

Furthermore, while cellular technologies have been planned and deployed mainly for cellular communication, a TBS network is exclusively dedicated to positioning. Thus, the TBS beacons can be placed with the goal of minimizing the DOP of the network (see Section 2.5.2 of this book), increasing the positioning accuracy.

On the other hand, the main disadvantage is that it requires the deployment and maintenance of a completely new beacon infrastructure. There is only one TBS network to date, which is the Metropolitan Beacon System (MBS) deployed by the company NextNav in the United States. MBS uses a network of beacons transmitting spread-spectrum signals combining CDMA and TDMA and using gold-codes (similar to GPS) on the 920-MHz band, using licenses that cover up to 95% of U.S. cities [14, 15]. Furthermore, MBS also uses atmospheric pressure reference stations and barometric measurements for height estimation in order to obtain accurate 3D locations. Barometric sensor positioning will be covered in Chapter 12 of this book.

In 2013, the Communications Security, Reliability, and Interoperability Council of the United States conducted an indoor location test bed, whose report was published by the FCC in [16]. The results of the report showed that MBS is able to achieve the FCC requirements for indoor location.

There are currently no TBS deployments outside the United States.

9.6 CONCLUSION

This chapter introduced the noncellular network positioning technologies most widely used for 5G and LTE location-based services. From the well-known Wi-Fi and Bluetooth technologies to the Terrestrial Beacon System, all the technologies presented here have advantages and disadvantages.

Among the advantages, one of the most important is that these technologies enable indoor positioning systems in scenarios where neither GNSS nor cellular network-based positioning have succeeded. On the other hand, the main disadvantage of Wi-Fi and BTLE technologies is that they are not under the control of the network operator, so their availability and accuracy cannot be ensured, making them a difficult candidate for emergency services. Nonetheless, they are a great candidate for commercial applications. In reality, Wi-Fi and BTLE-based positioning offers very accurate results, as most readers have likely already noticed when using a delivery service or a navigation application on their smartphones.

Current Wi-Fi and BTLE positioning implementations are based on power measurements (RSSI) and database matching algorithms. However, Wi-Fi RTT has the potential to become a game-changer in the near future.

All together, GNSS as seen in Chapters 6 and 7, OTDOA/UTDOA and ECID as seen in Chapter 8, and the technologies seen here form the positioning ecosystem in LTE. The next chapter will focus on the positioning ecosystem of 5G, partly formed by updated versions of the LTE techniques, and partly formed by new additions, such as angular-based technologies.

References

[1] Wi-Fi Alliance™, https://www.wi-fi.org/who-we-are/history, accessed on September 2019.

[2] Leichtman Research Group, *83% of U.S. Households Get an Internet Service at Home*, press release, 21st of December, 2018.

[3] Parks Associates, *Staking a Claim in the Connected Home: Service Provider Solutions*, white paper, 2018.

[4] IEEE 802.11-2016, *IEEE Standard for Information Technology–Telecommunications and Information Exchange between Systems; Local and Metropolitan Area Networks–Specific Requirements - Part 11: Wireless LAN MAC and PHY Specifications*, December, 2016.

[5] National Emergency Address Database, CTIA, http://www.911nead.org/, accessed on September 2019.

[6] Zegeye, W. K., et al, "WiFi RSS Fingerprinting Indoor Localization for Mobile Devices," in *Proc. IEEE 7th Annual Ubiquitous Computing, Electronics & Mobile Communication Conference (UEMCON)* , New York City, Oct. 20 - 22, 2016, pp. 1-6.

[7] He, S., and Chan, S. -. G., "Wi-Fi Fingerprint-Based Indoor Positioning: Recent Advances and Comparisons," in *IEEE Communications Surveys and Tutorials*, Vol. 18, No. 1, pp. 466-490, 2016.

[8] Van Diggelen, F., Want, R., and Wang, W., *How to Achieve 1-Meter Accuracy in Android* GPS World, July 2018, https://www.gpsworld.com/how-to-achieve-1-meter-accuracy-in-android/, accessed on September 2019.

[9] IEEE 802.15.1-2005, *IEEE Standard for Telecommunications and Information Exchange Between Systems - LAN/MAN - Specific Requirements - Part 15: Wireless Medium Access Control (MAC) and Physical Layer (PHY) Specifications for Wireless Personal Area Networks (WPANs)*, June, 2005.

[10] The Bluetooth Special Interest Group, https://www.bluetooth.com/, accessed on September 2019.

[11] The Bluetooth SIG, *Bluetooth Specification Version 4.0*, June 2010.

[12] The Bluetooth SIG, *Bluetooth Specification Version 5.1*, January 2019.

[13] 3GPP TS 36.355, *LTE Positioning Protocol (LPP)*, V15.5.0, September, 2019.

[14] Pattabiraman, G., and Gates, C., "Terrestrial Beacons Bring Wide-Area Location Indoors," *GPS World* Vol. 27, pp 38-40, July 2016.

[15] Metropolitan Beacon System, Nextnav, http://www.nextnav.com/network#terre, accessed on September 2019.

[16] CSRIC Working Group 3, E9-1-1 Location Accuracy, *Indoor Location Test Bed Report*, March, 2013, https://transition.fcc.gov/bureaus/pshs/advisory/csric3/CSRIC_III_WG3_Report_March_%202013_ILTestBedReport.pdf.

Chapter 10

5G Positioning Technologies

10.1 INTRODUCTION

The previous chapters introduced all the positioning technologies used in legacy cellular networks up to LTE. Some of these technologies, more specifically the RAT-independent technologies (A-GNSS, Wi-Fi, BTLE, etc.), can be directly used in 5G NR without any extension or modification. On the other hand, the RAT-dependent technologies (OTDOA, ECID, etc.) are tightly coupled to the cellular network specification. Although positioning algorithms, such as trilateration, will still be basically the same as described in Chapter 2 of this book, the positioning measurements and other parameters, such as measurement resolution, will be specific to 5G NR. This chapter introduces the native 5G NR positioning technologies.

3GPP Release 15 first standardized the 5G network. However, the first version of the standards did not include any 5G NR native positioning technologies. In order to meet the regulatory requirements for emergency services, the 3GPP agreed to support RAT-independent positioning technologies and legacy LTE-based OTDOA and ECID measurements and postpone all further positioning-related standardization activities for Release 16. During Release 16, the 3GPP conducted a study on 5G NR positioning and identified a series of positioning candidates. The results of the study and simulations, as well as the selected positioning technologies, are detailed in TR 38.855 [1].

The 5G network positioning methods are partly based on some of the legacy technologies. For instance, ECID has been redefined for 5G, while OTDOA has become DL-TDOA. Other positioning technologies have been extended. The RTT measurement, part of ECID in LTE, was supporting timing measurements only to the active cell. In NR, the RTT to neighbor cells has been defined as well, in the multi-RTT method. Finally, there are a few positioning technologies that are completely new in 5G NR. Angle-based techniques, such as downlink angle of departure (DL-AoD), benefit from the 5G inherent massive MIMO and beamforming capabilities.

In total, six new or refarmed methodologies have been incorporated to TS 38.305 [2]: DL-TDOA, DL-AoD, Multi-RTT, NR E-CID, UL-TDOA and uplink angle of arrival (UL-AoA). The 3GPP has studied as well PDoA and carrier-phase based techniques. However, the complexity of such techniques compared to their performance did not justify their addition to the Release-16 specifications.

During the definition of the new positioning standards, the 3GPP has classified the different methods into three categories: the downlink-based methods, such as DL-TDOA and DL-AoD; the uplink-based methods, such as UL-TDOA or UL-AoA, and the downlink- and uplink-based solutions, such as multi-RTT. Nonetheless, the positioning algorithms behind downlink and uplink methods are largely identical. Thus, this chapter will classify the methods with regard to the type of positioning measurement they use. The first section will focus on pointing out the main differences between LTE and NR RANs, and how that affects the positioning functionality. The following sections will analyze the timing-based, signal power-based ,and angle-based positioning technologies. Afterward, the main sources of error affecting 5G-based positioning will be introduced. Finally, the chapter will give a summary of all the different technologies.

10.2 DIFFERENCES BETWEEN LTE AND NR

5G NR was introduced in Chapter 5 of this book, explaining its basic architecture and the main differences with respect to legacy cellular networks. This chapter will once again point out the main differences between LTE and NR, which can make an impact on the performance of positioning technologies.

One of the words that is often used to describe 5G NR is *flexibility*: 5G NR offers a plenitude of configuration possibilities for almost every imaginable parameter: multiple possible subcarrier spacing, a wide selection of possible bandwidths, different bandwidth parts, and so forth. The following are some of the 5G NR features with critical impact to positioning.

10.2.1 Frequency and Beamforming

The first and most obvious difference is the use of frequency ranges. The 3GPP has defined two frequency ranges for 5G. The FR1 goes up to 7.125 GHz, a bit higher than the LTE uppermost frequency. This range offers similar propagation conditions to LTE, but 5G supports much higher bandwidths (up to 100 MHz). The higher bandwidth may allow to achieve better measurement performance and better multipath resolution, improving the overall positioning accuracy.

The real change comes with the second frequency range, FR2 or mm-wave. This range starts at 24 GHz and has much higher propagation losses. The use of directive antennas is required to increase the link budget. The main advantage of FR2 is the spectrum availability, which allows much higher bandwidths of up to 400 MHz.

Due to the higher frequencies, massive MIMO and beamforming are key aspects of 5G. They are required to increase the antenna gain and compensate the propagation losses. For positioning, these features enable angle-based technologies. In FR2, where the different antenna beams are narrower, angle-based technologies may deliver very good positioning accuracy.

10.2.2 Numerology and NR Time Unit

In LTE, all the timing-based positioning measurements were defined with respect to the LTE sampling time, called T_s. Both the RSTD and the UE Rx-Tx time difference are reported in multiples of T_s. The basic reporting accuracy for RSTD is 1 T_s, while the enhanced reporting accuracy introduced with the LPP Release 13 updates is 0.5 T_s. Analogously, the measurement accuracy requirements for mobile devices were also defined as multiples of T_s. Thus, the LTE sampling time has greatly influenced the LTE positioning performance.

This LTE basic time unit corresponds to the sampling time for one OFDM symbol with 2048 points for the inverse fast Fourier transform (IFFT). Since the duration of the OFDM symbol is the inverse of SCS, the LTE sampling time is defined by Equation 10.1, where Δf represents the SCS and N_f the number of points in the Fourier transform.

$$T_s = \frac{1}{\Delta f \cdot N_f} = \frac{1}{15000 \cdot 2048} \approx 32.55ns \tag{10.1}$$

The physical layer of NR shares commonalities with LTE: a frame also lasts 10 ms and is divided into 10 subframes of 1 ms duration. However, the subframes in NR are divided into a number of slots, each containing 14 OFDM symbols. The number of the slots in a subframe is variable and depends on the SCS. 5G NR does not have a single SCS. In fact, five different values for the SCS are possible as of now: 15, 30, 60, 120, and 240 kHz. Furthermore, 480 kHz could be potentially

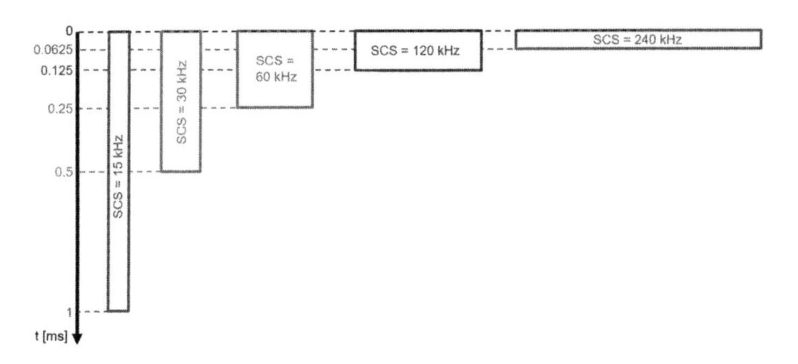

Figure 10.1 5G slot duration vs subcarrier spacing.

supported in the future. Each SCS defines a different slot duration, as shown in Figure 10.1.

If the NR basic time unit had been defined based on the SCS, this unit would have been different for each numerology, which is not practical. Instead, 5G defined the basic time unit, T_c, based on the maximum potentially supported SCS, 480 kHz, as shown in Equation 10.2. Also notice that the value of N_f is doubled for 5G NR:

$$T_c = \frac{1}{\Delta f_{max} \cdot N_f} = \frac{1}{480000 \cdot 4096} \approx 0.5ns \tag{10.2}$$

In consequence, there is a constant relation between the LTE and the 5G NR time units, $k = \frac{T_s}{T_c} = 64$. That means if the same reporting resolution of one time unit is used for 5G NR, the results could potentially be 64 times more accurate than for LTE. As seen during Chapter 8 of this book, 1 T_s in LTE was equivalent to a little under 10 m of distance, taking into account the signal propagation speed. For 5G, 1 T_c is equivalent to approximately 15 cm.

10.2.3 UE-Based Positioning

In LTE, all RAT-dependent positioning methods support UE-assisted mode only. Thus, the UE performs the required positioning measurements and reports them back to the network for the calculation of the UE position. UE-based mode, where the UE receives the necessary data to calculate its own position, is not supported. The main reason for this restriction is that the UE is not provided the base station coordinates, which are needed for the calculation. LPP did not foresee any assistance data element to provide this information.

For 5G NR positioning, LPP has been extended with the necessary assistance data elements and the downlink-based positioning methods (DL-TDOA and DL-AoD) support UE-based positioning. This feature has advantages such as reducing the network load, since the position calculation can be taken up by the UE, and also reducing the positioning session latency, since the UE can collect all information and calculate its location without further message exchange with the network. Furthermore, combined with the assistance data broadcast via the posSIB (see Chapter 14 of this book), it enables a fully passive positioning. Both the positioning signals and the required assistance data are broadcast periodically. Hence, the mobile device can perform the position calculation autonomously without additional signaling exchange with the network.

10.3 TIMING-BASED TECHNOLOGIES

This section covers the 5G NR positioning technologies based on signal timing measurements. The positioning algorithms that use these measurements are multilateration and trilateration, as explained in Chapter 2 of this book. These algorithms are well known and widely used in other technologies, such as GNSS or LORAN. Even in cellular network positioning, such methodologies have been used since GSM times (see Chapter 8). Thus, the methods introduced in this section are an evolution or upgrade of the methods previously used in LTE and other RATs.

10.3.1 DL-TDOA

DL-TDOA is the successor of the LTE's OTDOA technology. Its working principle is basically the same as described in Chapter 2 of this book. The mobile device measures the elapsed time between the reception of positioning signals coming from different base stations.

Analogous to LTE, the mobile device can use any cell synchronization signal, such as PSS or SSS. These signals are designed to be easily received within the base station coverage area. However, they are also designed not to interfere with neighbor base stations. Thus, the mobile device will likely not see many PSS / SSS from neighbor base stations. Therefore, the 3GPP, in TS 38.211 [3], has defined the PRS for 5G NR as well. The PRS is specifically designed to allow the mobile device to see multiple PRSs from multiple neighbor cells. Hence, parameters such as scheduling, power, and signal bandwidth are configured in a way that the signal can be decoded from afar, while minimizing the interference in the normal operation of other base stations.

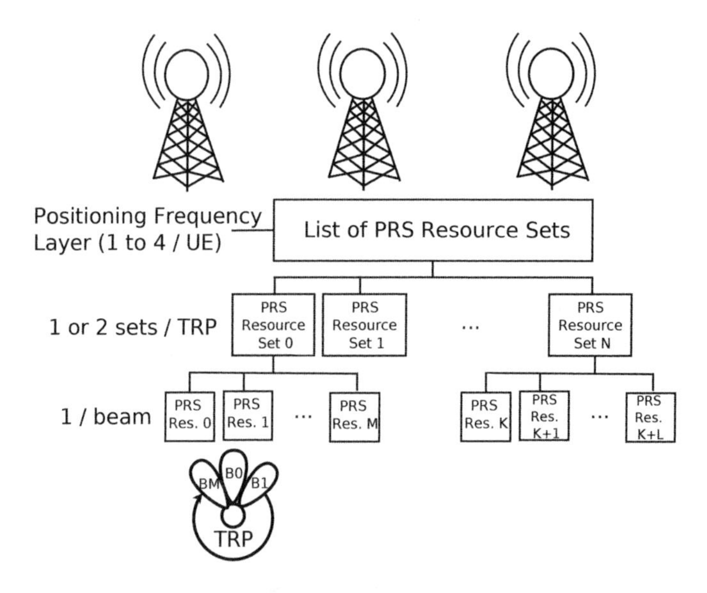

Figure 10.2 Configuration of the PRS resources in 5G.

The main difference between the LTE PRS and the NR PRS is the flexibility in the configuration of the signal. 5G NR organizes the PRS in DL PRS resource sets. In order to better understand the structure of the PRS resources, they are graphically represented in Figure 10.2.

The PRS resource sets are summarized in a list in the positioning frequency layer [3]. All the PRS resources within the same positioning frequency layer share some common parameters (e.g., the SCS or the bandwidth). A mobile device receives up to four positioning frequency layers. Each list contains multiple PRS resource sets, which can be one or two per transmission and reception point (TRP). Each TRP has one DL PRS resource for each of its beams, as seen in the lowest layer of Figure 10.2. A DL PRS resource is identified by a PRS ID. There are 4096 different PRS IDs.

Similar to LTE, the PRS can occupy any bandwidth up to the bandwidth of the NR carrier. However, the PRS bandwidth in NR is not configured individually for each PRS resource. Instead, it is given for the whole positioning frequency layer. Since NR supports very large bandwidths (up to 100 MHz in FR1 and 400 MHz in FR2), it is likely that the PRS signal will not be transmitted across the whole NR carrier, but in a portion of it. The lowest physical resource block (PRB) containing

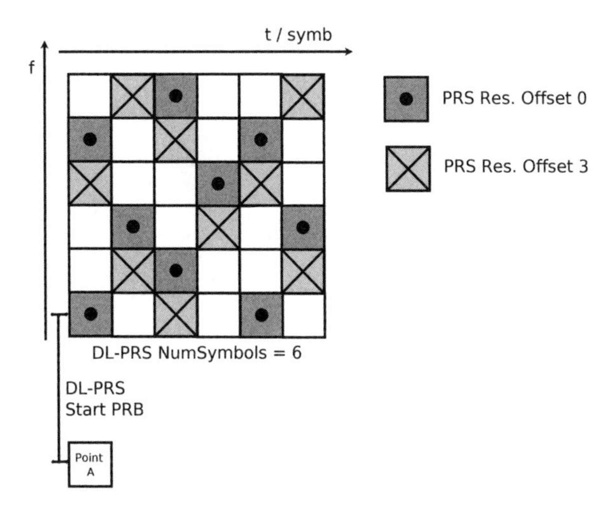

Figure 10.3 Example of PRS resource allocation for two PRS resources and comb size of 4.

PRS is defined by the parameter DL-PRS Start PRB. This parameter defines the offset to the Point A, a reference point used to determine the NR resource block grids.

The allocation of the PRS resources within the resource bandwidth follows a comb structure with a possible size of 2, 4, 6, or 12 carriers. The specific location of each PRS resource is defined by the resource offset. Figure 10.3 gives an allocation example with two PRS resources, with offsets 0 and 3, using a comb size of 4.

It is possible to mute specific PRS occasions following a muting pattern. 5G NR supports the configuration of two different muting patterns affecting a particular PRS resource. The first muting pattern is defined on the PRS resource set level and applies to all the resources within the set. The second muting pattern is specific for each PRS resource and can be used to mute individual resources. Furthermore, the PRS transmission can be switched on and off more dynamically than in LTE, based on the current positioning demands of the users of each gNB.

All the parameters related to the PRS configuration are provided to the mobile device as part of the DL-TDOA assistance data. Furthermore, the assistance data can also include the base station coordinates and other parameters needed by the mobile device to perform a UE-based location fix. 5G NR has inherited the possibility to broadcast the assistance data in the posSIBs, introduced to LTE in Release 15. This allows a more efficient use of the network resources when compared to on-demand

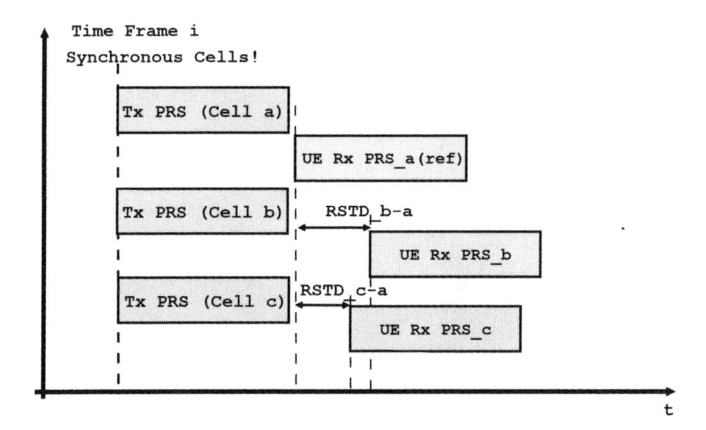

Figure 10.4 DL-TDOA RSTD measurement process.

transmissions, and also enables the mobile device to obtain a position fix passively, without the need for any additional signaling. Nonetheless, dedicated assistance data transmissions as part of the positioning session are still supported.

10.3.1.1 RSTD Measurements

DL-TDOA positioning is based on RSTD measurements, defined in TS 38.215 [5]. The procedure to obtain an RSTD measurement is identical to OTDOA (LTE), seen in Chapter 8 of this book. Nonetheless, it is briefly recalled here.

The RSTD measurement is defined as the difference in the time of arrival of the PRS from a neighbor base station compared to the time of arrival of the PRS from the reference base station. This measurement is illustrated in Figure 10.4. The mobile device decodes the PRS from multiple base stations. It selects one (typically the serving cell) as reference, PRS_a in the figure. Afterward, it computes the time difference between the reception of all other PRS (PRS_b and PRS_c) with respect to the reference. This results in the two RSTD measurements shown in Figure 10.4.

The RSTD in LTE is reported in multiples of T_s (approximately 32 ns). However, in 5G NR, the RSTD will be reported in multiples of T_c, which is about 0.5 ns. At the time of the writing, the 3GPP has not yet decided the granularity of the RSTD reports for 5G NR (see TS 38.133 [4]). Parameters such as the SCS, the measurement bandwidth, and the frequency range will influence the achievable RSTD resolution. The discussions move in the range of 1 to 4 T_c of measurement granularity for FR2 and 4 to 16 T_c of measurement granularity for FR1, which uses narrower bandwidths and smaller SCS. In order to give an idea of how this translates to positioning accuracy, 1 T_c reporting granularity translates into the uncertainty in the RSTD measurement seen in Equation 10.3. A measurement resolution of 4 T_c would translate into approximately 60 cm of uncertainty, and 16 T_c is equivalent to an uncertainty of 2.4 m. For a quick comparison, the U_{RSTD} for LTE is approximately 4.88 m:

$$U_{RSTD} = v_p \cdot 1 \cdot T_c \approx 15cm \tag{10.3}$$

Figure 10.5 compares the positioning uncertainty of a location fix with OTDOA and DL-TDOA in a function of the RSTD measurement resolution. As can be seen, in all cases the positioning uncertainty (shaded area in the image formed by the intersection of all the dashed lines) is smaller for DL-TDOA. Furthermore, for measurement resolutions of 4 T_c and 1 T_c, the positioning uncertainty is not even visible in the image, since the total uncertainty area is a few square centimeters.

In conclusion, DL-TDOA has the potential to achieve a much better accuracy than its LTE counterpart. However, as was already the case for OTDOA, DL-TDOA is very adversely affected by errors in the base station synchronization. Taking as an example the 3GPP study in [1], it can be seen that DL-TDOA is not able to achieve the desired positioning accuracy if the base station synchronization error is 50 ns based on the simulation results.

10.3.2 UL-TDOA

UL-TDOA is based on the same working principle as DL-TDOA, but the time difference measurements are done by the network of base stations. Each gNB or TRP measures the timing of the uplink signal of the mobile device and provides this information to the LMF of the 5G NR network. An important difference with the counterparts in legacy RATs is that the uplink timing measurements are not based on the random access burst, but in the UE UL sounding reference signal (SRS). The UL SRS, defined in TS 38.211 [3], has been extended for positioning, for instance by increasing the possible number of consecutive OFDM symbols.

The UL-TDOA measurement procedure is illustrated in Figure 10.6. Upon receiving a positioning request from the network, the serving base station determines the necessary UL SRS resources and sends the SRS configuration to the mobile device. The mobile device transmits the configured SRS to each neighbor gNB. The neighbor gNBs measure the time of arrival of the corresponding SRS signal and

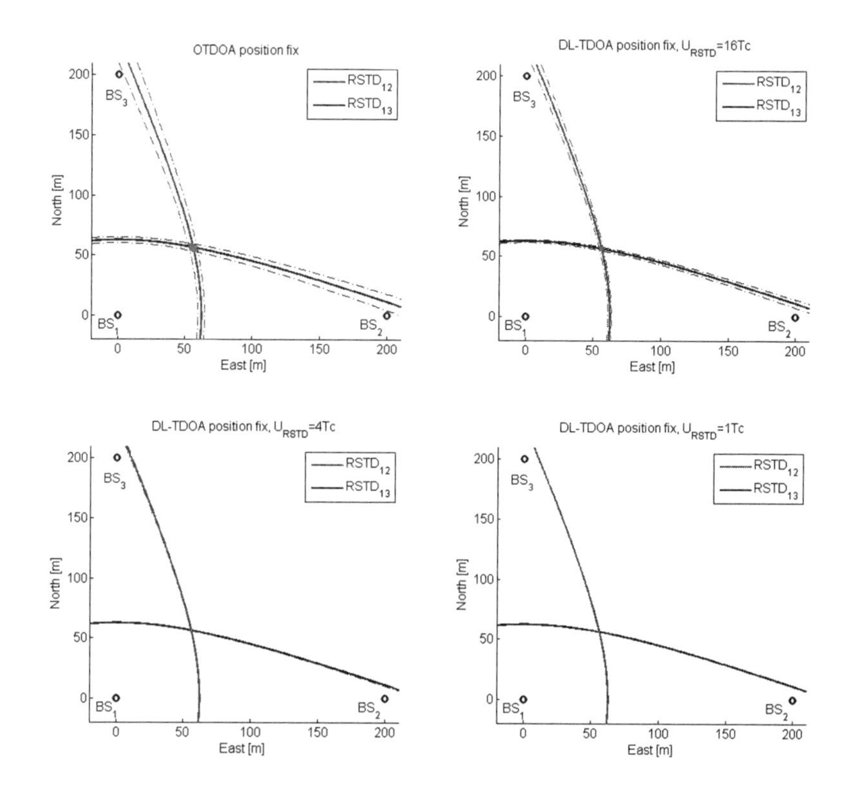

Figure 10.5 RSTD measurement uncertainty for OTDOA and DL-TDOA (depending on measurement resolution).

transmit this information to the LMF, which calculates the time difference measurements as shown in the figure.

Since the SRS may need to be received in different directions and at different distances, the UE may need to send multiple SRSs in order to reach all base stations. For example, in Figure 10.6, the mobile device sends three SRS for reaching four base stations. The procedure that the mobile device needs to follow to send the positioning SRS is clarified in TS 38.214 [6].

From the mobile device perspective, the advantage of UL-TDOA with respect to DL-TDOA is the simplicity, since it does not require any specific positioning measurement. The mobile device's only responsibility is to transmit the UL SRS. On the other hand, the network implementation is more complex and it requires specific

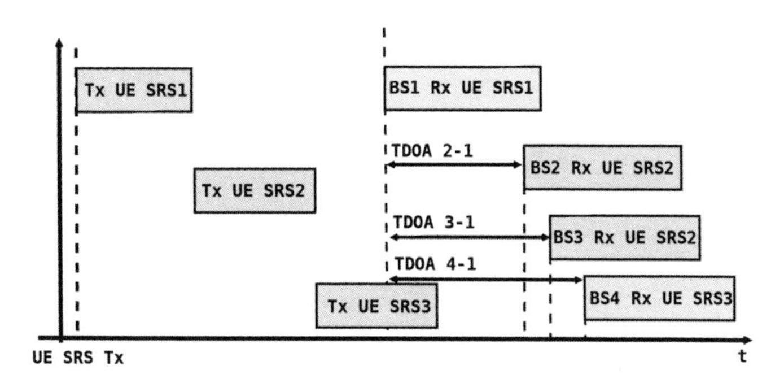

Figure 10.6 UL-TDOA measurement process.

measurement functionality. Since the measurement takes place in the network, it does not support a UE-based mode. Furthermore, the mobile device does not get any information on the calculated position, so this method is not useful for commercial location services that require the device to know its own position.

10.3.3 Multi-RTT

The main disadvantage of the RTT measurement as specified in LTE ECID is that it only supports the UE Rx-Tx measurement with respect to the serving base station. The distance to the neighbor base stations needed for a trilateration algorithm can only be estimated based on power measurements. As has been seen in Chapters 2 and 8 of this book, power measurements are not as accurate as timing measurements, and the resulting position does not meet the expected requirements.

This problem has been solved in 5G NR by introducing a new positioning method called multi-RTT. This methodology defines a procedure to perform RTT measurements additionally on neighbor base stations, enabling the trilateration algorithm using only timing measurements. The 3GPP decided to keep both methods separately for NR: NR ECID refers only to power measurements to neighbor cells, while multi-RTT refers to the procedure using multiple RTT measurements both to serving and neighbor base stations.

The RTT measurement procedure to the serving cell is mostly identical to LTE. Prior to the RTT measurement, the mobile device needs to adjust its uplink timing following the timing advance adjustment procedure. As in LTE, the gNB expects the uplink and the downlink transmissions to be synchronized so that the gNB Rx-Tx time (difference between transmission and reception times) is zero. In

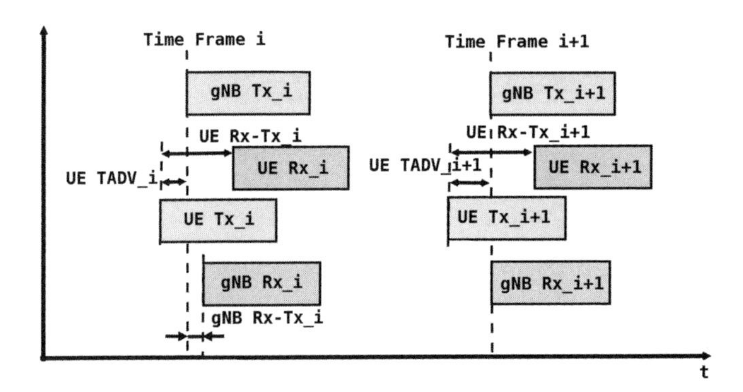

Figure 10.7 RTT measurement process to the serving base station.

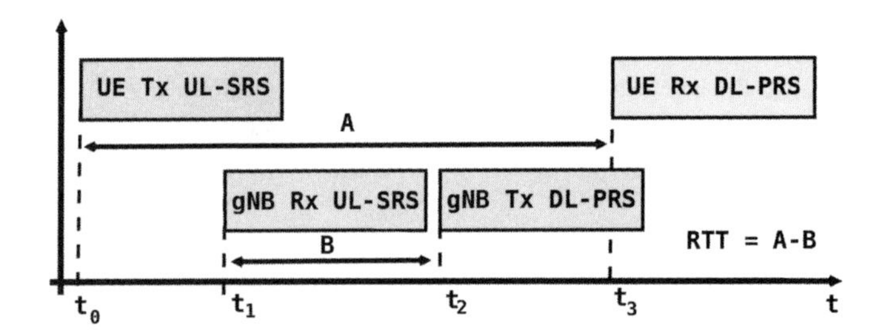

Figure 10.8 RTT measurement process to neighbor base stations.

order to achieve this requirement, the mobile device needs to transmit its uplink *ahead* of time, so that it arrives to the base station at the required time instant, after taking into account the propagation delay.

The serving cell RTT measurement process is explained in Figure 10.7. In the first time frame, i, the gNB measures its Rx-Tx time difference and sees it is different from zero. Thus, it sends a timing advance adjustment command to the mobile device. In time frame $i + 1$, the uplink timing of the mobile device is corrected and the gNB Rx-Tx time difference is zero. Hence, the UE Rx-Tx time difference is exactly the RTT.

The RTT measurement to the neighbor base stations is slightly different than the RTT measurement to the serving base station, as seen in Figure 10.8. Similar to UL-TDOA, the UE sends an UL SRS signal at time instant t_0. The neighbor gNB receives this UL-SRS at time instant t_1, and transmits a downlink PRS at time instant t_2. The UE receives the PRS at time t_3. The following measurements are reported to the LMF:

- A: The UE reports t_3 - t_0, the elapsed time between the transmission of the SRS and the reception of the PRS.
- B: The gNB reports t_2 - t_1, the elapsed time between the reception of the SRS and the transmission of the corresponding PRS.

The LMF can calculate the RTT by subtracting A-B, as shown in Equation 10.4.

$$RTT = (t_3 - t_0) - (t_2 - t_1) \tag{10.4}$$

At the time of the writing, the 3GPP has not yet finalized the discussion on the RTT measurement reporting granularity, which will be captured in TS 38.133 [4]. Nonetheless, similar values as for the RSTD measurements are being proposed (i.e., between 1 and 4 T_c for FR2 and between 4 and 16 T_c for FR1). Thus, the corresponding RTT measurement uncertainty (U_{RTT}) will oscillate between 15 cm and 2.4 m. In LTE, the reporting resolution for RTT was set to 2 T_s, which translates to approximately 9.76 m. Figure 10.9 shows the uncertainty in the calculated position based on the potential reporting granularity candidates. The positioning uncertainty is represented by the shaded area, which is only visible in the case of the 16 T_c measurement resolution. For the other two resolutions, the resulting uncertainty is so small that it is not visible in the figure. For comparison, the upper right figure shows the LTE ECID positioning uncertainty. Since LTE ECID supported only one RTT measurement, the positioning measurements to the neighbor base stations are done assuming power measurements. The figure shows only a qualitative estimation of the accuracy of power measurements compared to the RTT.

The main advantage of multi-RTT with respect to DL-TDOA is that the positioning accuracy does not depend on the base station synchronization. However, it increases the network load due to the additional uplink signaling generated.

10.4 SIGNAL POWER-BASED TECHNOLOGIES

In addition to the timing-based measurements, the mobile device can also report power measurements for positioning. Power measurements have the advantage that they do not require any additional features in the mobile device and are readily

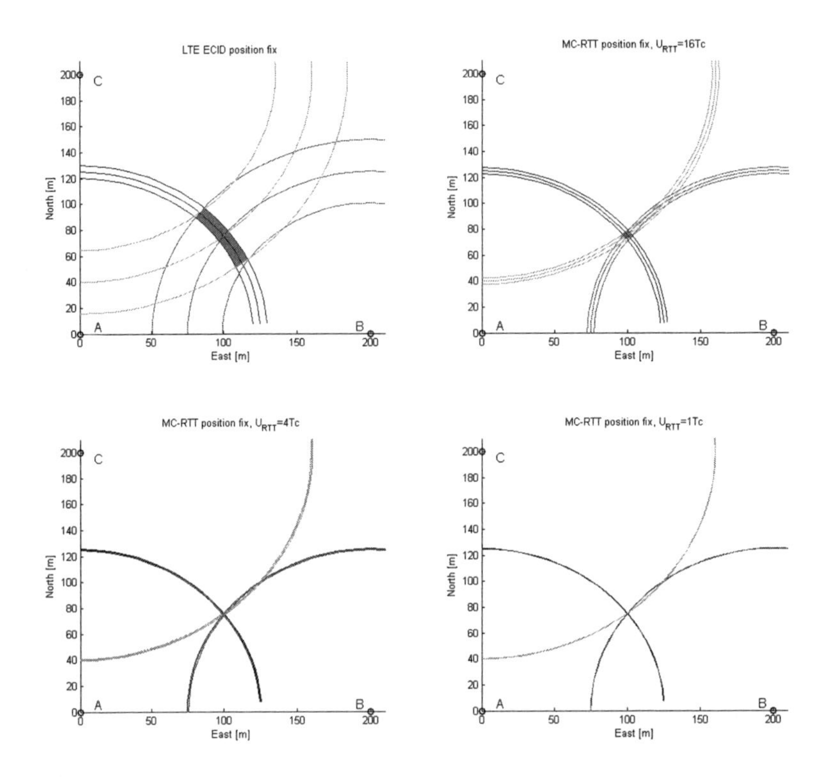

Figure 10.9 Multi-RTT positioning uncertainty (depending on RTT measurement resolution).

available. On the other hand, the distance calculation based on power measurements is inherently much more inaccurate than the distance calculation based on timing measurements. Thus, power measurements for positioning are only used as a last resort in order to achieve a low-latency rough location estimate.

10.4.1 NR ECID

The NR ECID method is significantly different to what was called ECID in LTE. First and foremost, NR ECID as defined in TS 38.305 [4] is based solely on power measurements. The RTT measurement is no longer part of the NR ECID method, and is performed in the context of a multi-RTT positioning fix. A second major difference is that for NR ECID, the mobile device is not expected to perform any specific measurement. It should instead report whatever information is already available at the time of the request.

The power measurements used for NR ECID are divided in two groups: power measurements based on the synchronization signal (SS), and power measurements based on the channel state information reference signal (CSI-RS).

10.4.1.1 Synchronization Signal Power Measurements

The synchronization signal reference signal received power (SS-RSRP) and the synchronization signal reference signal received quality (SS-RSRQ), defined in TS 38.215 [5], are the NR counterparts of the LTE RSRP and RSRQ measurements. These measurements have been used as a complement of the ECID RTT already present in LTE times. The same measurements are also required for radio resource management (RRM) procedures. Thus, they are typically available at the mobile device.

The SS-RSRP measurement reports are defined in TS 38.133 [4] with 1-dB resolution, while the SS-RSRQ reporting is defined with a resolution of 0.5 dB.

10.4.1.2 CSI-RS-Based Power Measurements

The CSI-RSRP and CSI-RSRQ measurements use the CSI-RS instead of the SS. Apart from that, the reporting resolution and all other aspects related to positioning are identical as for the SS-based measurements described in the previous section.

10.4.2 PRS RSRP

5G NR has also defined the procedure to measure and report the received power for the PRS signal, PRS-RSRP (see TS 38.215 [5]). A fundamental difference with the other power measurements described in this section is that the PRS-RSRP is used only for positioning and it must be explicitly requested. Thus, it is not part of the NR ECID measurement set.

The mobile device can be configured to report up to eight PRS-RSRP measurements on different PRS resources from the same cell. The measurement is useful to fine tune angle-based technologies by determining which beams the mobile device sees with higher power than others. Furthermore, it can also be used to help in mitigating multipath effects.

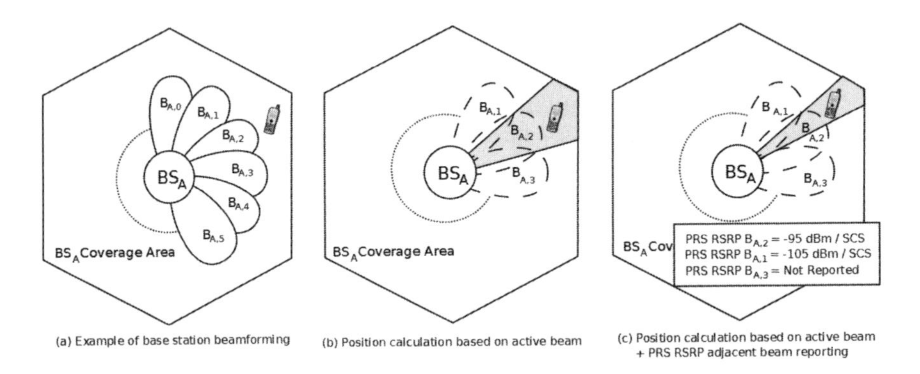

Figure 10.10 Angle-based positioning scenarios with one base station.

10.5 ANGLE-BASED TECHNOLOGIES

Although angle-based positioning technologies have been around for a while, it is with 5G NR, and particularly with FR2, when they can reach their full potential. Angle-based positioning technologies rely on finding the direction in which the object to locate is with respect to a reference point. In 5G NR, with the beamforming capabilities, this feature comes basically out-of-the-box, since each beam of the base station is associated with a beam index. The mobile device can easily differentiate between the different beams.

Beamforming is also supported at the mobile device using smaller antenna arrays. A mobile device (especially FR2 capable devices) consists of multiple patch antennas in different directions to scan the whole space. However, these beams are not very useful for positioning since they are not linked to a specific direction. The mobile device can be in the pocket, or held in the hand horizontally or vertically next to the ear for talking, and so forth. Since the mobile device can move and rotate freely, it is impossible at the network to differentiate which beam points in which direction. On the other hand, the base station beams are fixed. Thus, the 3GPP angular positioning technologies are always linked to base station beams: DL-AoD and UL-AoA.

Before going into the details of each of the technologies, it is worthwhile to list the different positioning use cases of angular techniques. Figure 10.10 (a) shows the concept of base station beamforming. When the mobile device completes the network registration, it gets connected to an active beam. By using this information, the network is able to determine an area where the mobile device can be located. In

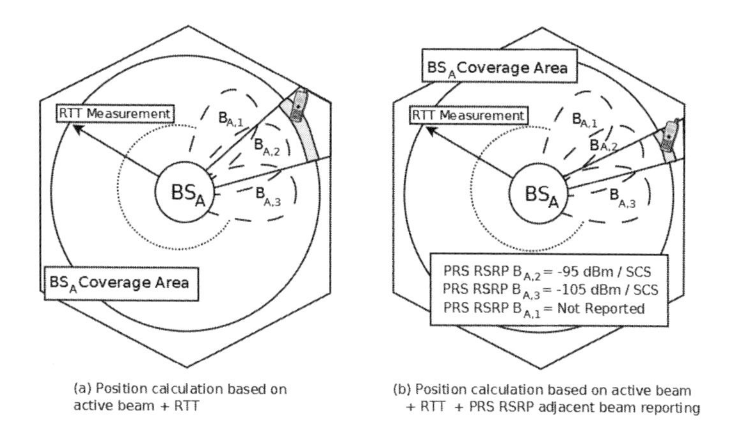

Figure 10.11 Angle-based positioning scenarios including RTT.

Figure 10.10 (b), the mobile device is connected to beam $B_{A,2}$. Thus, it is potentially located in the shaded area in the image. The positioning accuracy of this basic angular technique depends mainly on the beamwidth of the base station beams and on the radius of the coverage area of the base station. The beamwidth of the base station beams is an implementation detail, and it can be different for different base station manufacturers. In general, an educated guess is that FR1 beams will be wider than FR2 beams and that their width will range approximately between 10°and 45°.

A further refinement of angular positioning technologies would be to use the PRS RSRP measurement (see Section 10.4.2) of adjacent beams of the base station. Based on the power reported for each beam, it is possible to refine the position calculation by further delimiting the area of potential locations. Figure 10.10 (c) gives an example where the beam $B_{A,1}$ is reported with approximately 10 dB less power than the active beam ($B_{A,2}$). The other adjacent beam, $B_{A,3}$, is not reported at all, meaning that the mobile device probably cannot receive any PRS signal from this beam. Thus, the area of potential locations for the mobile device can be narrowed to the shaded area in the figure.

The next possible improvement is to include RTT or timing advance (TA or TADV) measurements. This gives an estimate of the distance between the base station and the target device, which results in a positioning circle around the base station. In Figure 10.11 (a), the RTT measurement is further combined with the active beam, obtaining a relatively small area of potential locations, shaded in the figure. In Figure 10.11 (b), by overlaying the PRS RSRP measurements, this area is approximately halved.

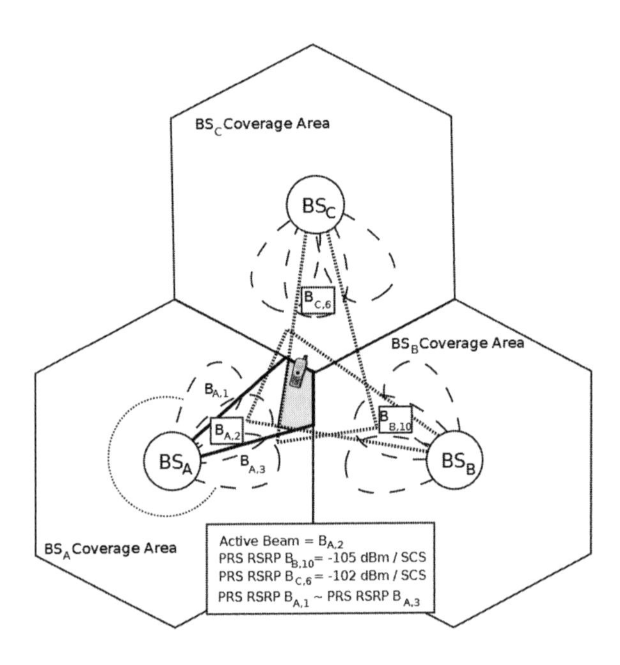

Figure 10.12 Angle-based positioning scenarios with multiple cells.

Finally, the mobile device could also be seeing different beams from multiple base stations. An example is shown in Figure 10.12. Important to note is that the PRS signal, as was mentioned in Section 10.3.1.1, is designed to reach further than the cell boundaries. Thus, the mobile device can still see PRS signals from beams $B_{B,10}$ and $B_{C,6}$ even though it is outside of the base station's B and C coverage areas.

10.5.1 DL-AoD

DL-AoD is a positioning method based on measuring the downlink PRS signals from different beams received by the mobile device [2]. The measurement is performed by the mobile and reported back to the network. As with DL-TDOA, it also supports UE-based positioning, provided that the network sends the base station's location and the beam information in the corresponding assistance data, including the direction in which each beam is transmitted. If the assistance data is broadcast using the positioning SIBs, this positioning method can be performed completely on the UE side without any specific uplink transmission.

10.5.2 UL-AoA

UL-AoA is the opposite method to DL-AoD. The base station identifies (using its beamforming capabilities) from which direction the uplink signals of the mobile device are coming. Since the measurement is performed on the base station side, it does not support any UE-based mode.

10.6 OTHER POSITIONING CANDIDATES

Apart from the technologies explained so far, the 3GPP has also studied other positioning candidates in TR 38.855 [1]. These positioning candidates were not included as part of the specification in Release 16 mainly due to the technical complexity of the required methodologies. However, since they may be included in the future, they will be briefly introduced in this section.

10.6.1 Carrier-Phase-Based Positioning

One of the additional methods proposed was based on carrier-phase measurements of the PRS signal. The positioning concept is similar to the high-accuracy GNSS enhancements seen in Chapter 7 of this book. In order to enable accurate carrier phase measurements, one of the propositions [7] was to transmit a dedicate carrier-phase PRS (C-PRS) as a sinusoidal wave using the guard band of the regular 5G signal.

The mobile device would receive both the standard PRS and the C-PRS signal, and use the first one for the normal RTT or TDOA measurements, while the latter is used only for the carrier-phase calculation.

Carrier-phase-based positioning offers (in theory) an accuracy comparable to high-accuracy GNSS methods such as RTK. On the other hand, the carrier-phase measurements in the mobile device involve high computational complexity. Furthermore, the synchronization requirements for the entire base station network are much tighter than the current operational requirements. Thus, this technology has been discarded for Release-16 5G NR positioning.

10.6.2 PDoA Positioning

A fairly similar method also proposed for high accuracy positioning is PDoA [8]. Similar to carrier-phase based positioning, PDoA is based on calculating the phase of the received PRS. However, it uses these measurements to compute the difference in the phases received for multiple subcarriers of the PRS.

Although the theoretical performance of PDoA methods is also high, the technical complexity caused it to follow the same fate as carrier-phase-based positioning.

10.6.3 Hybrid Positioning

5G NR RAT-dependent positioning technologies, especially DL-TDOA and multi-RTT, are suitable for hybrid positioning solutions (i.e., the combination of RAT-dependent and RAT-independent technologies).

The most promising combination in the hybrid area is between DL-TDOA, multi-RTT and GNSS, making use of the WLS algorithm presented (for LTE) in Section 8.9 of this book. The DL-TDOA and multi-RTT equations can be linearized in the same way that the OTDOA and ECID equations were linearized in LTE. Thus, whether the RAT-dependent measurements come from LTE or from NR is transparent to the hybrid positioning algorithm.

10.7 SOURCES OF ERROR IN 5G NR-BASED POSITIONING

This section will describe the major sources of error in 5G NR positioning. Since most of the sources have already been described for LTE, WCDMA, or GSM networks in Section 8.10 of this book, this section will focus mainly on the differences with respect to the legacy networks.

10.7.1 Network Synchronization

The network synchronization remains as one of the biggest sources of error for DL-TDOA positioning. This error occurs when the local time reference of the base station is not fully synchronized to the network clock, as shown in Figure 10.13. For TDD cells with less than 3 km of radius, the ITU [9] has defined a limit of up to ± 1.5 μs of recommended synchronization error. Since the synchronization error can be positive or negative, the total synchronization error between two TDD base stations can be up to 3 μs.

The effect of this synchronization error in TDOA-based positioning has been studied in detail in Section 8.10.1 of this book. However, in order to get a rough idea, 3 μs of synchronization error can translate to 900 m of error in

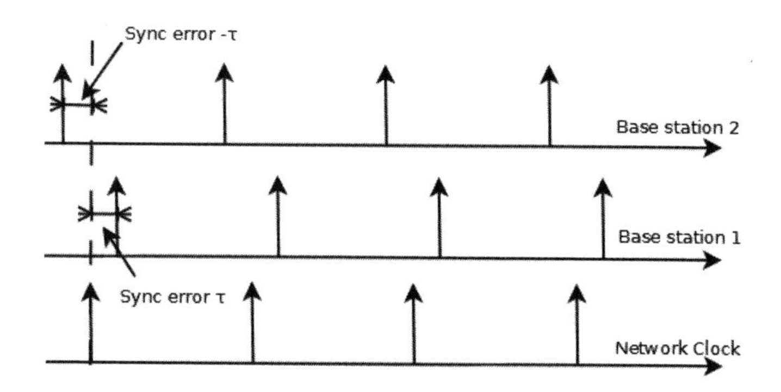

Figure 10.13 Base station synchronization error.

the range measurement based on RSTD. And this applies only to TDD networks, since there is no ITU recommendation for FDD. Thus, it is obvious that DL-TDOA positioning requires a network synchronization that is orders of magnitude better than the conventional network operation requirements. In TR 38.855 [1], the 3GPP studied the DL-TDOA positioning performance under two different network synchronization conditions: ideal network synchronization (i.e., 0 ns of error) and 50-ns synchronization error. In the latter case, the performance of DL-TDOA already could not meet some of the most stringent requirements.

So, once it is established that DL-TDOA requires a highly synchronized network, the next question is how to achieve that. The most straightforward way would be to equip every base station with an atomic clock synchronized to GNSS time. However, this implies a large cost in network deployment and maintenance. Apart from the rubidium-based oscillators used in atomic clocks, there are other types of oscillators that can be used. Typical oscillators include temperature-controlled crystal oscillators (TCXO) and oven-controlled crystal oscillators (OCXO). The TCXO can achieve up to ±100-ns accuracy, while OCXO can get up to ±50 ns. More information on the accuracy of different oscillators can be found in [10]. Nonetheless, this approach would also incur increased maintenance and deployment costs, and it does not achieve the accuracy level required to meet all 5G NR commercial positioning requirements.

An additional possibility without the need to deploy expensive additional hardware is to measure the base station synchronization error and communicate this error to the mobile device as part of the assistance data. This can be achieved for instance by taking advantage of another of the 5G NR features, IAB. IAB aims

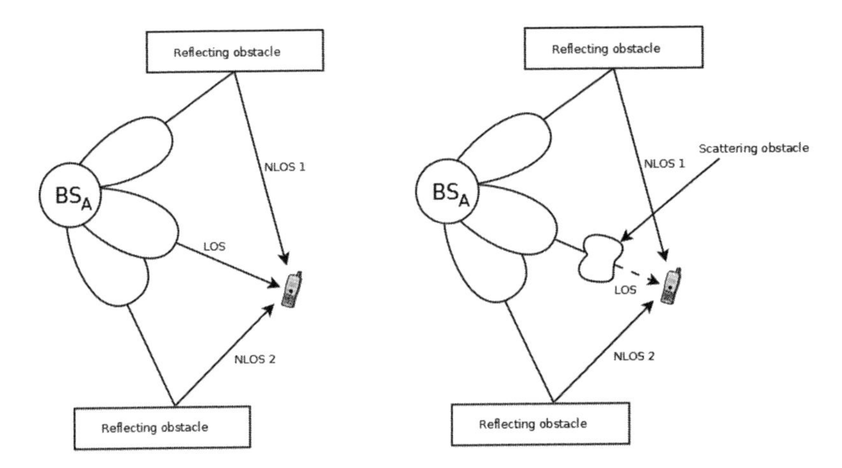

Figure 10.14 Multipath propagation.

to define an air interface to replace the fiber optics used for base-station-to-base-station communication. The main motivator for IAB is to decrease the deployment cost of the very dense base station network that will be needed for 5G NR mm-wave communications. However, this can also be used to increase the DL-TDOA positioning performance. If the base stations, using the IAB interface, listen to the PRS signal from neighbor base stations, the ToA of the signal can be used to calculate a distance between the pair of base stations. Assuming that the real distance between the base stations is known, the difference between the real and the calculated distances can be used to estimate the time synchronization error between them.

The network could maintain a database of synchronization errors between each pair of base stations and send this information to the mobile device in order to subtract the synchronization error from the positioning algorithm. The assistance data messages defined for DL-TDOA already foresee a placeholder to provide this timing error correction factor.

10.7.2 Multipath Propagation

Multipath propagation is another error source that can severely affect 5G NR positioning. Multipath occurs when the transmitted signal encounters obstacles that create reflections, diffraction, scattering, and other effects. The mobile device may receive some of these reflected or diffracted signals (typically known as NLoS

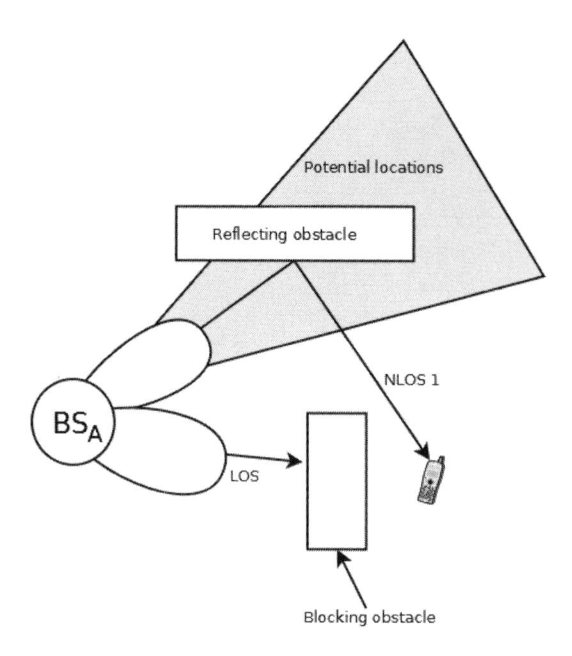

Figure 10.15 Error caused by multipath in DL-AoD positioning.

paths) in addition to the LoS or direct path signal, as shown in the left part of Figure 10.14. Furthermore, the LoS path may be affected by scattering, causing an additional delay, or even be blocked completely, as seen in the right part of Figure 10.14.

The effect of multipath in positioning depends on the type of measurements that are being performed. For timing-based measurements, if an NLoS path is selected for the measurement, the travel time of the signal is of course longer than the LoS path, adding an error to the reported measurement. A similar effect can be observed for the power-based measurements, since the NLoS components suffer higher attenuation. For the angle-based measurements, the effect can be even more devastating: if the wrong beam is selected, the positioning algorithm will think that the device is in a completely wrong direction, as shown in Figure 10.15. The LoS signal is blocked by an obstacle, and the base station considers that the reflected beam NLOS1 points to the direction where the mobile device is located. This results in the potential locations represented by the shaded area in the figure.

The 3GPP has studied the multipath effect and the potential solutions as part of the Release 16 positioning study. For DL-TDOA, one of the mitigation techniques proposed by 3GPP is to allow the configuration of up to four different DL PRS RSTD measurements for each base station pair. The measurements can be performed on different DL PRS resources or even resource sets. They are combined afterward in one single measurement report. This method can help in determining the signal path arriving first to the mobile device, which is more likely to be the LoS path.

Further multipath mitigation techniques are in scope of 5G NR positioning studies in the releases to come, as mentioned in TR 38.855 [1].

10.7.3 Other Sources

Apart from the two sources described, which will play a major role in the overall 5G NR positioning performance, there are other sources of error:

- The geometry of the base station constellation will affect the DOP and contribute to positioning performance. This has been analyzed in Chapter 8 of this book.
- Measurement quantification also plays a role in measurement accuracy. However, this role in 5G NR will be typically very small when compared to the other error sources. This error was analyzed for each positioning method separately throughout the chapter.
- The measurement accuracy of the mobile device will also influence the positioning performance for all methods that require any kind of UE measurements. Measurement accuracy will be limited by the test requirements defined in TS 38.133 [4], since all mobile devices would need to meet those requirements in order to be certified for sale. At the time of the writing, these accuracy requirements have not been finalized yet for Release 16.
- Positioning can also be affected by location database errors, which include errors in the base station geographic coordinates, in the antenna position, and so forth.

10.8 CONCLUSION

This chapter introduced the positioning technologies defined for 5G NR in the 3GPP Release 16 timeframe. There are six standardized methods: DL-TDOA, DL-AoD, UL-TDOA, UL-AoA, NR ECID, and multi-RTT.

The time difference based methods, DL-TDOA and UL-TDOA, have promising results but only if they are used in a highly synchronized network. The different options to achieve this network synchronization, both by hardware and software corrections, have been introduced briefly. On the other hand, multi-RTT technology does not require tight network synchronization, making it a very attractive alternative. The downside is a slight increase of the network load due to the additional UL traffic when compared to DL-TDOA.

Angle-based technologies have also shown very good performance and are an interesting positioning method, especially in FR2 deployments, due to the narrower beams required to improve the mm-wave link budget. However, their accuracy will depend on the width of the base station beams. Furthermore, angle-based technologies will be severely affected by multipath.

NR-ECID, which is based only on already available power measurements, may not be the most accurate option. Nonetheless, it has other advantages such as low computational effort on the mobile device side and very low latency.

Nevertheless, the 5G NR standardization work for positioning does not conclude with Release 16. The commercial requirements, as will be seen in Chapter 11 of this book, have not yet been fully covered by the existing technologies. Further studies are also required in multipath mitigation and other positioning candidates, such as carrier-phase based positioning or PDoA, could still come in future releases.

With regard to the hybridization and RAT-independent technologies, a tighter integration with RTK and RTK-PPP could also be studied in the future. Furthermore, topics such as GNSS reliability and integrity become very relevant, given that 5G NR positioning will also address safety-critical applications such as V2X or UAV control. Moreover, for IIoT applications, other KPIs such as low energy consumption take precedence over high positioning accuracy.

The 3GPP has already started a new positioning study item for Release 17, aiming to address some of the open questions. Thus, it is expected that 5G NR positioning will continue to grow and improve in the years to come.

References

[1] 3GPP TR 38.855, *Study on NR Positioning Support*, V16.0.0, March 2019.

[2] 3GPP TS 38.305, *NG-RAN; Stage 2 Functional Specification of User Equipment (UE) Positioning in NG-RAN*, V15.6.0, April 2020.

[3] 3GPP TS 38.211, *NR; Physical Channels and Modulation*, V16.1.0, April 2020.

[4] 3GPP TS 38.133, *NR; Physical Channels and Modulation*, V16.2.0, January 2020.

[5] 3GPP TS 38.215, *NR; Physical Layer Measurements*, V16.0.1, January 2020.

[6] 3GPP TS 38.214, *NR; Physical Layer Procedures for Data*, V16.1.0, April 2020.

[7] 3GPP R1-1901980, *Further Discussion of NR RAT-Dependent DL Positioning*, CATT, February 2019.

[8] 3GPP R1-1903346, *Discussions on DL Only Based Positioning*, LG, February 2019.

[9] ITU-T G.8271.1/Y.1366.1, *Network Limits for Time Synchronization in Packet Networks*, Recommendation, October 2017.

[10] Meinberg Funkuhren GmbH & Co. KG Website, https://www.meinbergglobal.com/english/specs /gpsopt.htm, accessed on February 2020.

Chapter 11

Comparison of the Positioning Technologies

11.1 INTRODUCTION

The previous chapters introduced all the different positioning technologies and their working principles. This chapter will take a look at the same technologies in a more practical manner. It will give an overview on the various aspects that play a role in the mobile ecosystem and ultimately determine whether a location technology is successful in the market or not.

The first part of the chapter summarizes the results of the 3GPP NR Positioning Study on RAT-dependent positioning technologies, which focused mainly on achievable positioning accuracy in different real-life scenarios. The results of this study are briefly explained and complemented with similar performance data for RAT-independent technologies.

The second part of the chapter focuses on other metrics beyond the scope of original 3GPP the study (e.g., battery consumption, network complexity). These additional metrics also play a major role when deciding network deployment and important aspects such as scalability, maintenance, and operational costs. A qualitative comparison on the strengths and weaknesses of each technology is given.

Finally, a summarized overview of all technologies is given, so that the reader can compare them and identify which are best suited for each particular use case.

11.2 PRIMARY METRIC: ACCURACY

The most important performance indicator when it comes to location-based services is of course positioning accuracy, which was defined in Chapter 2 of this book. Accuracy is the most critical factor in deciding whether a positioning technology is suited for a specific use case or not. Thus, this will be the primary metric used to compare the different methods.

11.2.1 Statistics

As positioning accuracy is a statistical variable, it can be expressed by a cumulative distribution function (CDF). The CDF of the horizontal error is typically given as a graph. However, in many cases, it is sufficient to evaluate the graph only at certain values (e.g., at 50%, 67%, or 90%) to obtain meaningful performance indicators. Typically the CDF of the circular error probability (CEP) is given.

If no percentage is given, CEP normally refers to CEP_{50} (i.e., 50% of the measurements are within a circle of said radius). All the studies presented in this section will refer to the CEP_{50}, unless explicitly specified.

The 3GPP 5G positioning study [1] for Rel-16 compares the position errors achieved by every RAT-dependent technology with a set of performance requirements. These requirements match the potential requirements of TR 22.862 [2].

- Regulatory requirements:
  - Horizontal CEP_{80}: $<$ 50 m
  - Vertical CEP_{80}: $<$ 5 m
  - End-to-end latency and TTFF: $<$ 30 s
- Commercial indoor requirements:
  - Horizontal CEP_{80}: $<$ 3 m
  - Vertical CEP_{80}: $<$ 3 m
  - End-to-end latency and TTFF: $<$ 1 s
- Commercial outdoor requirements:
  - Horizontal CEP_{80}: $<$ 10 m

- Vertical CEP_{80}: < 3 m
- End-to-end latency and TTFF: < 1 s

As already discussed in Chapter 5 of this book, the Rel-16 positioning requirements do not cover some of the most demanding positioning use cases. TR 22.862 states that "Requirements related to higher accuracy positioning have not been confirmed. Further work on positioning requirements is needed." In particular, the commercial services positioning requirements specified in TS 22.261 [3] are more demanding than those investigated in study TR 38.855. This has raised concerns in a part of the industry (see also [4]). Nonetheless, a Rel-17 study is already ongoing, with the aim to further improve positioning accuracy. The current plan for this study and the corresponding work item is to be completed by the end of Q2 2021.

11.2.2 RAT-Dependent Technologies: 5G Positioning Study

Instead of complex analytical evaluations, 3GPP often uses simulations to compare technologies. Several common scenarios are defined and various 3GPP members submit the results of their simulations. Multiple scenarios are necessary, as the technologies vary drastically depending on the environment they are used in. Some deliver best performance outdoors with open-sky view (e.g., in rural areas), while others rely on a dense network of transmitters. In this section, the suitability of RAT-dependent technologies in different scenarios is analyzed and their accuracy is presented. The results in this section are based on a summary of the studies under TR 38.855 [1].

The 3GPP has evaluated positioning performance for three scenarios. Each of the scenarios depicts a real-life deployment possibility with different sites or base stations. The studies are conducted at different frequencies. The different deployment scenarios are shown in Figure 11.1, Figure 11.2, and Figure 11.3. There are two types of users: vehicles and pedestrians. The percentages under each figure represent the proportion of pedestrians versus vehicles simulated for each of the scenarios.

- Indoor office, with frequencies of 2, 4, and 30 GHz; 12 sites, and a distance between sites of 20 m. It is shown in Figure 11.1.
- Urban micro, with frequencies of 2, 4, and 30 GHz; 7 or 19 sites (3 sectors each), and a distance between sites of 200 m. It is shown in Figure 11.2.

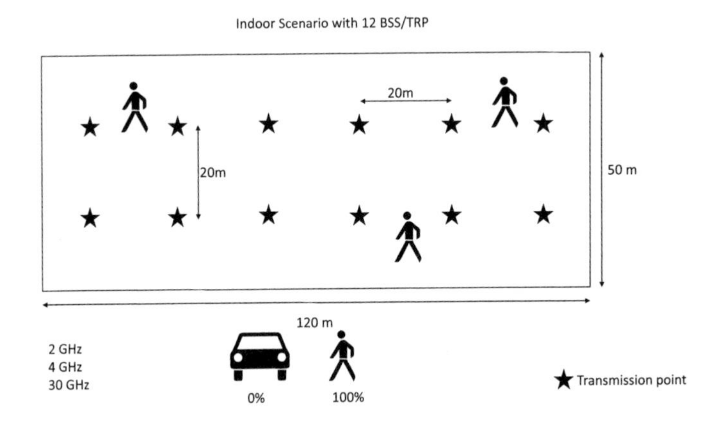

Figure 11.1 The indoor open office scenario investigated in the NR positioning study.

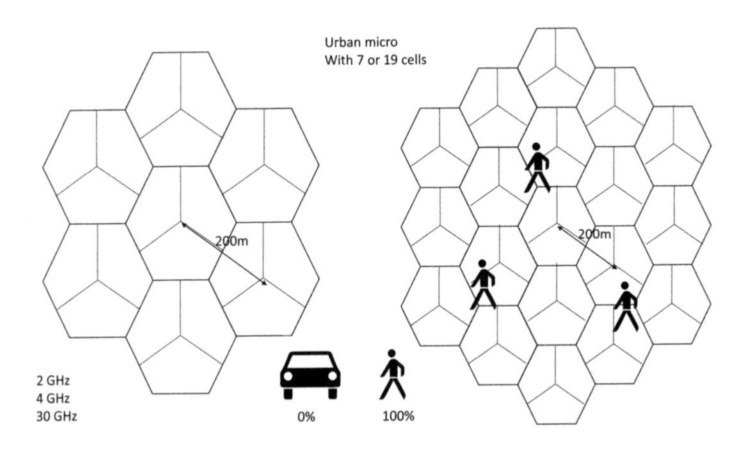

Figure 11.2 The urban micro cell scenario investigated in the NR positioning study.

- Urban macro, with frequencies of 2 and 4 GHz, 7 or 19 sites (3 sectors each) and a 500 m distance between sites. It is shown in Figure 11.3.

These scenarios are a subset of the generic scenarios used for 5G physical layer performance evaluation in TR 38.802 [5]. Some of the scenarios in TR

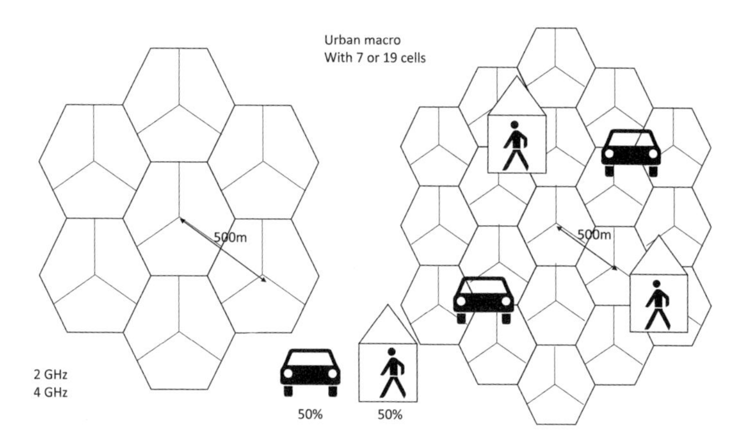

Figure 11.3 The urban macro cell scenario investigated in the NR positioning study.

38.802 were not evaluated for positioning performance, notably the rural scenario, which has significantly larger intersite distances of 1732 m and 5000 m and lower frequencies.

As the base station antenna configuration is essential for the performance of angle-based methods, it was exactly specified for each scenario by a set of seven numbers following the 3GPP nomenclature [6]: (M, N, P, Mg, Ng, Mp, Np). Each of the seven elements is defined as follows:

- M: Number of vertical antenna elements within a panel, on one polarization
- N: Number of horizontal antenna elements within a panel, on one polarization
- P: Number of polarizations
- Mg: Number of panels in a column
- Ng: Number of panels in a row
- Mp: Number of vertical transceiver units (TXRUs) within a panel, on one polarization
- Np: Number of horizontal TXRUs within a panel, on one polarization

Based on these seven parameters, TR 38.855 defines the gNB antenna configuration for each of the scenarios under consideration as follows:

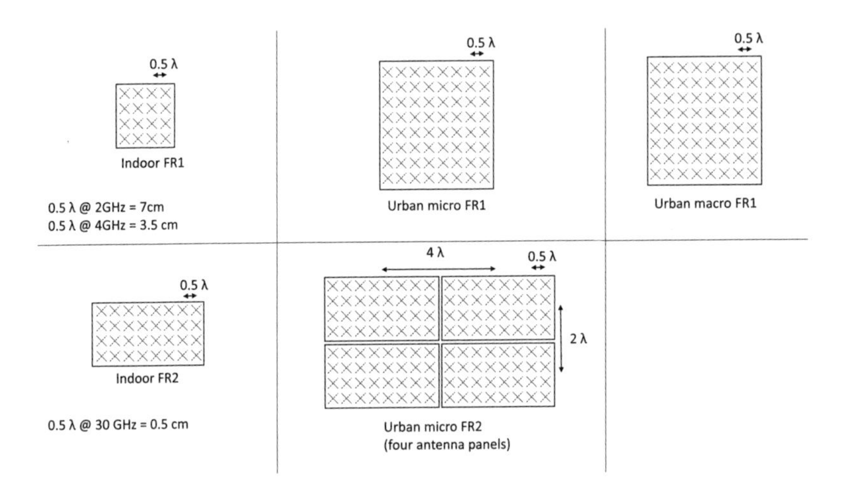

Figure 11.4 Overview over the BS antenna arrays assumed for the 3GPP NR positioning study.

- Urban macro, < 6 GHz: (M, N, P, Mg, Ng) = (8, 8, 2, 1, 1)
- Urban micro, < 6 GHz: (M, N, P, Mg, Ng) = (8, 8, 2, 1, 1)
- Indoor, < 6 GHz: (M, N, P, Mg, Ng) = (4, 4, 2, 1, 1)
- Urban micro, > 24 GHz: (M, N, P, Mg, Ng) = (4, 8, 2, 2, 2)
- Indoor, > 24 GHz: (M, N, P, Mg, Ng) = (4, 8, 2, 1, 1)

As these numbers are hard to visualize, Figure 11.4 depicts the simulated base station antenna arrays. To sum up, for frequencies below 6 GHz indoors, a massive MIMO antenna with 16 antenna elements was assumed, while for the urban macro and micro scenarios, a 64-antenna element array was assumed. For frequencies above 24 GHz indoors, a 32-element array antenna was assumed, while for the urban micro an antenna consisting of four 32-element array antennas was assumed. The urban macro scenario was not investigated for frequencies above 24 GHz.

Table 11.1 details the results of the 3GPP study for the various RAT-dependent technologies. It is worth noting that for the TDoA methods, the 3GPP considered two different synchronization errors between base stations; that is, 0 ns, or no error, and 50 ns. As seen in Chapter 10 of this book, the synchronization between base stations plays a very critical role in the accuracy of TDOA.

Table 11.1
Study Results

Scenario	Indoor Reg.	Indoor Com.	Micro Reg.	Micro Com.	Macro Reg.	Macro Com.
DL-TDOA 50 ns	12/12	0/6	12/12	0/6	12/12	0/6
DL-TDOA 0 ns	12/12	12/12	12/12	12/12	12/12	7/12
DL-AoD FR2	1/1	1/1	1/1	-	-	-
DL-AoD+ZoD	-	-	0/1	-	-	-
DL-AoD+ZoD+TA	-	-	1/1	-	-	-
DL-TDOA+AoD 50 ns	1/1	0/1	-	-	-	-
DL-TDOA+AoD 0 ns	1/1	1/1	-	-	-	-
UL-TDOA 50 ns	8/8	1/4	6/6	0/4	3/5	0/4
UL-TDOA 0 ns	8/8	4/4	6/6	4/6	3/5	3/5
UL-AoA	3/3	3/3	1/1	0/1	0/1	-
Multi-RTT	5/5	5/5	5/5	5/5	2/2[1]	0/2
SC-RTT+UL-AoA	1/1	1/1	1/1	1/1	1/1	0/1

Note 1: only one out of two studies showed compliance for indoor UEs
Note 2: "Reg." stands for regulatory and "Com." for commercial requirements.

For each technology/scenario combination, Table 11.1 gives the number of studies conducted and the number of studies that found that the requirements can be met. For instance, for UL-TDOA with 0-ns sync error, five studies analyzed the urban macro scenario and three of these came to the conclusion that the commercial requirements can be met. In most cases, all studies had similar findings, but as in the example above there is also some dissent visible. From this evaluation it becomes obvious that a 12/12 result (12 studies conducted, all stating that requirement can be met) gives more confidence than a 1/1 result.

Several conclusions can be drawn from Table 11.1 and the studies summarized under TR 38.855:

- Regulatory requirements can be met with most technologies, even with 50-ns base station synchronization uncertainty.
- AoD and AoA technologies fulfill the commercial criteria as defined in TR 38.855 for the indoor scenario.
- Assuming 50-ns sync error:
  - Only multi-RTT and serving cell round-trip time (SC-RTT)+AoA fulfill the commercial criteria as defined in TR 38.855 for urban micro.

- No technology fulfills the commercial criteria for the urban macro scenario.

11.2.3 RAT-Independent Technologies

Technologies that do not directly utilize the NR radio link for position measurements are usually not considered by 3GPP RAN1 for studies. Therefore, the author had to rely on non-3GPP studies for the accuracy of other technologies. Hence, the results in this section may not be one-to-one comparable to the results of the study TR 38.855 used for RAT-dependent technologies.

11.2.3.1 GNSS and Hybrid GNSS+TDOA

Even though usually considered to be out of scope for RAN1 studies, three 3GPP members still submitted a joint analysis on the hybrid use of GNSS+TDoA [7]. The results of this study were not considered in TR 38.855, as no GNSS scenarios were agreed upon in advance. The GNSS scenario was based on dual-frequency L1+L5 with Galileo, GPS, GLONASS, and BeiDou.

The hybrid simulation indicated that GNSS+TDoA (without base station sync error) would fulfill the horizontal commercial requirements for outdoor usage. Furthermore, even GNSS alone would fulfill the horizontal requirements (see Table 11.2).

Table 11.2

GNSS and HYBRID GNSS Accuracy

Scenario	Micro H_{err}	Macro H_{err}	Micro V_{err}	Macro V_{err}
GNSS	7.19 m	3.72 m	4.55 m	5.48 m
GNSS, TDOA 0 ns	1.46 m	1.24 m	3.76 m	4.41 m
GNSS, TDOA 50 ns	6.61 m	3.71 m	4.53 m	4.98 m

11.2.3.2 Barometric

Based on the FCC's request for vertical caller location capabilities, the organization of North American network operators known as CTIA founded a test-bed company. This company conducted an extensive study on barometric pressure performance and summarized the results in a report [8]. Two vendors for barometric solutions were evaluated with several commercial-grade mobile phones. Both indoor and outdoor performance was analyzed, also including the effects of air-conditioned, sealed buildings. The vertical position error is summarized in Table 11.3 (see as well [8], Section 11.2).

Table 11.3

Barometric Vertical Accuracy

	NextNav	Polaris Wireless
Vertical error (80% of cases)	1.8 m	4.8 m

Both vendors fulfill the regulatory requirement and one vendor demonstrates that it is possible to achieve the commercial requirements as well. Further studies on z-axis accuracy are ongoing at the time of writing (Stage Za and Stage Zb [9]), but were not available for evaluation.

11.2.3.3 WLAN

The European Project HELP112 conducted an analysis on the performance of WLAN location vs. A-GNSS and network location (cell ID) using the AML solution, seen in Chapter 3 of this book [10]. The average location error was given instead of the 80% percentile that was given in the 3GPP and FCC studies. Therefore the data in this study is not one-to-one comparable to the other studies in this chapter. Nevertheless, it gives an idea about the relative performance of WLAN location in comparison to A-GNSS. Data was taken from a live network trial conducted in four European countries and is summarized in Table 11.4.

Table 11.4

WLAN, GNSS and Cell ID Performance in Field Tests, Average Accuracy

Country	Cell-ID	GNSS	WLAN
Austria	1550 m	6 m	20 m
Italy	1377 m	28 m	24 m
Lithuania	5506 m	21 m	35 m
United Kingdom	1983 m	14 m	24 m

11.3 ADDITIONAL METRICS

While 3GPP has spent significant effort on the study of positioning accuracy in various scenarios, other aspects were not analyzed as part of the Release 16 studies. Due to the limited number of studies available, these deployment considerations are only qualitatively analyzed in this chapter. The 3GPP Release 17 study item on positioning technologies is targeting a deeper analysis on some of the metrics mentioned here, as for instance the energy and battery consumption.

11.3.1 Time-to-First-Fix

The TTFF measure is self-explanatory. The TTFF measures the time elapsed from the moment when a positioning measurement is requested until the moment when the positioning results are available in the E-SLMC or the LMF. This includes not only the measurement time, but also the latency of the network. The classification of the technologies in this chapter is based on the following categories for TTFF:

++ Immediately available (if UE registered in network)

+ <1 s (commercial service requirement)

- <30 s (regulatory service requirement)

-- >30 s

Only a few technologies are able to meet a (cold start) TTFF of less than 1 s. However, all technologies are able to meet the regulatory service requirements of 30 s (see HELP 112 [10], FCC test beds). Once the initial fix is available, further fixes can typically be achieved in less than 1 s, meeting the commercial requirements.

11.3.2 UE Energy Consumption

The technologies vary significantly in the energy required by the mobile device per fix. While some do not add any (additional) energy consumption, as they use measurements that occur anyway during the regular operation of a mobile device or are performed on the network side, others involve significant signal processing in the UE with the sole purpose of acquiring a position fix (e.g., GPS).

With the reduction of structure size on silicon, energy consumption is reduced with every new chip generation. Even though absolute power figures will be outdated soon due to advancements in technology, the following figures might be helpful to get an idea of how the technologies perform relative to each other. The following estimates are based on the same generation of devices.

11.3.2.1 GNSS Technologies

Values from a data-sheet [11] and a presentation [12] of two energy-efficient chipsets are given in Table 11.5 for the comparison of energy consumption.

Table 11.5

Energy Consumption of Two Efficient GNSS Chipsets

Parameter	u-blox M8M	Sony CXD5603GF
Power: Acquisition (1 GNSS)	57 mW	11 mW
Power: Tracking (1 GNSS)	14.4 mW	6 mW
Energy: Estimate unassisted cold fix (30 s acquisition)	1710 mWs	330 mWs
Energy: Estimate assisted fix (6 s acquisition)	342 mWs	66 mWs

In unassisted mode, assuming a good sky view, the acquisition typically takes 30 s. In assisted mode, the typical "on" time is a fraction; here 6 seconds are assumed. Energy consumption scales down proportionally if the energy for the transmission of the assistance data is neglected. However, in the case of an open-sky view, the energy used to transfer the assistance data might even exceed the savings, meaning that standalone GNSS under optimal conditions could use less power than A-GNSS. Nonetheless, under obstructed sky view, a cold start may take significantly more than the 30 s mentioned above. In that case, A-GNSS will use less energy and achieve a higher yield than the standalone GNSS mode.

11.3.2.2 Cellular Technologies

As there are no studies available on the energy consumption of TDOA-based positioning to the author's knowledge, a coarse analytical estimate is derived in this section to make an educated guess about the energy consumption. For the evaluation of the energy consumption of cellular devices, Table 11.6 is used. It originates from Table 7.1.7.4-1 in [13], a 3GPP Technical Report on the NB-IOT technology, which is the most power-efficient member of the 3GPP technology family.

For DL-only technologies, no transmit operation is necessary (assuming the UE-based mode, which has been incorporated in 5G). Therefore, power consumption is assumed to be similar to a normal receive mode with 60 mW when the UE

Table 11.6

Energy Consumption of a Typical Cellular NB-IOT Modem in Different Operation Modes

Operating mode	Power (mW)	Notes
Transmit (+23 dBm) integrated PA	500	+23 dBm with 45% PA efficiency for class B (including Tx/Rx switch insertion loss) plus 60 mW for other circuitry.
Transmit (+23 dBm) external PA	460	+23 dBm with 50% PA efficiency for class B (including Tx/Rx switch insertion loss) plus 60 mW for other circuitry.
Receive synchronization (PSS/SSS)	70	Accounts for more complex digital processing during synchronization.
Receive normal (non-PSS/SSS)	60	Includes digital mixing/decimation to single 15-kHz subchannel, and subsequent demodulation of this subchannel.
Sleep	3	Corresponds to maintaining accurate timing by keeping RF frequency reference active.
Standby	0.015	Common assumption.

needs to explicitly measure DL signals (for monitoring DL-TDOA PRS). The duty cycle for DL and UL technologies depends on the network parameters. In this example, according to the study in TR 38.855 ([1], Table 8.1.1.1-1), nine positioning occasions are used per fix, each with a duration of 1 ms. One position fix ideally adds only 9 ms of additional time in receive mode for a UE, which would lead to approximately 0.54 mW of additional energy consumption.

UL-TDOA requires the UE to transmit a SRS. Assuming the more efficient power amplifier from Table 11.6, the power is in the order of 460 mW if the UE transmits at full power. As multiple neighbor stations need to receive the signal for positioning, it is a reasonable assumption that the UE will typically transmit SRS for positioning at full power. According to TR 38.855 ([1], Table 8.2.1.1-2), four occasions of 4 symbols length each were used for a position fix. Assuming a subcarrier spacing of 15 kHz, a symbol has a length of approximately 0.0714 ms. This leads to a duration of $4 * 0.0714$ ms ≈ 0.286 ms, per occasion and a total of $4 * 0.286$ ms ≈ 1.14 ms of total transmission time. This results in a total energy of 1.14 ms $*460$ mW ≈ 0.52 mWs for sending the SRS signal.

Multi-RTT requires transmission of SRS and reception of PRS. The measurement time assumed here is of the same length as in the DL-TDOA and UL-TDOA

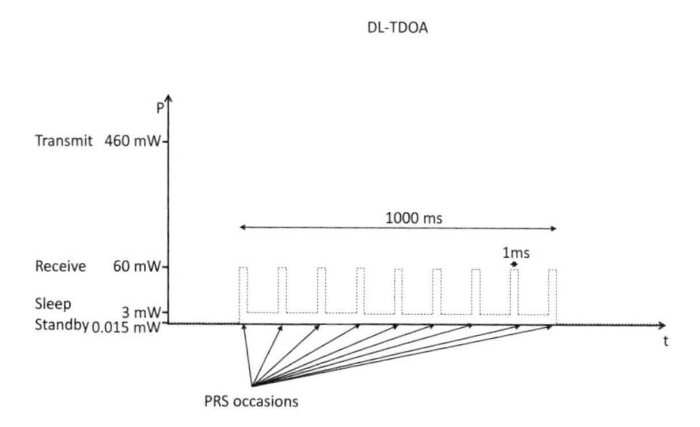

Figure 11.5 Energy consumption of DL-TDOA.

examples (1 s). For this estimate, we add the energy for transmission of SRS and reception of PRS.

For all three technologies, DL-TDOA, UL-TDOA, and multi-RTT, the UE must maintain an accurate timing in between the PRS and SRS occasions. This means that it must remain at least in sleep mode for the whole measurement time and cannot go into the lower power standby mode. Assuming a sleep time of roughly 1 s in all three cases, we obtain 1 s $*3$ mW $=$ 3 mWs (neglecting that the sleep time is actually a little bit shorter, as the UE is in receive or transmit mode for a few milliseconds).

Table 11.7

Coarse Estimate for Energy per Cellular Positioning Fix

	DL-TDOA	UL-TDOA	Multi-RTT
Energy per fix	3.54 mWs	3.52 mWs	4.06 mWs

Even though this analytical estimate, which is summarized in Table 11.7, is rather an educated guess and the numbers are highly dependent on actual network configuration and UE implementation, it can be assumed that DL-TDOA, UL-TDOA and multi-RTT add only a little extra energy consumption to the regular

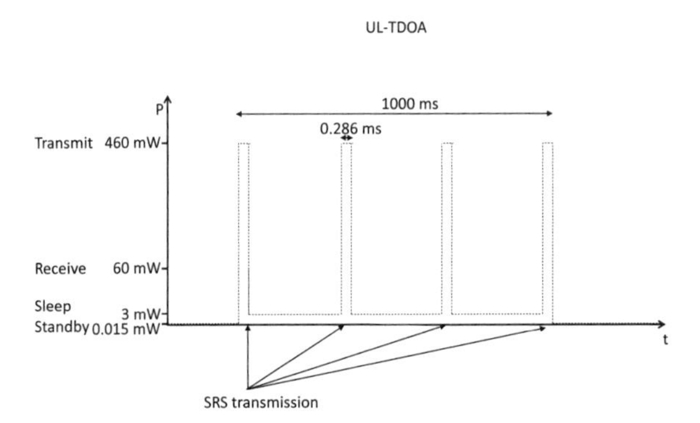

Figure 11.6 Energy consumption of UL-TDOA.

modem operation. They typically require significantly less energy than an (A-) GNSS fix.

11.3.2.3 WLAN Positioning

RSSI-based WLAN positioning requires a scan for AP beacons over the available bandwidth (around 80 MHz at 2.4 GHz and around 600 MHz at 5 GHz in the United States). Furthermore, a WLAN client typically sends a probe request on each channel to invoke a timely response instead of waiting for the beacon signal. According to a white paper [14] of a company specializing in WLAN location, a low-power optimized device's energy consumption is in the order of 50-100 mWs per scan. Of course, it is possible to use this method only opportunistically, when a WLAN scan is necessary for other purposes, but to achieve a timely result, it may still be necessary to trigger a scan for the sole purpose of positioning.

Time-of-flight-based WLAN positioning requires the exchange of multiple messages with multiple APs. The energy consumption per fix can be assumed to be higher than for RSSI-based WLAN.

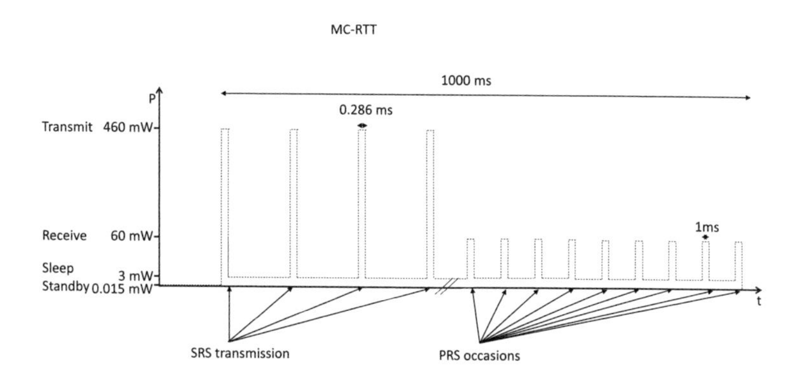

Figure 11.7 Energy consumption of multi-RTT.

11.3.2.4 IMU

Using an IMU for navigation requires the continuous usage of three gyroscopes and three accelerometers at high rates (e.g., 200 Hz). According to a presentation [15], these sensors incur a significant power drain of around 90 mA. Multiplied by a typical Li-Ion voltage of 4.2 V, one obtains 378 mW of continuous power consumption. Even though those figures are from 2009 and might have improved in the meantime, IMU navigation would be very energy consuming compared to other methods, especially if only occasional fixes are required.

11.3.3 Network Load

Another important aspect to take into account when comparing the different positioning methods is the traffic load they impose on the network. Technologies that depend on bidirectional communication for positioning measurements increase the network load with each additional positioning request. This imposes a capacity limit in terms of location server load and radio link usage. Technologies that use only information that is available anyway (AoD, AoA, ECID) do not impose an additional limit other than the general limit of regular simultaneous operation of devices of the network.

On the other hand, passive positioning relies solely on the DL and does not require data transmission (in case the location is only required in the UE). Thus, it does not have such a capacity limit. Besides the scalability advantage of passive location methods, they might also enhance the performance in the case of low signal power levels: even though the signal might not be strong enough for a communication link, it might still be useful for a measurement. Another advantage is the inherent low latency, as no communication is involved.

A prominent example of passive positioning is the standalone GPS and its GNSS siblings, which can cope with any number of users, as each user only listens passively. A-GNSS requires the assistance data exchange, which was a bidirectional communication up to Rel-15. However, Release 15 introduced assistance data broadcasting, so currently A-GNSS can also be used passively. TDOA methods can also benefit from the broadcasting possibilities. However, the base station coordinates need to be broadcasted as well. Nonetheless, UE-based DL-TDOA is already supported in 5G as of Release 16. Incidentally, cell ID could also be used in a passive way with this information.

However, for cellular network positioning technologies, it is worth noting that the network capacity is also reduced by the periodical transmission of specific positioning signals such as the PRS. Some methods rely on the continuous transmission of such signals, while others only require these signals on demand, while the location function is actually used.

11.3.4 UE Complexity

In general it is desirable to reduce the UE complexity for a given functionality. Methods like Cell ID or ECID add no or very little extra UE complexity (mainly software extensions), as the measurements are still needed for the regular UE operation. The transmission of SRS signals for UL-TDOA is also no additional implementation effort (unless 3GPP decides to introduce a special positioning SRS signal). On the other hand, DL-TDOA typically relies on measuring additional positioning signals (PRS), which increases the UE's SW complexity. If DL-TDOA is used in multiple frequency bands (interfrequency), it also requires calibration of the RF front-ends and further synchronization effort.

Finally, GNSS adds the most complexity, as it requires one or two dedicated antennas, RF front-ends and hardware correlators. Despite this, in the smartphone market, GNSS is currently an undisputed part of each position solution due its accuracy. However, for the extremely price- and power-sensitive IoT market, some implementations try to avoid the usage of GNSS.

11.3.5 Base Station Complexity

As gNBs (the 5G RAN) constitute the vast majority of a network operator's hardware investment, a low impact of positioning in this component is very important.

Most RAT-independent technologies use the mobile network mainly for data transfer and do not have any impact on gNB complexity (WLAN, barometric pressure, coarse time A-GNSS). For highly optimized Fine Time A-GNSS (with SIB9), the gNB needs to transmit additional SIBs, slightly increasing the complexity.

DL-TDOA increases the complexity further, as PRS signals need to be scheduled, potentially only on-demand of the location server. UL-TDOA and multi-RTT incur the highest complexity addition in the gNB, as neighbor gNBs need to be extended to receive signals from UEs, which are not served by them and hence are not closely synchronized to these gNBs. UL-AoA and DL-AoD introduce no significant complexity to (massive MIMO) gNBs, as the angular information is required for beamsteering during normal operation.

Positioning methods may also require a network synchronization level much tighter than required for any other 5G service (DL-TDOA and UL-TDOA). This might require very expensive hardware, increase the cabling effort of cell sites, or incur continuous effort for maintenance. Obviously, this is undesirable from a network operator point of view, as it raises the cost.

RAT-dependent technologies also require knowledge of the exact antenna location (vertical and horizontal) and orientation at each cell site. This may complicate the network roll-out process or add additional cost for measuring each site afterward.

11.4 TECHNOLOGY COMPARISON

For a compact overview, each of the metrics introduced so far is mapped to a simple grading scheme. Each technology is then evaluated against this scheme in the next section. Effort refers to the amount of work and the costs associated with the deployment and maintenance of each technology.

- gNB complexity:

  ++ Possible in all 3GPP-compliant gNBs

  + Low extra effort (mainly core affected)

  - Higher effort (RAN impacted)

    -- Very high effort (e.g., special care must be taken during installation or operation, e.g., sync)

- UE complexity:

    ++ Possible in all 3GPP-compliant UEs

    + Low extra effort (mainly software affected)

    - Higher effort (physical layer impact)

    -- Very high effort (e.g., extra RF path, calibration)

- Network capacity usage:

    ++ No network capacity needed in addition to messages that are exchanged anyway

    + Negligible usage of network capacity; only needed during positioning

    - Noticeable usage of network capacity; only needed during positioning

    -- Permanent usage of network capacity, even if no position session is ongoing

- Accuracy (horizontal CEP, typically):

    ++ <10 m (commercial requirements)

    + <50 m (regulatory requirements)

    - <500 m

    -- >500 m

- Battery drain per cold start fix (additional to regular UE operation):

    ++ 0 mWs

    + < 10 mWs

    - < 100 mWs

-- > 100 mWs

- TTFF (e.g., position available in E-SMLC/LMF):

  ++ Immediately available (if UE registered in network)

  + <1 s (commercial service requirement)

  - <30 s (regulatory service requirement)

  -- >30 s

11.4.1 Technology Matrix

This section's main element is Table 11.8, which gives a high-level overview over the technologies discussed so far in this chapter. Due to the size limitation, some abbreviations were used in Table 11.8:

- Avail.: Availability of the technology
  - O: Outdoor coverage
  - I: Indoor coverage
  - U: Urban coverage
  - R: Rural coverage
- Bat.: Battery impact
- Acc.: (Position) accuracy
- TTFF: time-to-first-fix
- Netw. capac.: Network capacity reduction due to position technology

11.4.2 Notes per Technology

This section summarizes all the different positioning technologies with a brief description of their strengths and weaknesses.

Table 11.8

Overall Deployment Considerations for Each Technology Discussed

	Netw. capac.	Passive	Bat.	Acc.	TTFF	Cmplx. UE	Cmplx. gNB	Avail.
Cell ID	++	Yes	++	−−	++	++	++	R/U/I/O
eCID	+	Yes	++	−	+	++	++	R/U/I/O
eCID RxTx	+	No	++	−	+	+	−	R/U/I/O
IMU	++	Yes	−−	−−	++	−−	++	R/U/I/O
Barometer	+	No	+	++	+	−	++	R/U/I/O
A-GNSS	−	Yes	−−	++	−	−−	++	R/U/O
WLAN RSSI	+	Yes	−−	+	+	+	++	U/I
WLAN ToF	+	No	−−	++	+	−	++	U/I
UL-AoA	++	No	++	++	+	++	−	U/I/O
DL-AoD	++	No	++	++	+	++	−	U/I/O
UL-TDOA	−	No	+	+	+	++	−−	U/I/O
DL-TDOA	−−	Yes	+	+	+	−	−−	U/I/O
Multi-RTT	−−	No	+	++	+	−	−−	U/I/O

11.4.2.1 Cell ID

Cell ID requires no more than a database with the latitude and longitude of each cell site. If an operator stores this information anyway, this method causes virtually no extra expenses. As the Cell ID is necessary for the normal operation of a mobile, it adds no extra complexity on the device side.

This technology can be considered the baseline for mobile positioning, since it is virtually always available. However, the accuracy equals the cell size and is therefore not sufficient for commercial or regulatory requirements according to TR 38.855.

11.4.2.2 Enhanced cell ID

The basic cell ID can be enhanced by signal power and quality measurements of neighbor cells. These are readily available without any additional positioning measurements for all 3GPP-compliant devices. Usage of the UE-RxTx time difference measurement with the serving cell involves additional base station capabilities. Generally, the impact on the device and the network is still low. A location server is required to perform a location fix. The accuracy is still typically not sufficient for commercial and regulatory applications, since it uses only one timing measurement and it relies on power measurements to complete the equation system. Its enhancement, multi-cell RTT, offers much better accuracy.

11.4.2.3 UL-AoA, DL-AoD

UL-AoA and DL-AoD only deliver good results if beamforming (i.e., massive MIMO) is used. Its accuracy improves with more antenna array elements, and therefore the best results can be expected with mid- and especially high-frequency bands. In contrast, multipath propagation will severely impact its accuracy.

11.4.2.4 UL-TDOA

The requirement for UL-TDOA is a time-measurement unit, typically colocated or integrated in the base station software. Its accuracy depends on network synchronization (see also the next section, DL-TDOA). On the UE side, no extra complexity is needed, as regular sounding reference signals can be used, which are necessary for other network functionality. Due to the limited power of the UE's UL signal, the number of neighbor cells that can be used for a measurement is typically smaller than the number of neighbor cell signals visible to the UE in the DL-TDOA case. The network capacity is impacted, as neighbor cells should ideally not schedule any data transmission of other UEs while one UE transmits SRS for positioning.

11.4.2.5 DL-TDOA

Broadcasting of PRS by the base station reduces the available cell bandwidth for other purposes. While PRS had to be transmitted constantly in 4G, regardless of whether a UE used them, 5G can also only send PRS on demand. This limits the network capacity impact. In 4G, according to [16], Table 5-2, the capacity decrease is between 0.04% and 3.75% depending on the configuration.

As PRS signals are exclusively used for positioning purposes, supporting DL-TDOA increases the UE complexity and not all UEs support this method (e.g., Release 15 devices do not support NR-PRS). Supporting DL-TDOA on multiple frequency bands increases the complexity in the UE even further, as it requires precise calibration of RF group delays.

One of the main drawbacks of DL-TDOA is that it requires a highly synchronized network. A synchronization error of 50 ns renders this technology already useless for the commercial requirements as defined in TR 38.855. Nonetheless, 5G features, such as the IAB, can help improve the synchronization of the network without the need for additional expensive hardware.

11.4.2.6 Multi-RTT

Multi-RTT (RxTx measurements to multiple base stations) can be considered a combination of UL-TDOA and DL-TDOA from a complexity point of view. The UE complexity is similar to DL-TDOA and the UL-power limits the performance in sparse network configurations with few neighbors like in UL-TDOA. The advantage of this scheme is its independence from precise network synchronization. While UL- and DL-TDOA's performance is impacted by small errors in the nanosecond range, multi-RTT's performance is not limited by the network's synchronization.

The downside to multi-RTT is the need to send UL transmissions to multiple base stations, increasing the load of the network.

11.4.2.7 WLAN RSSI

Similar to E-CID, this method relies on power measurements of neighbor WLAN APs. A database of WLAN AP locations is required, which might not be available at a network operator. Such a database is typically provided by specialized third parties or mobile operating system vendors. Newer WLAN APs can broadcast their coordinates, but this approach requires professional installation of access points. The performance of WLAN RSSI is typically better than cellular ECID because of the massively higher number of WLAN APs compared to cellular network sites as of 2019. As RSSI is a power measurement, it is inherently less accurate than time-based approaches (e.g., GNSS, WLAN time-of-flight, or multi-RTT).

11.4.2.8 WLAN Time-of-Flight

WLAN time-of-flight (ToF) utilizes IEEE 802.11-2016 Fine-Time-Measurements [17] to estimate the round-trip time between a UE and a WLAN AP. It is similar in its principle to multi-RTT and can reach very good accuracy (1-2 m, see [18]). However, the WLAN APs need to support this protocol. As of 2019, only a handful of AP models support WLAN RTT measurements. Therefore, its usability depends on the proliferation of newer WLAN APs. Being a RAT-independent technology, the impact on the base station is minor.

11.4.2.9 A-GNSS

The top-IP variant (SUPL) of A-GNSS is particularly attractive for network operators, as it does not involve any network modification. The actual location server may be located on the internet and maintained by a third party (OS vendor, GNSS chipset vendor). Alternatively, the network operator may run its own server, but this server may remain independent of the mobile network and hence be easier to maintain.

If assisted GNSS data is transferred via the control plane protocols, the MME/AMF and the location server in the core are involved. The location server needs continuous data from an A-GNSS monitoring network or needs a GNSS receiver. If an operator chooses to implement coarse-time assistance, only the location server needs a fairly tight sync, but the rest of the network may remain not synchronized. For fine-time assistance, a relatively well-synchronized (10 μs) network is required and the base stations need to report their time offsets to the location server.

On the device side, significant effort is necessary, for example, a dedicated GNSS antenna(s), RF-frontend(s) and GNSS correlators on silicon, either in a dedicated GNSS chip or added to the SoC.

11.4.2.10 IMU

Inertial-navigation theoretically holds the promise of being an infrastructureless, ubiquitous positioning method. In practice, 2019's state-of-the art micro electro mechanical systems (MEMS) gyroscopes and accelerometers are orders of magnitude away of being usable for navigation. Due to their drift, a location will move away in a few seconds. However, they can be used to keep the position during a short outage of other positioning methods (e.g., in a tunnel).

11.4.2.11 Barometric Pressure

Limited to the vertical axis by its nature, barometric pressure sensing can achieve good accuracy even with low-cost sensors. It is important that the barometric sensors in the mobile devices are appropriately calibrated, and a network of reference pressure sensors is required to compensate the effects of the weather. Being RAT-independent, the impact on the cellular network is low in terms of complexity and capacity.

11.5 CONCLUSION

The manifold aspects of positioning are expressed by the various metrics against which each technology can be measured. The evaluation and comparison of the planned 5G positioning technologies shows that there is (still) no one-size-fits-all technology. Like in 4G, depending on the actual task, different technologies may be suitable. It is necessary to select the appropriate technology for the task, either optimizing accuracy, yield, battery consumption, or time-to-fix. In many cases, it is also beneficial to use two methods, as they are complementary to each other, such as WLAN-RSSI and GNSS. While GNSS performs best outdoors in open-sky scenarios, it is useless in deep indoor scenarios. On the other hand, WLAN is most useful in indoor scenarios with many APs in the vicinity and not useful in rural areas. Furthermore, the hybridization of several technologies, such as SC-RTT and UL-AoA, seems promising to reach the demanding positioning requirements laid out for 5G.

It must be noted that the higher levels of commercial requirements (< 0.3 m horizontal and vertical) in TS 22.261 [3] are most likely not achievable in all scenarios by the technologies introduced in 5G Rel-16.

For rural and suburban environments, high-accuracy GNSS technologies such as RTK and its variants might achieve this accuracy. For dense urban and indoor scenarios, there is still a performance gap that might demand more advanced technologies, which might be introduced with future 3GPP releases. Several advanced RAT-dependent technologies with the potential to achieve the required accuracy were proposed but not included in Release 16 due to its complexity and the tight timelines. A few examples were seen in Chapter 10 of this book, as for instance carrier phase based positioning. 5G positioning is still a work in progress, and positioning enhancements beyond Release 16 will surely come in the near future.

References

[1] 3GPP TR 38.855, *Study on NR Positioning Support*, V16.0.0, March, 2019.

[2] 3GPP TR 22.862, *Feasibility Study on New Services and Markets Technology Enablers for Critical Communications; Stage 1*, V14.1.0, October, 2016.

[3] 3GPP TS 22.261, *Service Requirements for Next Generation New Services and Markets*, V16.9.0, September, 2019.

[4] 3GPP RAN1 R1-1902190, *Considerations on RAT Independent and Hybrid Positioning for NR*, February, 2019.

[5] 3GPP TR 38.802, *Study on New Radio Access Technology Physical Layer Aspects*, V14.2.0, September, 2017.

[6] 3GPP TR 38.901, *Study on Channel Model for Frequencies from 0.5 to 100 GHz*, V16.0.0, October, 2019.

[7] 3GPP RAN1 R1-1902549, *TP on Hybrid Positioning and GNSS Enhancements for TR 38.855*, February, 2019.

[8] 9-1-1 Location Technologies Test Bed, LLC, *Report on Stage Z*, 2018.

[9] 9-1-1 Location Technologies Test Bed Seeks Vendor Participants for 2020 Stage Zb Campaign http://www.911locationtestbed.org/Stage_zb.html, accessed on December, 2019.

[10] European Commission, "Deliverable D5.1 Final Report on Task 4 End-to-end Pilots" of *Pilot Project on the Design, Implementation and Execution of the Transfer of GNSS Data During an E112 Call to the PSAP*, July, 2017.

[11] NEO-M8, *u-blox M8 Concurrent GNSS Modules Data Sheet*, UBX-15031086 - R05, January, 2020.

[12] Kazukuni, T., et al, "Sony's CXD5603GF The Lowest Power Consumption GNSS Chip in the IoT Tracker Market," *Proc. 31st International Technical Meeting of the Satellite Division of the Institute of Navigation (ION GNSS+ 2018)*, Miami, Florida, September 24-28, 2018, pp. 515-537.

[13] 3GPP TR 45.820, *Cellular System Support for Ultra-Low Complexity and Low Throughput Internet of Things (CIoT)*, V13.1.0, December, 2015.

[14] Skyhook, *Wi-Fi Scanning, Power Consumption and Bandwidth for LPWAN*, https://www.skyhook.com/wi-fi-scanning-power-consumption-lpwan, accessed on December, 2019.

[15] Sharkey, J., *Coding for Life - Battery Life, That Is*, 2009, https://de.scribd.com/document/16917367/Coding-for-Life-Battery-Life-That-Is, accessed on December, 2019.

[16] Fischer, S., *Observed Time Difference of Arrival (OTDOA) Positioning in 3GPP LTE*, Qualcomm Technologies Inc., June, 2014.

[17] IEEE Std 802.11-2016 *Part 11: Wireless LAN Medium Access Control (MAC) and Physical Layer (PHY) Specifications*, December, 2016.

[18] Wi-Fi Location: Ranging with RTT https://developer.android.com/guide/topics/connectivity/wifi-rtt, accessed on December, 2019.

Chapter 12

Other Positioning Technologies: Sensors

12.1 INTRODUCTION

As discussed in earlier chapters, in particular Chapters 3 to 5, the need to accurately locate a wireless device has mandated the usage of technologies beyond GNSS. As shown in Chapter 4, a multitude of technologies are combined together in order to provide a navigation system the ability to accurately estimate its location and safely navigate its environment.

In the previous chapters of this book we focused on GNSS and its enhancements, cellular network-based positioning technologies such as OTDOA or ECID and noncellular network technologies, such as Wi-Fi or Bluetooth. However, there is one fundamental part of positioning systems, especially for navigation applications, which is still missing: sensors.

There are numerous sensor systems involved in location-based services. Each of them accomplishes a certain function in localization, environment recognition, and navigation. This chapter will start with the typical sensors that can be found in a mobile phone and are already used for positioning today: the IMU and the barometer. The second part of the chapter deals with other sensor systems that play a minor role in smartphones but are very important to new technologies, such as V2X and UAV, as seen in Chapters 4 and 5 of this book. These sensors include, among others, ultrasonic sensors, radar, and lidar. The chapter will end with an overview of sensor fusion (i.e., how to combine the input from multiple sensors and other technologies for positioning and navigation).

This chapter will allow the reader to understand the complexity of modern positioning systems and the need to integrate multiple technologies together in the next generation 5G NR systems. However, some of the topics, especially sensor

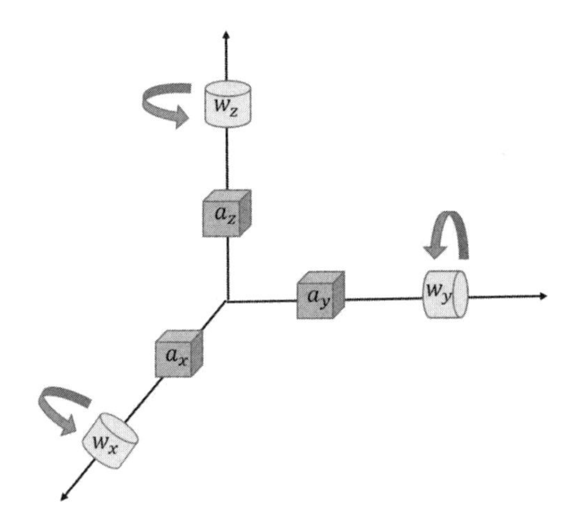

Figure 12.1 Description of an inertial measurement unit.

fusion, are described only at an introductory level, as they are not in the scope of this book. The reference section contains useful resources for a deeper understanding of the topics and technologies described here.

12.2 IMU

Currently, mobile devices come equipped with all kind of sensors, including accelerometers, barometers, magnetometers, gyroscopes, and proximity sensors. However, two in particular are significant for positioning systems: the IMU, which comprises accelerometers and gyroscopes and will be described in this section, and the barometer, which will be described in the next section.

The IMU is a module that combines at least three accelerometers and three gyroscopes to measure the acceleration and the turn rate of a device. A gyroscope provides information on the orientation of the device with respect to an axis by generating an electrical signal proportional to the angular velocity in that axis. Analogously, an accelerometer provides information on the acceleration on its corresponding axis by generating an electrical signal proportional to this acceleration. As shown in Figure 12.1, the IMU provides motion information using gyroscopes

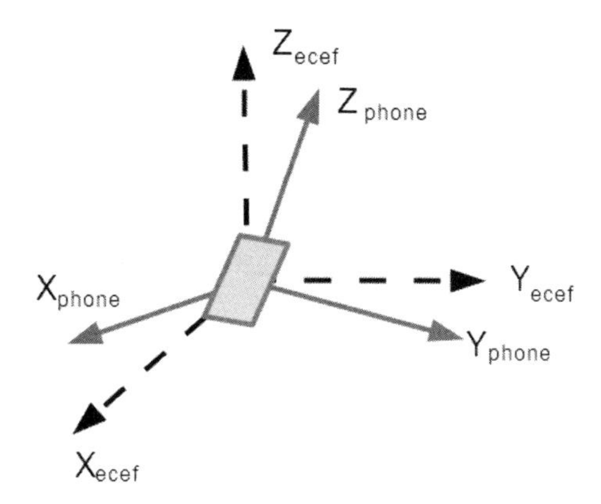

Figure 12.2 Comparison of the phone coordinate system vs the absolute coordinate system.

and accelerometers in a 3D Cartesian coordinate system with respect to the frame in which the IMU is mounted.

An IMU module provides information to augment GNSS positioning or any other positioning technology for navigation applications (it is of no use for static positioning, as the accelerometers and gyroscopes only measure movement). Nonetheless, the location estimation based on IMUs is no trivial task, as explained in [1], due to the multiple error sources that affect the IMU, including noise, misalignment errors, and drifts. High-accuracy IMUs are very expensive (in the range of thousands of dollars) and out of scope for cellular network positioning. Furthermore, when it comes to mobile phones and other handheld devices, the device is not attached to a fixed coordinate frame, so it becomes very complicated to give absolute estimates on the location (e.g., the mobile phone user can have the device in a hand, or inside a pocket, and in every possible orientation). This issue can be better understood with Figure 12.2. The phone has its own coordinate system, x_{phone}, y_{phone}, and z_{phone}. The relation between this coordinate system and the absolute ECEF coordinates (see Section 2.2.1 of this book), represented by the dashed arrows in the figure, is usually not known and is also not static.

Apart from the orientation uncertainty described before, the following errors typically affect an IMU:

- Sensor noise: depending on the type of sensor, there are different types of noise that cause errors in an IMU. Noise is generated in a sensor due to thermal effects and electrical effects. The most common types of noise affecting the IMUs are thermal noise and flicker noise.
  - Thermal noise is the main source of errors in electronic sensors and is caused by the thermal agitation of the electrical carriers present in semiconductors. This noise can be modeled as a white noise with a Gaussian distribution probability.
  - Flicker noise affects electronic devices at lower frequencies and is caused due to impurities in the semiconductor structure. This noise is also known as pink noise or 1/f noise. This noise can be represented by a power spectral density that is inversely proportional to the frequency.

- Misalignment errors are caused due to misalignment of the computational frame when compared to the frame of the mounted sensor. There are two types of misalignment error seen in sensors:
  - Package to frame misalignment errors appear when the body frame of the IMU is not mounted perfectly aligned to the measurement axis. As shown in the left part of Figure 12.3, this causes the vector components of the acceleration to be offset by a few degrees with respect to the body frame.
  - Interaxis frame alignment errors are similar to package to frame errors but are due to nonorthogonal orientation of the sensors in the different axis (i.e., not all the angles between the x, y, and z axis are 90°). Interaxis misalignment are typical production errors.
- Cross-axis coupling errors are caused by a part of the signal sensed in one of the axes being coupled to the other axis. This is usually caused due to electromagnetic signal coupling between orthogonal axes. As shown in Figure 12.3, this error causes the acceleration measured in one axis to influence the value measured in the other axis.

Some of the error sources, in particular the misalignment, can be calibrated and partially or completely removed. Nonetheless, the measurement accuracy of commercial IMUs does not suffice to become a standalone positioning method,

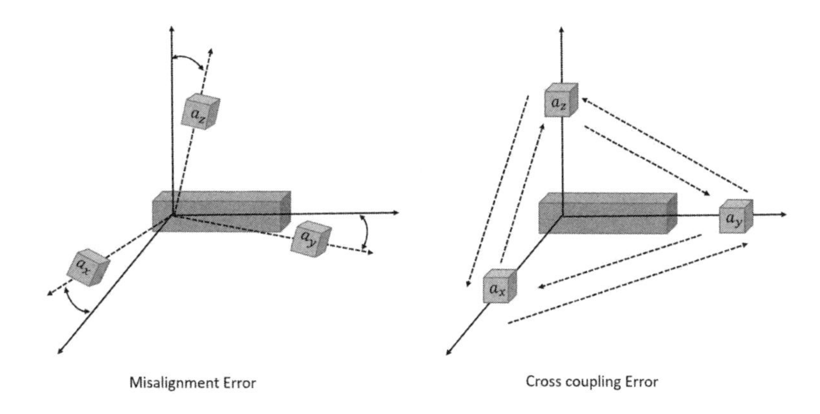

Figure 12.3 Misalignment and cross-coupling errors in an IMU.

especially for mobile phone positioning, where the typical IMUs available are low cost. Even with the phone in a static position, for instance lying on a table, the IMU keeps registering accelerations and turn rates. Furthermore, there is the influence of gravity, with a constant acceleration g towards the center of the Earth. If gravity would always affect the same axis, it could be easily filtered. However, a mobile phone can be held and rotated in multiple directions, so the gravity contribution could spread over multiple axes and the distribution is constantly changing. For all these reasons, any algorithm that takes the raw IMU measurements and calculates the distance and direction by the conventional ways is doomed to fail. The errors accumulate over time and the positioning calculation quickly diverges.

Nonetheless, IMUs can be used as a complimentary system in navigation to overcome temporary unavailability of the other positioning technologies (e.g., GNSS signal loss inside a tunnel), or as part of a sensor fusion algorithm, which will be seen later in this chapter.

12.3 BAROMETER

Another one of the sensors that can be typically found in a smartphone is the barometer, used to measure the atmospheric pressure. The atmospheric pressure can be used to estimate the altitude over sea level using the formula in Equation 12.1, where p_0 is the average sea level pressure, with a value of 101325 Pa (Pascal);

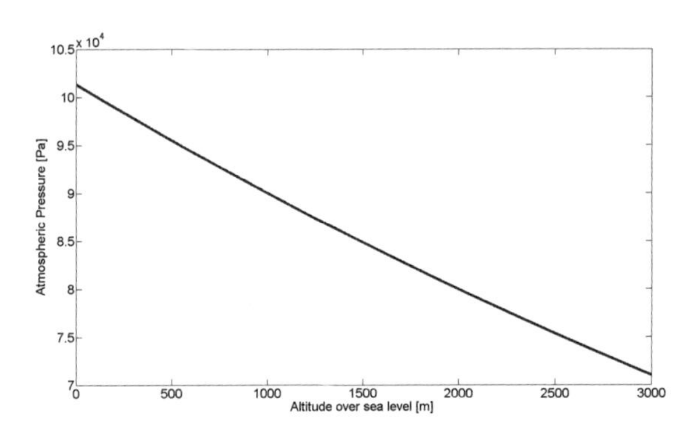

Figure 12.4 Atmospheric pressure in a function of the altitude over sea level.

g is the gravitational acceleration, with a value of 9.81 m/s^2; M is the Molar mass of air, with a value of 0.0289644 Kg/mol; R_0 is the universal gas constant, with a value of 8.31447 J/(mol·K); and T_0 is the sea level standard temperature, with a value of 288.16 K.

$$p \approx p_0 \cdot e^{-\frac{g \cdot M \cdot h}{R_0 \cdot T_0}}. \tag{12.1}$$

Thus, atmospheric pressure measurements are being used for altitude determination. The FCC and the 3GPP have set a vertical accuracy requirement of ± 3 m for indoor positioning, which is considered to be the standard height separation between consecutive floors in a building. Let us see if the barometer can be used to achieve that accuracy. Considering an altitude range from 0 to 3,000 m over sea level, the approximated atmospheric pressure in a function of the altitude is shown in Figure 12.4. The variation follows an inverse exponential, so relation altitude variation versus pressure variation is not constant. However, for an altitude of around 500 m over sea level, each meter of difference accounts for approximately 10 Pa. We will take as an example a typical mid- to high-range mobile phone pressure sensor, the Bosch BMP280, which can be bought for approximately 3 to 4 dollars. According to the data sheet [2], it has an absolute accuracy of $\approx \pm$ 1 hPa (i.e., 100 Pa). Translating that to meters would result in an accuracy of $\approx \pm$ 9 to 10 m, which is a very good accuracy for most applications. However, for indoor positioning, 9 m of error can be the equivalent of an error up to three floors and does not meet the regulatory requirements. Typically, pressure sensors offer much better relative accuracy than absolute accuracy. That means they can be used very accurately to calculate vertical movements, but not so accurately to calculate the absolute vertical position. Looking at the data sheet of the sensor mentioned above, it can be seen that it has a relative accuracy of $\approx \pm$ 0.12 hPa, which results in approximately 1 m.

It is also worth noting that the atmospheric pressure in Equation 12.1 is an approximation and the actual value is highly affected by local temperature and weather conditions. Hence, the actual atmospheric pressure value needs to be corrected depending on the current climatic conditions at the time and place of the measurement. This correction cannot be performed so easily on the mobile device. In order to compensate for these effects, a reference station placed somewhere near the actual device location can be used. The mobile device measures and reports the uncompensated barometric pressure (UBP) to the location server. The location server, with the help of the reference station, compensates the UBP measurement to estimate the actual altitude.

Both barometers and IMU are sensors that can be typically found in current smartphones. However, Chapters 4 and 5 of this book have already shown that, starting with LTE and most prominently in 5G NR, there are plenty of new positioning use cases involving other devices such as cars or drones. Thus, the next sections will focus in other sensors that are not used for mobile phones, but can be useful for these new applications.

12.4 RADAR

The first experiments showing that radio waves could be reflected from solid objects were carried out by Heinrich Hertz in the last decades of the nineteenth century. However, it was not until 1904 that Christian Hülsmeyer obtained a patent on using this phenomenon to detect the presence of metallic objects and estimate the distance to them. He called his invention "telemobiloscope." Starting in the 1930s, radio detection and ranging (RADAR) found its usage in a variety of military, aerospace, and civilian applications. Radar applications are manifold: to locate targets on the ground, air, and at the sea, to track targets, objects, and missiles; in air traffic control applications for monitoring and regulating air traffic; coordinating airport ground functioning; and in aircraft landing systems. Airplane radars are used to detect other airplanes and monitor weather conditions. Meteorological radars are used for weather forecasting and to detect precipitation and storms. Marine radars are used to detect and locate ships and navigate through obstacles or near the shore, where there might be submerged objects such as rocks.

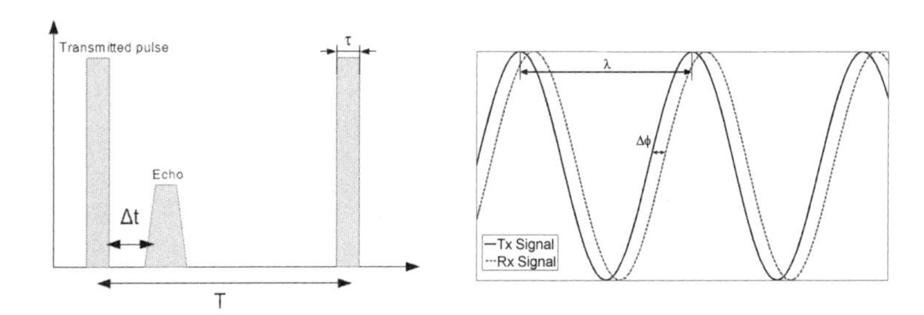

Figure 12.5 Example of pulsed (left) and continous (right) wave radars.

Experiments with radars in the automotive sector started in the early 1990s and the first high-end cars equipped with radars were seen in the early 2000s. The requirements of the automotive industry have driven the development of small, compact radars systems that can be deployed on commercial vehicles for applications such as adaptive speed control and collision detection.

A radar transmits an electromagnetic signal and receives the echoed signal. The time taken for the transmitted signal to be received back (ToF) is the basic principle used by most radars to find the distance to the target object. The electromagnetic signal is reflected in many directions by the target and the angle at which the reflected signal is received determines the location of the target relative to the transmitter. The Doppler shift in the signal determines the motion of the target relative to the transmitter.

Radars can be bistatic, if the receiver and the transmitter are placed at different locations, or monostatic, if the receiver and the transmitter are part of the same unit. Depending on the target application, there are numerous types of radars [3]:

- Pulsed radars transmit a series of short pulses of a very strong signal with silent periods in between the transmissions where the echo signal can be received, as shown in Figure 12.5. The echo is received Δt after the transmission, representing the ToF measurement. Thus, the distance to the target, also called range, is defined by Equation 12.2, where c represents the propagation speed of the signal, in this case the speed of light. The range of the radar is limited by the periodicity of the pulsed signal, T. If the object is beyond the maximum range, the system would not be able to distinguish to which of the transmitted pulses the reflected echo corresponds, leading to an ambiguity in the solution and probably a false detection.

$$R = \frac{c \cdot \Delta t}{2} \tag{12.2}$$

Pulsed radars find extensive usage in military, air traffic control, and remote sensing applications. They are very effective in measuring the distance to the target.

- Continuous wave (CW) radars transmit high-power unmodulated continuous-wave signals [4], as shown in Figure 12.5. In order to distinguish between the transmitted and received signals, spatial separation is used (i.e., bistatic radars). The radar transmits a sinusoidal wave of a certain wavelength, λ, which gets reflected in the target and is received; see the dashed line in the figure. Based on the phase difference between transmitted and received signals, $\Delta\phi$, the distance to the object, R, can be calculated as shown in Equation 12.3.

$$\Delta\phi = -2 \cdot \pi \frac{2 \cdot R}{\lambda} \tag{12.3}$$

CW radars are not very effective for measuring the distance to an object, since they have an ambiguity if the phase shift in the received signal is bigger than 2π. However, the are very effective for measuring the speed of a vehicle. Thus, they are largely used for traffic control.

- Frequency-modulated continuous-wave (FMCW) radars combine both advantages of the radar systems seen before by being able to accurately measure range and velocity. This is achieved by modulating the transmitted signal in frequency [5, 6]. This type of radar transmits a signal with continuously increasing frequency, called chirp, as shown in Figure 12.6. The duration of the signal is called chirp time, t_c in the figure, and determines the maximum range of the radar. The chirp bandwidth is defined as $BW = f_{c2} - f_{c1}$ and determines the resolution of the radar. The frequency difference between transmitted and received signals is called beat frequency, B_t and is used, together with the starting frequency f_{c1}, to calculate an intermediate frequency as shown in Figure 12.6. The distance to the target, R, can be calculated by using Equation 12.4. The speed calculation relies on the Doppler effect, the same as for CW radars.

$$B_t = (f_{c2} - f_{c1}) \cdot \frac{2R}{c} \tag{12.4}$$

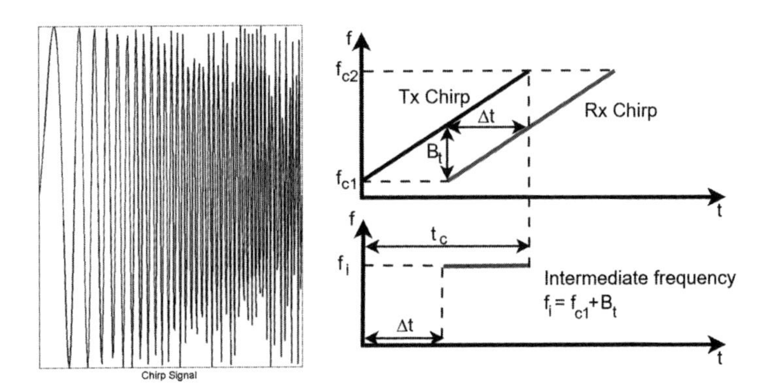

Figure 12.6 Chirp signal for a FMCW radar.

FMCW radars are very popular in commercial applications, especially in the automotive area, due to their small size and operation flexibility. Since range measurements are not dependent on the received power, the received signal can be amplified without losing any significant information. These radars are also robust against interference from other radars operating in the same area, as their working principle does not depend on the amplitude of the received signal.

12.4.1 Automotive Radars

The need for radars in the automotive area has led to the development of small-size and low-cost, high-resolution radars. As shown in Chapter 4, they are an integral part of driver assistance systems such as emergency braking and adaptive speed control. Single-chip radar systems based on silicon germanium technology are the basis for these kind of radars. They typically operate in the W-band on 24-GHz, 77-GHz, and 79-GHz frequency ranges, due to the over 500 MHz of bandwidth availability on 77 and 79 GHz (see IEEE radar frequency ranges in [7]). Typically, most of the automotive applications use FMCW radars that allows accurate measurement of both range and velocity.

Figure 12.7 shows the three types of radar systems typically installed in a car and their applications:

- Long range radar (LRR) has a high transmit power and directivity ($\approx$55 dBm effective isotropic radiated power (EIRP)) and a maximum range of around

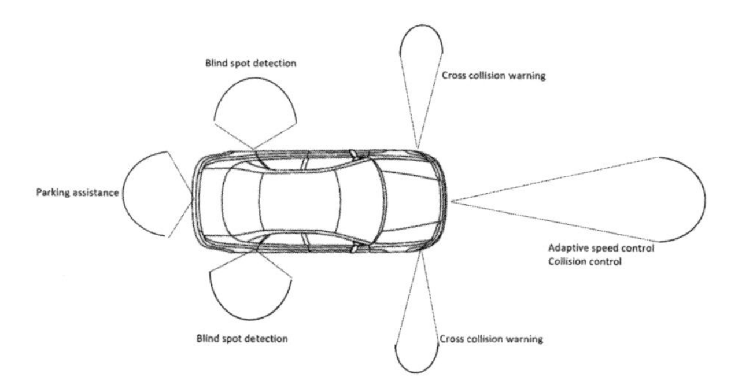

Figure 12.7 Representation of the radar systems in a car.

250 m with a chirp bandwidth below 1 GHz. As a result, this radar does not have a high-range resolution and angular resolution and cannot distinguish between two objects close to each other. It is used for anti-collision and adaptive speed control applications.

- Medium range radar (MRR) has lower transmits power than a LRR and operates typically up to 100 m. It has a chirp bandwidth of around 4 GHz, which enables better range resolution and angular resolution.
- Short-range radar (SRR) serves applications such as park assistance and operates up to 50 m. It has a much better range resolution than the other two types. Ultrasonic sensors serve similar purpose. SRRs may be combined together with MRR systems.

12.5 LIDAR

The light detection and ranging (LIDAR) technology was initially deployed in the space industry and is used for applications such as remote sensing or topographic information collection, such as information on tree growth. The first usage of lidar was in 1971 during the Apollo 15 mission to map the surface of the moon. It was later used in remote sensing applications such as mapping of vegetation and forest coverage. Aircraft-based lidars are very effective in civil engineering applications to create models of terrain, buildings, and mountains. As explained in Chapter 3 and Chapter 4, the advent of assisted driving and autonomous driving has pushed lidar technology in the realm of consumer electronics. These applications come hand

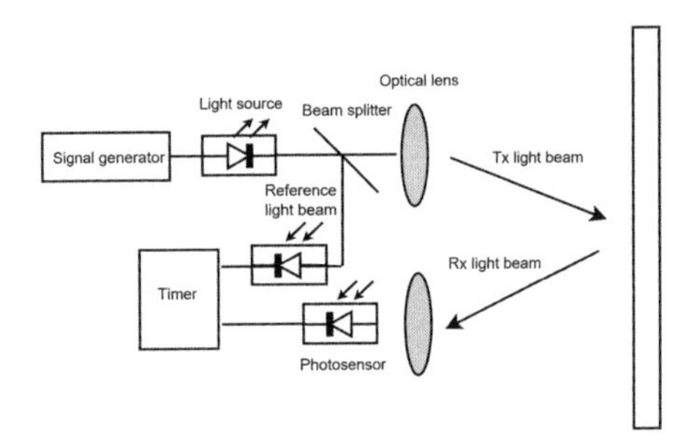

Figure 12.8 Principle of operation of a lidar.

in hand with a need for producing robust and cost-effective solutions, driving the innovation in the lidar field.

Lidar uses laser signals generated on 905-nm and 1500-nm wavelengths [8].

- The 905-nm wavelength is closer to the visible light spectrum and using high intensity could have adverse effects on the human retina. Hence, lidar on this wavelength is limited in range and intensity. However, it has relatively lower absorption loss due to the humidity in the atmosphere.
- The 1500-nm wavelength can be operated at a higher intensity than 905 nm due to lower negative health effects. However, it has lower performance in rainy and snowy conditions due to the higher absorption caused by the air humidity.

The principle of operation of a lidar is shown in Figure 12.8. The lidar generates a signal that is emitted by a high-power laser source. The light beam is sent through a splitter, which collects a part of the generated light beam and feeds it back as a reference for the receiver. A system of optical lenses are used to transmit the light beam. The reflected beam is received and routed through the optical lenses to a photodetector, which converts the detected photons back to electrical signals. Using the receiving and reference signals, the time measurement circuitry calculates the ToF of the light beam.

Lidar is similar to radar except for the fact that the wavelengths of the signals are in the nanometer range compared to the centimeter- and millimeter-range wavelengths used for radar. As a result, lidar has much higher resolution than radar and it is more effective at distinguishing between two objects close to each other. Thus, lidars are especially useful for object recognition and categorization.

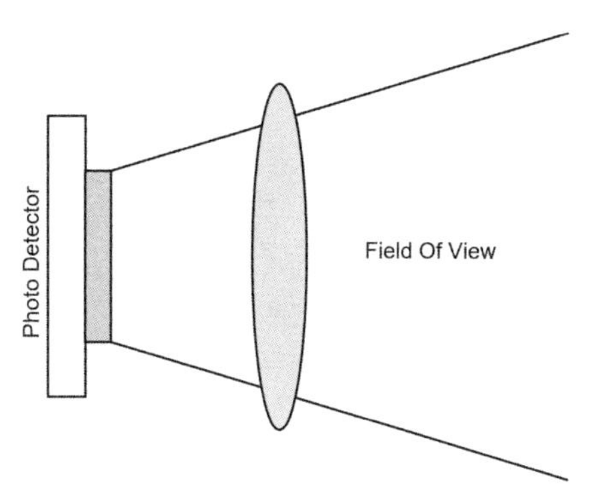

Figure 12.9 Field of view of a lidar.

Although lidars, radars, and camera systems overlap in multiple areas, lidars are much more effective in generating 3D models of the environment. On the other hand, they are much more adversely affected by bad weather conditions and interference from other light sources. Furthermore, lidars do not work well at short distances, where the short propagation time damages the system resolution.

12.5.1 Field of View

The field of view (FOV) of a lidar represents the area of observation where a lidar can detect and measure targets, as shown in Figure 12.9 [9]. The aperture of the optical lens is one of the most important factors to determine the FOV for a static lidar. The larger the aperture and the larger the surface area of the photo detector, the greater the FOV of the lidar. However, larger apertures result in lower resolution at higher distances. Thus, the choice of aperture size is an important aspect in lidar design.

Different techniques can be used to increase the FOV of a lidar while maintaining a small aperture of the optical lens. Some possibilities are to scan the area with a moving laser beam or to use a flash laser, which works like the flash of a photo camera lighting up the whole scene. Area scanning lidars send laser pulses continuously in multiple directions and the receiver collects the reflections from all of those directions. There are various types of lidar implementing the scanning functionality, for instance mechanical lidars or optical phased array lidars.

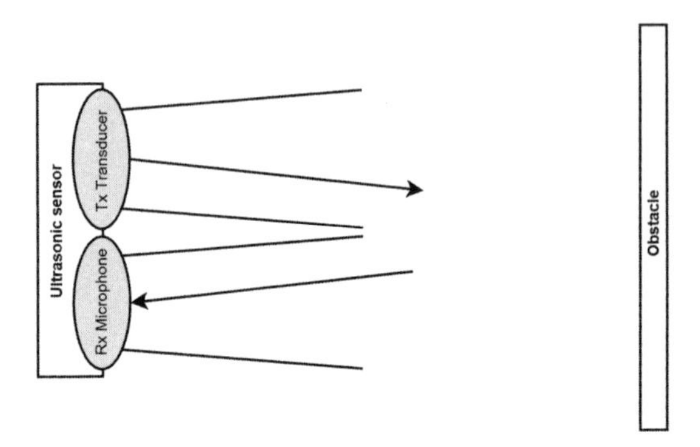

Figure 12.10 Description of an ultrasonic sensor.

12.6 ULTRASONIC

Ultrasonic sensors fall into the category of proximity sensors. They are very reliable sensors for the detection of objects that are very close. Ultrasonic sensors work on the principle of sending high-frequency sound waves pulses out and receiving the reflected signal. These sensors are used typically in automotive areas for parking assistance and detection of nearby obstacles at lower speeds.

Ultrasonic sensors, as shown in Figure 12.10, send out sound pulses that get reflected by any nearby object and are received again. The transmitter and receiver in an ultrasonic sensor are made from transducers, which during transmission convert electric current into sound waves and on reception sound waves into electric signals. Depending on the sensor, the frequency of the electric pulses sent varies between 20 kHz and 4 MHz. The higher the frequency, the better the resolution of the sensor but also the higher the cost. The ToF measurements are used to calculate the distance to the target. The two most popular sensor types used for ultrasonic sensors are piezoelectric transducers and capacitive transducers.

The distance to the object can be calculated using the formula in Equation 12.5, where t is the ToF of the ultrasonic signal, v_s represents the speed of sound, and k is a constant value depending on the sensor's geometry.

$$R = t \cdot v_s \cdot k \tag{12.5}$$

The speed of sound is affected by the air temperature and humidity. For distance measurements at nominal temperatures, the effects of humidity can be neglected in practice. Hence, the speed of sound can be approximated as $v_s \approx$

$20.055 \cdot \sqrt{T}$, where T is the temperature expressed in Kelvin. In a temperature range of 0°to 40°Celsius, the speed of sound changes between 330 to 360 m/s.

Ultrasonic sensors are affected by the following factors:

- Ultrasonic sensors are effective in a range from few centimeters to a few meters and the accuracy of the sensors reduces considerably over a few meters.
- They are affected by crosstalk from other ultrasonic sensors nearby.
- The shape of the reflecting object can affect the strength and direction of the reflecting echo, which has an influence on the performance of the sensor.
- The field of observation or angular range of the sensor is typically up to 30 °.

Ultrasonic sensors are especially robust against bad weather, dust, moisture, and other lighting issues, which affect lidar and radar as will be seen later in this book. Additionally, they are easy to produce and hence are very cost effective sensors.

12.7 SENSOR FUSION

This chapter has introduced multiple sensors that can be typically found in different cellular network related positioning applications. It has been seen that each of these sensors are affected by external and internal perturbations such as noise, interference, bias, and coupling. In order to reduce the influence of these error sources, one of the most powerful techniques is sensor fusion.

Sensor fusion consists of combining data from multiple different sources in a way that the global uncertainty is reduced. Reducing the uncertainty results not only in more accurate results, but also increases the reliability and resilience on the system. For safety-critical applications, such as V2X, this is a very important topic. On a typical positioning scenario, sensor fusion can potentially combine data from technologies such disparate as IMUs, radar, lidar, GNSS, cellular network technologies, or Wi-Fi, as shown in Figure 12.11. These measurements, together with the last known status, are the inputs to the sensor fusion algorithm, which calculates the new status based on these inputs. The status includes not only the positioning information (e.g., latitude, longitude, and altitude), but also other important information such as velocity and bearing.

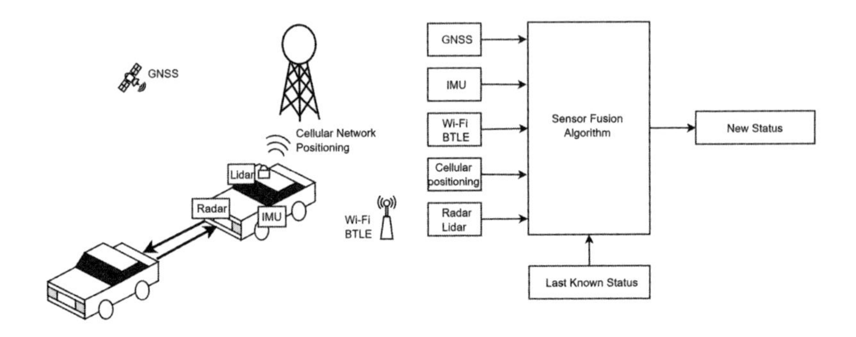

Figure 12.11 Sensor fusion scenario.

A sensor fusion algorithm can be implemented by defining a Bayesian network to model the positioning scenario and applying probabilistic tools such as the Kalman filter or the particle filter. Sensor fusion is a very complex topic that deserves a book just for itself. This book will just give a small introduction to sensor fusion so that the reader is able to obtain a basic understanding of the concept. A much deeper overview of sensor fusion can be found in [10, 11].

12.7.1 Sensor Fusion Applications

This section will cover some of the traditional sensor fusion applications in an informative way without giving too many details. Please refer to [10, 11] for a more detailed explanation.

12.7.1.1 Dead Reckoning

Dead reckoning, also known as deduced reckoning, is a process widely used in positioning systems to determine the current position based on the previous known position and motion information. This technique has been used for centuries in ships to estimate their current location based on the ship speed, direction (estimated using the compass), and time elapsed.

In modern-day automotive systems, dead reckoning plays an important role in estimating a location when GNSS signals are limited or unavailable. For example, a car traveling through a tunnel can use dead reckoning to estimate the location based on the last known GNSS fix and interpolate with IMU, magnetometers and odometer measurements. Dead reckoning can also be used to provide interpolated

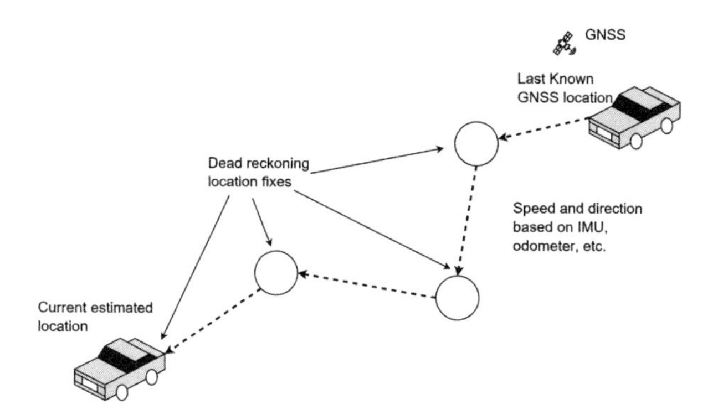

Figure 12.12 Dead reckoning scenario.

positions between infrequent GNSS fixes. IMU-based dead reckoning without any additional positioning source is typically not used, since the sensor noise causes the location to drift significantly from the ground truth.

An example of dead reckoning for a car is given in Figure 12.12. The GNSS signal is lost, and the navigation system of the car continues to calculate the current position based on dead reckoning using an IMU, an odometer, and other available measurements. Every location is estimated from the previous location plus the information about the motion of the car, obtained by measuring the speed, turn rates, and elapsed time.

12.7.1.2 Map Matching

Map matching is the process of mapping a set of measured coordinates from a measurement source to a map. The raw data collected from GNSS receivers or sensors (e.g., an IMU) is affected by multiple error sources such as multipath, biases, and sensor noise. Map-matching algorithms map these measured points to the closest segments of the road, enhancing the accuracy of the vehicle position.

Figure 12.13 shows a scenario similar to the dead reckoning example. In this case, the GNSS fix is not necessarily lost. Based on all available inputs (e.g., GNSS, IMU), the positioning algorithm estimates the location of the vehicle, represented by a circle in the figure. Afterwards, it compares this position against a known map. The updated location fix is represented by an inverted square.

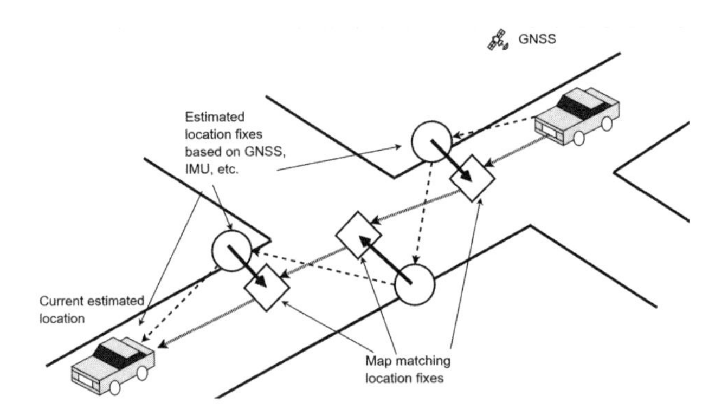

Figure 12.13 Map-matching scenario.

12.7.1.3 SLAM

In the map-matching algorithm, shown in the previous section, the map of the environment is known. This is the usual case in the automotive field, since the navigation systems in cars are already equipped with digital maps of roads. However, in other scenarios (e.g., emergency services after a natural disaster), an accurate, up-to-date map of the environment may not be available. Simultaneous localization and mapping (SLAM) addresses this problem.

SLAM can be seen as a combination of dead reckoning and map matching, as shown in Figure 12.14. If the map of a certain area is unknown, the SLAM algorithm calculates the current position of the vehicle based on the last known position and the motion information, as was done in dead reckoning. If the map of a certain area is already known, this information is used to refine the position calculation by map matching. In both cases, the resulting position is used in the mapping function to incrementally improve the map. SLAM greatly benefits from crowdsourcing to create and maintain accurate maps of the environment.

12.7.2 Bayesian Networks

A Bayesian network is a graphical model to represent how the different variables in a certain probabilistic problem interact with each other. The model is typically depicted as a directed acyclic graph. It is a very powerful tool that helps in the translation from a real-life scenario to a sensor fusion algorithm, and eases the implementation of a Kalman or particle filter.

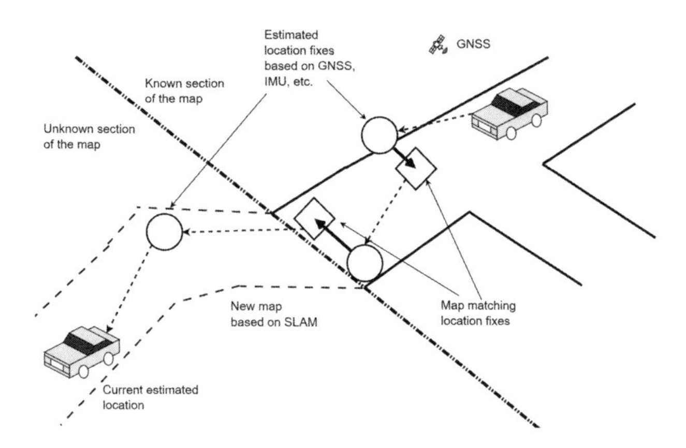

Figure 12.14 SLAM scenario.

A positioning scenario is better represented by a dynamic Bayesian network (DBN), in which the position of a certain object at a time instant k depends on the previous position at $k - 1$ and a series of variables and inputs. An exemplary model of a DBN for a positioning problem is shown in Figure 12.15. This example will be used throughout this section to help understanding the concepts explained here.

The DBN in Figure 12.15 represents a series of nodes connected between them during two consecutive time stamps, k-1 and k. The different nodes are:

- The *state* vector, X, contains all the relevant variables that need to be calculated to locate a certain object. This information can be, for instance, the estimated position of the object at a certain time instance. The state at time k is influenced by the previous state at k-1, by the motion vector applied, U, at time instant k and by the map, M.

- The *motion* vector, U, contains information about the motion applied to the object (e.g., the acceleration and the turn rate). The motion vector is influenced by the previous motion, due to inertia. For example, a car circulating at 100 km / h cannot be suddenly stopped from one moment to the next.

- The *map*, M, contains information about the environment, such as the the map of the roads, the layout of a building or the location of nearby objects, trees, traffic lights, and so forth. The map has an influence over the state vector, as it can potentially apply some constrains to the position of the object. For example, a car is likely to be on the road and a person cannot

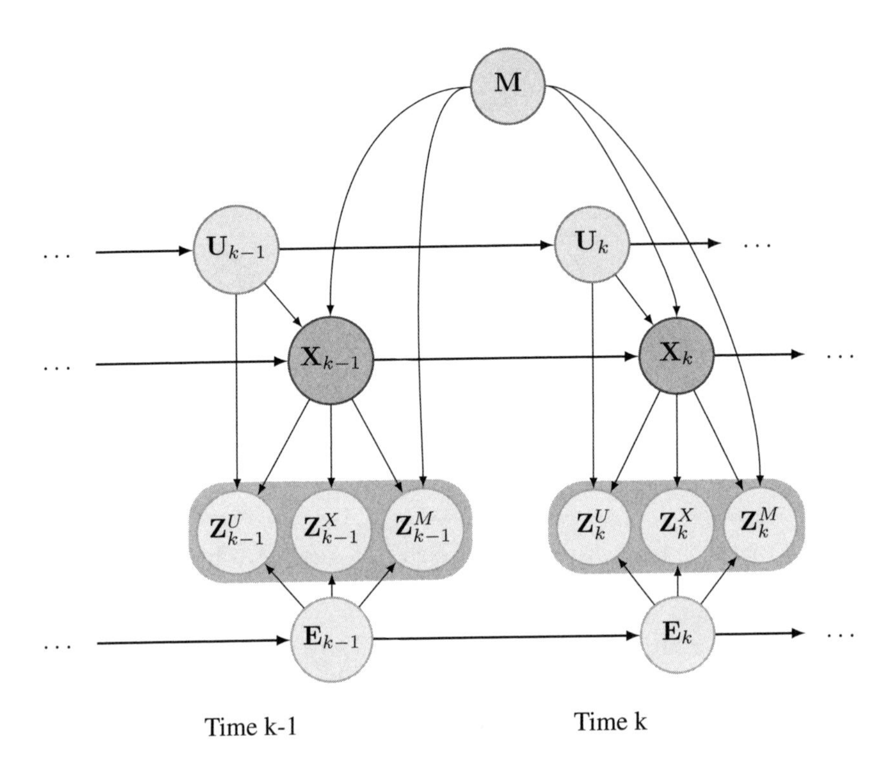

Figure 12.15 Example of a dynamic Bayesian network for a positioning scenario.

walk through walls. The map can be known, in which case we will be talking about localization or navigation, or unknown, in which case we have a SLAM problem.

- The *measurements* vector, Z, also called observations, contains the different measurements collected by all the involved sensors. In this example, it is split into three components, the motion measurements, such as the typical IMU measurements; the state measurements, which are direct observations of the state of the object (for example, GNSS directly measures the position of the device); and the map measurements, for instance a lidar measurement to a nearby object.

- The *error* vector, E, collects the errors associated to each of the measurements. Some errors are sporadic, but some others are cumulative, such as a drift. Thus, the error vector can be affected by the previous error.

12.7.3 Bayesian Probability

The system in Figure 12.15 evolves with time. The Markov assumption postulates that a state is complete if the knowledge of the present state is sufficient to determine the conditional probability distribution of the future states. If the Markov assumption is fulfilled, the system is called a Markov process. What this means is that the state X_k depends only on the previous state, X_{k-1}, the motion vector U_k and the map, in case this is known. Equation 12.6, called the system model, represents the probability of a state X_k in a function of its dependencies;

$$p(\mathrm{X}_k|\mathrm{X}_{k-1}, \mathrm{U}_k, \mathrm{M}). \tag{12.6}$$

The measurement model represents the probability of obtaining a particular set of measurements Z_k based on the rest of the variables involved in the system. For the example in Figure 12.15, the measurement model is shown in Equation 12.7. The measurements are affected by the state, by the motion (IMU measurements), and by the map (camera-based measurements, lidar, etc.). Both measurement model and system model are called posteriors in probability.

$$p(\mathrm{Z}_k|\mathrm{X}_k, \mathrm{U}_k, \mathrm{E}_k, \mathrm{M}). \tag{12.7}$$

Based on all the measurements available, it is possible to maintain a *belief* of the state and motion of the object to locate. The belief is represented by Equation 12.8, where the notation has been simplified from X_k, U_k to XU_k:

$$bel(\mathrm{XU}_k) = p(\mathrm{XU}_k|\mathrm{Z}_{1:k}, M). \tag{12.8}$$

Finally, the last probability function to be introduced is the *prediction*, shown in Equation 12.9, which is the estimation of the expected belief at a time instant k, based on all previous measurements, without the knowledge of the measurements at time k;

$$\overline{bel}(\mathrm{XU}_k) = p(\mathrm{XU}_k|\mathrm{Z}_{1:k-1}, M). \tag{12.9}$$

The computation of the belief from the prediction by incorporating the measurements is called measurement update or correction. A solution to calculate the

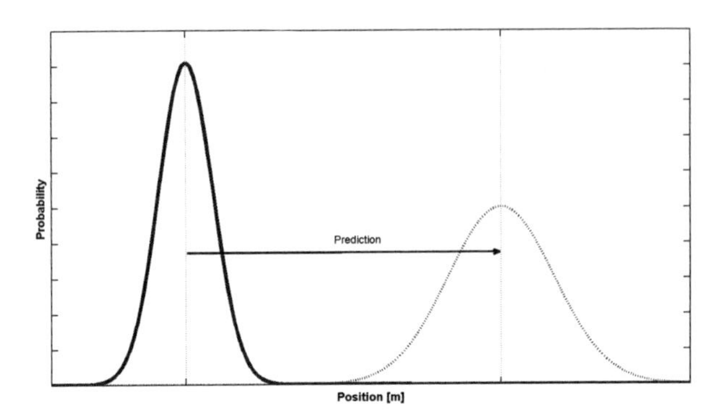

Figure 12.16 Prediction step for a Kalman filter.

belief from the measurements is a Bayesian filter. The Bayesian filter is a two-step iterative process. At a time instant k, the process does the following:

1. Predict the state X_k based on all the previous knowledge.
2. Correct the prediction with the new available measurement information Z_k.

12.7.4 Kalman Filter

The Kalman filter, developed by Kalman in 1960 [12], is a type of Bayesian filter that approaches the belief by a multivariate Gaussian distribution, $x \sim \mathcal{N}(\mu, \Sigma)$, where $\mu \in \Re^n$ stands for mean and has the same dimensionality as the state vector X and $\Sigma \in \mathbb{S}^n_{++}$ stands for covariance matrix and is symmetric and positive-semidefinite. Its dimension is the same as the state squared.

In Figure 12.16, the prediction step of a Kalman filter is explained. At a certain time instant k, the state of an object is represented by a Gaussian distribution with a certain mean and covariance. Based on the current knowledge of the position and the movement of the object (e.g., its direction and speed) the state of the object in $k + 1$ is predicted. For simplicity, this has been represented in Figure 12.16 as a horizontal movement. As can be seen, the Gaussian distribution has become wider. This is due to the prediction being based on previously estimated values, and not on actual evidence. Hence, the uncertainty of the actual state of the object has grown.

On the left side of Figure 12.17, the measurement for instant $k+1$ has become available. These measurements could be for instance a GNSS location fix or the results of the IMU. As can be seen, it is also a Gaussian distribution with a smaller

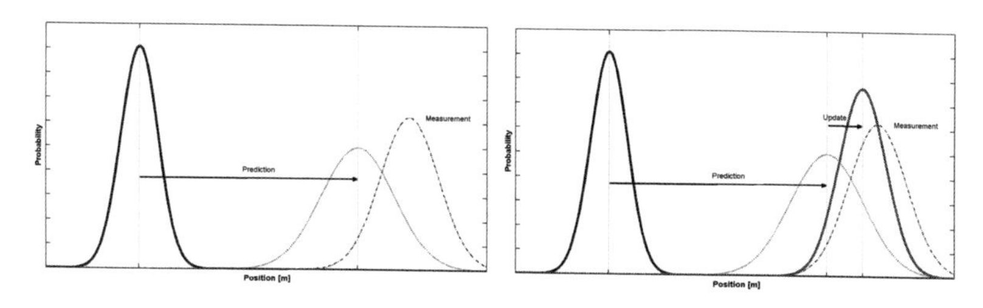

Figure 12.17 Measurement and update steps for a Kalman filter.

standard deviation than the prediction, because it is based on actual evidence. The uncertainty in the measurement comes from potential sensor errors. On the right side of Figure 12.17, the update has taken place using both the measurements and the prediction, resulting in a much more accurate solution for the state of the object.

This example has shown the basic working principle of a Kalman filter. The Kalman filter is an optimal estimator, meaning that it can always find the best solution. Nonetheless, Kalman filters present some limitations. The biggest of them is that they can only be applied to linear systems (i.e., systems where the predict and update steps can be performed by linear functions). To overcome this limitation, the extended Kalman filter (EKF) and the unscented Kalman filter (UKF) have been developed.

The EKF is an enhancement of the Kalman filter for nonlinear systems. It is based on approximating the nonlinear probability distribution of the belief by a Gaussian distribution. In order to do that, the nonlinear predict and update functions are approximated by the linear Taylor expansion. The EKF is no longer an optimal estimator. Its accuracy depends on the degree on similarity between the real functions and the Taylor approximation.

If the grade of nonlinearity of the predict and update functions is large, the performance of the EKF is generally poor. In such cases, the UKF offers better performance. The UKF relies on the unscented transformation technique for deterministic sampling by selecting a number of points around the mean of the probability distribution and propagating these points through the nonlinear functions.

The KF, EKF, and UKF can be combined and applied to the predict and update functions separately, depending on the nonlinearity properties of each. Together, they form a very powerful method for sensor fusion. Nonetheless, they are all parametric filters and only able to maintain a single belief at any point of time (i.e., only one available hypothesis for the state of the object).

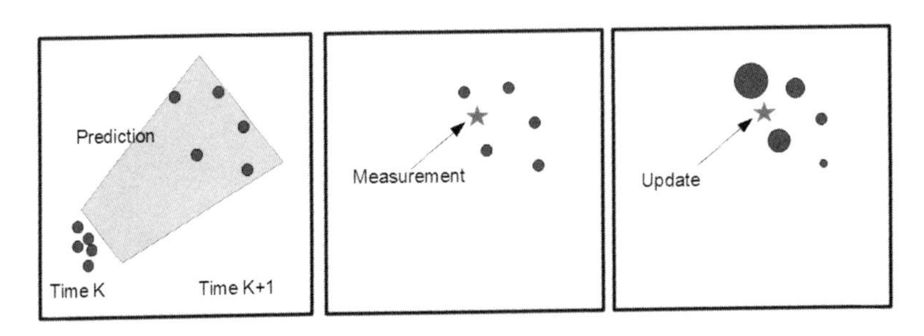

Figure 12.18 Prediction, measurement, and update steps for a particle filter.

12.7.5 Particle Filter

The particle filter (PF) is a nonparametric filter that approximates the probability distribution of the belief by a finite number of particles. Contrary to the KF and its enhancements, the PF is able to maintain multiple simultaneous hypothesis for the state of the object (each particle is a hypothesis).

Each particle i has an associated state x^i and a weight w^i that represents the likelihood of the x^i being the true state. During the prediction step, each particle is propagated differently based on the probability distribution of the prediction function. The update step is used to recalculate the weight of each particle based on the closeness to the received measurements.

This process is shown in Figure 12.18. The five particles are initially distributed around a certain position at time instant k. In the prediction step, the particles are propagated in different directions based on the prediction function. A measurement is received, represented by a star. This measurement is used during the update step to assign a weight to each particle, represented by its size in the figure.

In the example, the PF contains only five particles for simplicity, but typically the number of particles is much higher. The number of particles needs to be sufficient to adequately represent the probability distribution of the belief. The higher the non-linearity of this distribution, the more particles are needed. The particle filter is a much more powerful tool than the EKF / UKF for representing highly non-linear scenarios. On the other hand, the computational cost increases significantly with the number of particles.

12.8 CONCLUSION

This chapter introduced the sensor-based positioning technologies. Not all the sensors described here will be used for all applications. Typical mobile phone location services will probably focus on the use of low-cost IMU and atmospheric pressure sensors available in most consumer devices. On the other hand, safety-critical applications such as autonomous driving and unmanned aerial vehicles will probably implement a complex combination of radar, lidar, and other sensors. Thus, the sensor fusion algorithms briefly mentioned in the last part of the chapter will play a central role in the future of positioning.

Together with the previous chapters that presented A-GNSS, cellular, and noncellular network-based positioning technologies, the reader now has a complete picture of all the technologies involved in modern positioning applications. Thus, this chapter concludes the part of the book dedicated to explaining the different technologies. The last part of this book will now focus on detailing the different communication protocols necessary to perform a successful positioning session.

References

[1] Gomez de la Flor Palacios, A. *Inertial Measurement Unit Simulation for Location Based Services*, Master Thesis, University of Oviedo, July 2017.

[2] Bosch BMP280 datasheet, https://www.bosch-sensortec.com/bst/products/all_products/bmp280, accessed on September 2019.

[3] Scheer, J.A., Richards, M.A., Holm, W.A., and Melvin, W.L. *Principles of Modern Radar*, Scitech Publishing, 2013.

[4] Luebbert, U. *Target Position Estimation with a Continuous Wave Radar Network*, Cuvillier Verlag Goettingen, 2005.

[5] Komarov, I.V. and Smolskiy, S.M., *Fundamentals of Short-Range FM Radar*, Artech House, 2003.

[6] Jankiraman, M.. *FMCW Radar Design*, Artech House, 2018.

[7] 521-2002 *IEEE Standard Letter Designations for Radar- Frequency Bands*, IEEE, 2002.

[8] Dong, P. and Chen, Q. *LiDAR Remote Sensing and Applications*, CRC Press, 2018.

[9] Blaunstein, N., Arnon, S., Kopeika, N., and Zilberman, A. *Applied Aspects of Optical Communication and LIDAR*, CRC Press, 2009.

[10] Thrun, S., Burgard, W., and Fox, D. *Probabilistic Robotics*, MIT Press, 2005.

[11] Hall, D.L. and McMullen, S.A.H. *Mathematical Techniques in Multisensor Data Fusion*, Second Edition, Artech House, 2004.

[12] Kalman, R.E. "A New Approach to Linear Filtering and Prediction Problems," *Transaction of the ASME - Journal of Basic Engineering*, pp. 35–45, 1960.

Part III

Positioning Protocols

Chapter 13

Positioning Protocols in Cellular Networks

13.1 INTRODUCTION

Starting with the second generation of mobile cellular networks (GSM, C2K), the 3GPP, 3GPP2, and OMA specified network elements and protocols for caller location determination. The main drivers for this action were legal requirements (e.g., emergency caller location) and the desire to enable commercial location-based services. Each subsequent generation of mobile communication standards extended the capabilities. Over the years, accurate emergency caller location of mobile phones (outdoors) became the standard in the United States, with the FCC defining demanding E911 positioning requirements, as explained in Chapter 3 of this book. Furthermore, location information is used by a plethora of smartphone apps that also rely on the 3GPP, 3GPP2, and OMA protocols behind the scenes.

Colloquially, people tend to refer to "GPS" for location services, as if these two terms are equivalent. On the contrary, a mobile phone today uses in fact a heterogeneous mix of technologies to obtain an accurate location, GPS being only one of them. These technologies can be divided into RAT-dependent and RAT-independent groups, where the former uses the mobile network's RF signals to find the location and the latter uses signals from GNSS, Wi-Fi, Bluetooth, and so forth. Before the advent of smartphones around 2010, GNSS and Wi-Fi receivers were not widely deployed in mobiles. Simple RAT-dependent technologies like ECID played a key role in the early days and are still an important fallback in case no other technology is available.

Most location technologies have in common that they require some data exchange between the mobile and a location server in the network. The data exchanged includes UE capabilities, assistance data, and measurements. For this purpose, for (almost) every RAT standard, a location protocol to communicate with a mobile device was specified:

- GSM/GPRS: RRLP
- CDMA2000: TIA-801
- WCDMA: RRC (part of the overall WCDMA RRC)
- LTE: LPP
- NR: LPP

These protocols have been primarily defined for the control plane and are piggybacked in NAS or RRC. In the case of LTE, the NAS protocol TS 24.301 [1] provides a transparent container for LPP and LCS messages. Similarly, the NR NAS (TS 24.501, [2]) specifies an equivalent transparent container. Furthermore, most of these protocols can be transported over the user plane using the OMA SUPL standard.

This chapter will start with a general protocol description and some basics about data encoding and continue with the positioning protocols before LTE (i.e., for GSM, CDMA2000, and WCDMA). From there, it will also introduce user plane positioning and the OMA SUPL standard and end with some general aspects about user privacy and periodic sessions. The subsequent chapters will focus on the LTE and NR positioning protocols.

13.2 GENERAL PROTOCOL DESCRIPTION

Although there is a multitude of different positioning protocols depending on the cellular network, all share several similarities, such as the basic call flow or the different types of transactions. This section will explain some of these characteristics shared by all or the majority of the positioning standards.

13.2.1 Fundamental Transactions

A positioning session will always contain a combination of certain transactions. The basic blocks are the positioning capability exchange, the provision of assistance data, and the report of positioning measurements. Typically, for each of these transactions there is a request / provide message pair. However, in some cases the provide message may also be sent unsolicited. For example, the *LPP Provide Assistance Data* in C-Plane is typically sent by the network without a request from the UE side. This simple pattern covers all positioning technologies where UE cooperation is required: ECID, A-GNSS, DL-OTDOA/A-FLT, Wi-Fi, Barometric, Bluetooth, MBS, and is extensible for future technologies, like RTK/PPP. These fundamental transactions are used across all the different positioning protocols described in this book, with some exceptions. One of the transactions may be skipped in some of the protocols (e.g., the lack of capability exchange in earlier versions of the RRLP protocol). Furthermore, not all of the aforementioned technologies require assistance data. ECID and Bluetooth methods work without aiding data. For Wi-Fi and barometric, LPP specifies aiding data, but in actual implementations, it is not used as of today, as it provides little benefit in terms of TTFF or sensitivity. OTDOA, A-GNSS, and MBS in contrast benefit significantly from assistance data to reduce TTFF as well as energy consumption and to increase sensitivity.

On top of the above-mentioned messages, there is also an abort transaction and an error response for abnormal behavior. Additional to these transactions, if the bearer protocol is not reliable (e.g., the EPS NAS in C-Plane mode), an acknowledgment is requested for every single message.

13.2.2 Generic Call Flow

Based on the transactions defined in the previous section, the most generic positioning call flow is depicted in Figure 13.1.

Of course, this basic call flow will slightly differ for each of the protocols, as some of the transactions may be missing, some of the responses may come unsolicited, or the contents may be spread over multiple messages due to length limitations.

13.2.2.1 Transaction Initiator

An important aspect of the positioning session, which is not considered in the generic call flow in Figure 13.1, is how the positioning session is started. Which of the involved entities starts the request? The 3GPP in [3] foresees three possibilities:

- *MO-LR*: the UE starts the positioning session.
- *MT-LR*: the positioning is started by an external LCS Client.
- *NI-LR*: the network starts the location request in the context of an emergency call.

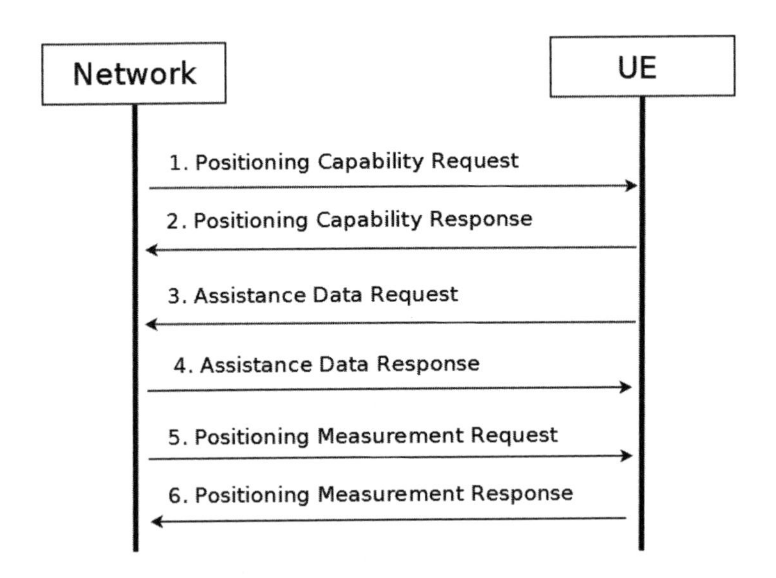

Figure 13.1 Basic positioning call flow.

The OMA has defined options equivalent to the 3GPP standards, but using different nomenclature. In the OMA positioning protocol for the U-Plane, SUPL (see Section 13.5), a MO-LR is called SET-Initiated session, where SET stands for SUPL-Enabled Terminal. The NI-LR is called Network Initiated session.

Typically, RRLP, RRC, or LPP C-Plane positioning sessions are initiated by the network. Nevertheless, a mobile may start a mobile originated location session via control plane by using the LCS protocol, which is explained in more detail in Section 13.6.1. In practice, most mobiles are configured to use the user-plane set-initiated session (see Section 13.5.4) rather than the MO-LR session when they require a location session.

13.2.3 Position Calculation Modes

All the protocols described in this book support at least two basic position calculation modes:

- UE-based: the calculation of the location coordinates takes place in the UE, potentially with the help of the assistance data provided by the network. The UE reports a calculated position.

- UE-assisted: the calculation of the location coordinates takes place on the network side. The UE provides measurements that are used for the calculation.

The nomenclature differs slightly depending on the protocol. In RRLP, these terms are referred as MS (mobile station) based or MS assisted, respectively. Furthermore, some protocols support additional methods, such as standalone (wherein the UE calculates its own GPS-based position without the help of any assistance data), but these are typically not used for emergency services.

Not all positioning technologies support both based and assisted modes; each has its own particularities. A-GNSS support both methods. Historically, UE-Assisted for A-GNSS mode was intended to reduce computational complexity in the GPS receiver (an aspect which became largely irrelevant due to Moore's law). For technologies like ECID, OTDOA, Wi-Fi, and Bluetooth, the UE-assisted mode has the advantage that the exact base station locations do not need to be transmitted to the mobiles. The precise locations may be considered a trade secret by some operators and therefore the UE-assisted mode helps to hide the network topology from competitors. There are occasional discussions in 3GPP on whether the base station location information should be added to LPP. However, this information has not yet been added for LTE LPP. Interestingly, in TIA-801, a UE-based mode for ECID and A-FLT were defined from the beginning. For NR LPP Release-16 and higher, which will be seen in Chapter 15 of this book, UE-based DL-TDOA is supported.

All RRLP, RRC, and LPP support multiconstellation, multifrequency A-GNSS for GPS, Glonass, BeiDou, Galileo, and QZSS. These A-GNSS measurements can be combined in the same transaction with other technologies such as ECID or OTDOA, but only if they are configured for the same position calculation mode. It is generally not possible to combine UE-based and UE-assisted measurements in the same transaction.

13.3 DATA ENCODING

The standardization of protocols aims at ensuring interoperability between different implementations; for example, a location server of vendor A and a location-enabled mobile of vendor B. Protocols make use of messages for information transfer between network nodes. Messages are structured to be efficient in terms of transmission bandwidth used and processing power required to encode and decode the messages. It is also very important that messages can be extended for future feature enhancement. Additionally, they should allow backward compatibility, where an

IEI	Information Element	Type/Reference	Presence	Format	Length
	Protocol discriminator	Protocol discriminator 9.2	M	V	1/2
	Security header type	Security header type 9.3.1	M	V	1/2
	CS service notification message identity	Message type 9.8	M	V	1
	Paging identity	Paging identity 9.9.3.25A	M	V	1
60	CLI	CLI 9.9.3.38	O	TLV	3-14
61	SS Code	SS Code 9.9.3.39	O	TV	2
62	LCS indicator	LCS indicator 9.9.3.40	O	TV	2
63	LCS client identity	LCS client identity 9.9.3.41	O	TLV	3-257

Figure 13.2 Definition of the CS Service Notification message in [1].

older version of a node can communicate seamlessly with a newer version of the node.

13.3.1 Tabular Encoding

Tabular encoding is one straightforward way of encoding messages that are represented in binary format. One method is to represent the subelements of the message (also known as information element (IE)) in type length value (TLV) format. TLV specifies message encoding in tables, which explicitly detail all bits of the message. This is the approach followed by 3GPP for NAS protocols, such as TS 24.301 [1], and by 3GPP2 for C.S0022-1 [4]. The TLV standard used by 3GPP is defined in TS 24.007 [5], including the different formats (V, LV, T, TV, TLV, LV-E, TLV-E) and the five categories of information elements (type 1, 2, 3, 4, and 6) used in all 3GPP TLV-based specifications.

An example of a TLV message definition can be found in Figure 13.2, taken from the CS Service Notification message in TS 24.301 [1]. The IEs in the message are placed sequentially as specified in the table. Both a transmitter and receiver follows this table to encode and decode the message contents. The column "Presence" indicates whether a certain IE is mandatory or optional. The column "Format" indicates which of the three parts (type, length, and value) will be sent for this IE. For mandatory IEs, the type is pre-defined and does not need to be transmitted. Optional IEs need to include at least the value and type, so that the receiver knows what to decode. The Length can be either predefined or variable, in which case it needs to be transmitted as well.

```
LCS indicator value
   Bits
8 7 6 5 4 3 2 1
0 0 0 0 0 0 0 0    Normal, unspecified in this version of the protocol.
0 0 0 0 0 0 0 1    MT-LR
0 0 0 0 0 0 1 0
   to  Normal, unspecified in this version of the protocol
1 1 1 1 1 1 1 1
```

Figure 13.3 LCS indicator IE value definition.

Taking a closer look at the LCS indicator from this example, it can be seen that it is a 2-octet value. The first of the octets contains the LCS indicator IE identification, used to indicate that this part of the message corresponds to a LCS indicator type. The second octet contains the value, defined in Figure 13.3. Up to Release 15, the LCS indicator has only one defined value, "MT-LR." All other possible values are left unspecified.

When extending the TLV scheme for a new feature, a new type is assigned. Older devices that do not know this type must skip the corresponding part, as they can derive the total length of the data with the length field. This approach looks simple at the beginning, but designing the specification in a consistent way and writing compliant encoders and decoders involves a lot of error-prone manual work.

One of the advantages of the TLV format is that the IEs are octet aligned, simplifying the encoding and decoding of messages. However, this format is not bandwidth efficient since it does not compress information below octet size.

13.3.2 ASN.1 Encoding

To make protocol standardization simpler and more consistent, the ITU-T and the ISO introduced the Abstract Syntax Notation One (ASN.1), which allows an unambiguous and extensible protocol specification. According to the ITU-T definition in [6], ASN.1 specifies elementary types like BIT STRING, BOOLEAN, or INTEGER and aggregations of these types: CHOICE, SEQUENCE, SEQUENCE OF, SET, or SET OF. Furthermore, it includes some more special types like UTCTIME. ASN.1 messages can be serialized into a sequence of bytes using ANS.1 Encoders. Similarly, decoders are used to process these messages. ASN.1 abstracts the message description from the message encoding. There are multiple message encoding formats that can be used, for example:

- Basic encoding rules (BER).

```
-- ASN1START

LPP-MessageBody ::= CHOICE {
    c1                      CHOICE {
        requestCapabilities         RequestCapabilities,
        provideCapabilities         ProvideCapabilities,
        requestAssistanceData       RequestAssistanceData,
        provideAssistanceData       ProvideAssistanceData,
        requestLocationInformation  RequestLocationInformation,
        provideLocationInformation  ProvideLocationInformation,
        abort                       Abort,
        error                       Error,
        spare7 NULL, spare6 NULL, spare5 NULL, spare4 NULL,
        spare3 NULL, spare2 NULL, spare1 NULL, spare0 NULL
    },
    messageClassExtension   SEQUENCE {}
}

-- ASN1STOP
```

Figure 13.4 Example of ASN.1 notation in the LPP message body [7].

- Packed encoding rules (PER).

The encoding format defines the octet structure of the underlying message. The PER format is used, for instance, for RRC message transmissions. This format is designed be bandwidth efficient by packing IEs and reducing the total number of octets needed compared to the BER format.

An example of ASN.1 notation can be found in Figure 13.4, taken from TS 36.355 [7]. This example shows the LPP message body defined for LPP. This message is a CHOICE, meaning that the content of the message can be different depending on what is selected for the item "c1." In particular, the LPP message body can contain any of *RequestCapabilities*, *ProvideCapabilities*, *RequestAssistance-Data*, *ProvideAssistanceData*, *RequestLocationInformation*, *ProvideLocationInformation*, *Abort* or *Error* messages.

If certain rules are followed, ASN.1 defined protocols are backward compatible (e.g., an older client can ignore new elements). For example, the latest RRLP version of 2019 is still compatible with the version from 2002, meaning that a legacy UE implementing this version should still work with a brand-new network element.

ASN.1 defines the operator "..." for addition of elements in future versions of the protocol while ensuring backward compatibility. An example can be seen in the *ProvideCapabilities-r9-IEs* in Figure 13.5. The Provide Capabilities message is a SEQUENCE, meaning that it contains a series of messages or containers. In

```
-- ASN1START

ProvideCapabilities ::= SEQUENCE {
    criticalExtensions      CHOICE {
        c1                      CHOICE {
            provideCapabilities-r9      ProvideCapabilities-r9-IEs,
            spare3 NULL, spare2 NULL, spare1 NULL
        },
        criticalExtensionsFuture    SEQUENCE {}
    }
}

ProvideCapabilities-r9-IEs ::= SEQUENCE {
    commonIEsProvideCapabilities        CommonIEsProvideCapabilities        OPTIONAL,
    a-gnss-ProvideCapabilities          A-GNSS-ProvideCapabilities          OPTIONAL,
    otdoa-ProvideCapabilities           OTDOA-ProvideCapabilities           OPTIONAL,
    ecid-ProvideCapabilities            ECID-ProvideCapabilities            OPTIONAL,
    epdu-ProvideCapabilities            EPDU-Sequence                       OPTIONAL,
    ...,
    [[  sensor-ProvideCapabilities-r13  Sensor-ProvideCapabilities-r13      OPTIONAL,
        tbs-ProvideCapabilities-r13     TBS-ProvideCapabilities-r13         OPTIONAL,
        wlan-ProvideCapabilities-r13    WLAN-ProvideCapabilities-r13        OPTIONAL,
        bt-ProvideCapabilities-r13      BT-ProvideCapabilities-r13          OPTIONAL
    ]]
}

-- ASN1STOP
```

Figure 13.5 Example of the usage of the "..." operator in the ProvideCapabilities LPP message [7].

this case, it contains again a CHOICE, but there is only one possibility to select, the *provideCapabilities-r9*. The container has spare values for future use and a potential *criticalExtensionsFuture*. The *ProvideCapabilities-r9-IEs* message is also a SEQUENCE, containing values for A-GNSS, OTDOA, and other positioning technologies seen in Chapters 6 and 8 of this book, plus the extension operator "...". In E-UTRA Release 13, 3GPP decided to incorporate new positioning technologies to LPP, such as sensor, Wi-Fi, and Bluetooth (explained in Chapter 9 of this book). The presence of the operator "..." allowed the extension of this message in a backward compatible way. This was used by the new elements added in Release 13 of this message, as seen by the "-r13" extensions (see Figure 13.5).

If the extension element is missing in a sequence, it cannot be extended in a backward-compatible way. An example of this is the *Ellipsoid-Point* sequence, which is used to describe a geographic shape to report WGS84 positioning coordinates. As it can be seen in Figure 13.6, the original element does not include the "..." operator, meaning that it cannot be extended. An extensible definition, including the "..." operator, is shown at the bottom of the image. Since the original *Ellipsoid-Point* cannot be extended, the 3GPP had to define new sequences in LPP Release 15 to include high accuracy parameters (e.g., see *HighAccuracyEllipsoid-PointWithUncertaintyEllipse* in [7]).

```
Ellipsoid-Point ::= SEQUENCE
{
  latitudeSign ENUMERATED {north, south},
  degreesLatitude INTEGER (0..8388607),
  degreesLongitude INTEGER (-8388608..8388607)
}
```

Non-extensible ASN.1 element

```
Ellipsoid-Point-Extensible ::= SEQUENCE
{
  latitudeSign ENUMERATED {north, south},
  degreesLatitude INTEGER (0..8388607),
  degreesLongitude INTEGER (-8388608..8388607),
  ...
}
```

Future-extensible ASN.1 element (notice the three dots at the end)

Figure 13.6 Original Ellipsoid Point sequence and example of a future-extensible definition [7].

There are several tool sets that can automatically generate java, C, C++, C#, or other languages' encoders and decoders from the ASN.1 message definitions. This speeds up development and also rules out human error to a large degree. The binary format on-the-wire depends on the configuration of the codecs. The encoding used for location protocols like RRC, RRLP, SUPL, and LPP is "BASIC-PER Unaligned Variant" [8].

13.3.3 HTTP2/JSON, OpenAPI 3.0.0, and YAML

Load balancing and dynamic deployment of 4G entities like an E-SLMC is not simple to achieve using traditional protocols. For that reason, the 5G core follows a different communication approach in general, such as for the interface between the AMF and the LMF. All interfaces within the 5G core are service-based (see TS 23.501 [9], TS 23.502 [10]), meaning they use a stateless HTTP2/JavaScript Object Notation (JSON) interface based on OpenAPI 3.0.0 syntax. This allows NFs like the AMF or the LMF to be deployed dynamically in virtual machines in a cloud and allows NFs of various vendors to easily interwork.

All 5G core internal services are specified in OpenAPI 3.0.0 specification in YAML Ain't Markup Language (YAML) format, which is included in a separate file with every 3GPP 5GC specification. An example of the OpenAPI 3.0.0 is shown in

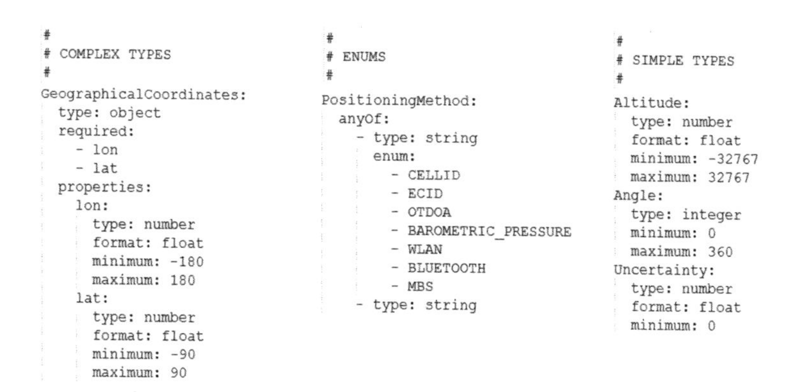

Figure 13.7 Example of the Open API definition for the 5G core network.

Figure 13.7, extracted from the NImf_Location API in TS 29.572 [11]. The figure shows three different schemas, corresponding to complex types, simple types, and enums.

Similar to the ASN.1 definition, the YAML code can be used to autogenerate encoders and decoders and removes error sources.

13.3.4 HTTP1/XML

The Mobile Location Protocol (MLP) for the Le interface (see [12]) is defined in xml and may be transported via Simple Object Access Protocol (SOAP), Wireless Session Protocol (WSP), or HTTP1. Later versions of the OMA specifications restricted the transport method to HTTP only.

13.4 C-PLANE LEGACY PROTOCOLS: 2G, 3G

This section will briefly describe the legacy positioning protocols used in 2G and 3G networks, pointing out the main differences to the LPP protocol used in 4G and 5G, which will be described in detail in the subsequent chapters.

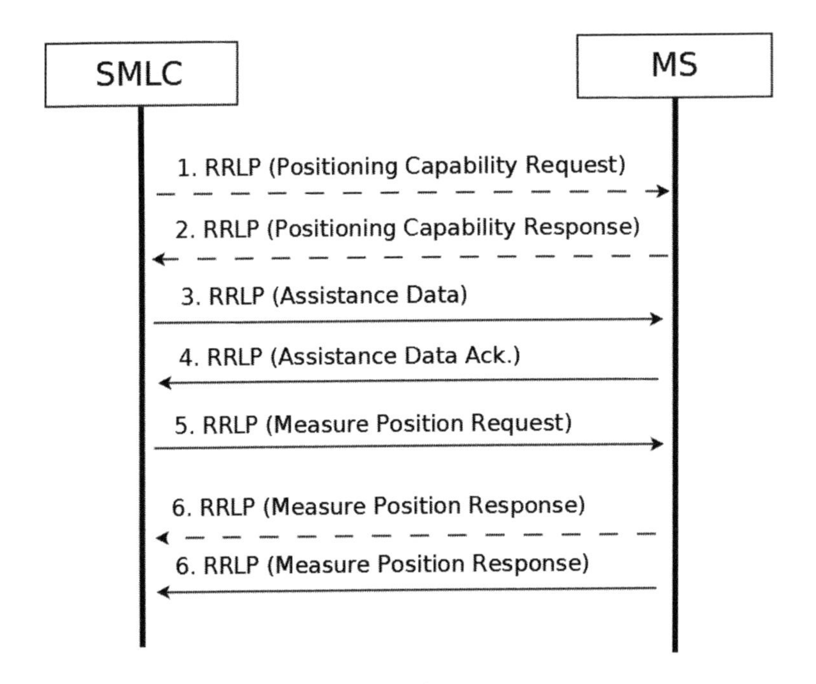

Figure 13.8 RRLP position session call flow.

13.4.1 GSM: 3GPP RRLP

RRLP was one of the first positioning protocols standardized, coming from the days of GSM [13]. The basic positioning session call flow for RRLP is shown in Figure 13.8.

The network initiates a positioning session with the Positioning Capability request. This part is in dashed lines because it is optional, as will be explained later. After the positioning capabilities have been exchanged, the network provides any needed assistance data, which is acknowledged by the MS. The network requests position measurements and the MS tries to perform the requested measurements and send a measurement response. The measurement report or assistance data may come in multiple messages due to message segmentation, as will be explained later in this section. At any point, the MS may send an RRLP Protocol Error message instead of the expected response to indicate that there has been a problem with the reception of the previous message from the network.

One of the main differences between LPP and RRLP is the initial lack of a capabilities transaction. Early versions did not include a capability exchange mechanism at all (i.e., steps 1 and 2 in Figure 13.8 did not take place). The UE's capabilities were conveyed to the server via RRC protocol. This concept was not problematic for C-Plane in GSM, as the location server in GSM (and WCDMA) is a part of the RAT. However, this was a problem when using RRLP with the OMA SUPL top-IP, as a SUPL server has no connection to the RAT and cannot query the capabilities. Consequently, SUPL was specified with its own capability exchange transaction that is only used in case RRLP is carried. After the SUPL specification was completed, RRLP was extended with a capabilities routine. To avoid ambiguous behavior in implementations, it was specified not to use the RRLP capability routine if used with SUPL.

RRLP is standardized in ASN.1. It supports MS-assisted and MS-based methods (known as UE-assisted and UE-based in LPP) and E-OTD and GNSS positioning technologies. The initial versions of RRLP were only designed for GPS. Later, it was extended for Galileo and then generalized for other GNSSs. The extension term GANSS refers to all GNSS except GPS. If a GANSS RRLP element is used without an explicit system identifier, it refers to Galileo. This differentiation between GPS and the rest of GNSS technologies is present in RRLP and also in the RRC protocol. The LPP was the first protocol to eliminate this distinction, meaning that the GNSS nomenclature changes between RRLP and LPP.

13.4.1.1 Segmentation

The maximum transmission unit (MTU) in GSM C-Plane might be exceeded by large RRLP messages. For that reason, RRLP allows segmentation. In practice, this problem can occur at two instances. The first is during the assistance data delivery procedure if the assistance data message is large. In this case the network may send assistance in multiple assistance data components and steps 3 and 4 in Figure 13.8 is repeated multiple times. The second case is during the position measurement response. The MS is allowed to separate the position message in a maximum of two separate parts of the same transaction, as shown in Figure 13.8 by the two steps labeled as 6.

Since the maximum transmission size of a RRLP-via-SUPL message is much higher, this segmentation is normally not used on U-Plane.

13.4.1.2 UE-Assisted Mode for GPS L1 C/A

While LPP encodes Doppler shifts in m/s and they need to be scaled to the actual GNSS signal's frequency, RRLP provides the values directly in Hz if used for GPS L1 C/A. Also the codePhase value definition is significantly different in RRLP for GPS L1 C/A when compared to other GANSSs or to LPP.

13.4.1.3 RRLP's ASN.1 Bug

Even though great care is taken to ensure compatibility, occasionally errors happen and lead to incompatibilities.

In RRLP specifications between Release 99 and Release 8, a formerly optional element (*referenceFrameMSB*) in the timing measurement sequence was accidentally changed to mandatory. This error was identified some years later and corrected with revisions (even old versions of the specification were changed retrospectively). Nonetheless, the mobiles produced in between might be affected by this issue. Luckily, this problem occurred in a protocol element rarely used.

Similarly, several Galileo elements were incompatibly changed in Release 12 of LPP, RRLP, and RRC (as the previous protocol versions were based on the draft ICD of Galileo that differed from the final Galileo ICD). For this reason, any entity implementing A-Galileo needs to use a protocol version of Release 12 or later to be compatible.

13.4.1.4 RRLP Summary

As RRLP evolved since the late 1990s, it was constantly extended and offers similar capabilities for A-GNSS as LPP, including support for GPS, Galileo, BeiDou, Glonass, and QZSS. Naturally, it does not support RAT-dependent technologies for LTE. Furthermore, it also lacks Wi-Fi, Bluetooth and barometric support. The requirement to maintain backward compatibility made the protocol more and more complex over the years. In the United States, RRLP does no longer play a significant role. However, many other regions of the world with less-strict emergency caller location requirements still utilize the protocol over both C-Plane and U-Plane.

13.4.2 WCDMA: 3GPP UTRAN RRC

WCDMA's RRC location part shares many commonalities with RRLP. In fact, most of its atomic message elements are almost exactly same as in RRLP. The main difference between GSM's RRLP, and WCDMA's RRC is the deep integration of the WCDMA location protocol within the overall RRC protocol in TS 25.331 [14]; there is no difference between general RRC measurements and specific location measurements. An A-GNSS location session is set up similar to a neighbor cell measurement for a handover. The UE positioning measurement request is integrated in the *Measurement Control* message and the UE positioning measurement results are integrated in the *Measurement Report* message. WCDMA RRC positioning supports WCDMA OTDOA, GPS, and GANSS technologies.

The deep integration of positioning within the RRC protocol may also have complicated and eventually prevented the use of WCDMA's RRC location part in SUPL. While SUPL 1.x and 2.x formally specify a container for RRC messages, RRC-over-SUPL was never implemented in commercial chipsets. Since WCDMA chipsets normally support GSM anyway, it was easier to reuse the RRLP protocol also in 3G-mode when using SUPL. Nevertheless, RRC was used by U.S. operators for WCDMA-based emergency calls and served its purpose for many years. While A-Glonass is supported by many chipsets in RRC, A-Galileo and A-BeiDou are usually only integrated in RRLP and LPP.

This deep integration of location routines in RRC was given up in LTE with the dedicated LPP protocol. 5G NR also uses the dedicated LPP protocol and split RRC from location procedures.

13.4.3 CDMA2000: 3GPP2 C.S0022 (TIA-801)

The 3GPP2 CS.0022 series ([4, 15–17]) has enabled location support in 3GPP2 cellular systems since 1999. It was specified for analog AMPS, IS-95 (cdmaOne), IS-2000 (CDMA2000), and EVDO.

According to the numbering scheme of 3GPP2, three revisions of the standard exist, which differ in feature support. The initial revision 0 is available in two versions; the higher version includes corrections and clarifications. The different versions and their additional features are shown in Figure 13.9.

The main positioning technologies used were cell-ID, A-GPS, and A-FLT (which is a ToA method based on the mobile network's base stations). Since the 3GPP2 networks were inherently synchronous to GPS time, they could provide fine-time assistance to the mobiles and the A-FLT measurements can be combined with GPS measurements for a hybrid location fix. The main mode used for 911 caller location was UE-assisted, but the protocol and most mobiles also support UE-based mode. Since foreign GNSS systems require a waiver for emergency call use, the

3GPP2 name	TIA name	Features	ASN.1 package	Published
C.S0022-0 v1.0	IS-801	Initial version	-	1999
C.S0022-0 v3.0	IS-801-1	Sat Health Info, Incompatible PilotPhaseMeasurements, PilotStrength, TotalRX Power, BaseID	-	2001
C.S0022-A v1.0	TIA-801-A	HRPD support, Several Extended Elements	-	2004
C.S0022-B v1.0, (planned for UWB)	-	Glonass, QZSS, Galileo, Beidou Modernized GPS, Extension element with ASN.1 encoding	C.R0022-B v1.0	2009

Figure 13.9 Versions of the CS.0022 specification.

use of Galileo, Glonass, and BeiDou was initially not required. When Galileo was granted the waiver, the CDMA systems were already largely phased out.

At the time of writing, most operators shut down their 3GPP2-compliant services. CDMA2000 is the last remaining standard of the family that is still in use as of 2020 for voice calls. In the foreseeable future, CDMA2000 will also be phased out and the TIA-801 location protocol will become obsolete.

13.5 SUPL: THE USER-PLANE LOCATION PROTOCOL

Based on the foundation of the 3GPP and 3GPP2 location protocols LPP, RRLP, and TIA-801, the OMA developed a standard (see [19]) that allows the transmission of these protocols over any IP data link. One driving force was to avoid the upgrade cost of legacy networks and to reduce signaling load by moving commercial location sessions to the U-Plane.

The network architecture and the difference in routing between C-Plane and U-Plane positioning sessions have been explained in detail in Chapters 3 to 5 of this book. The OMA SUPL protocol does not require any modifications of the core network or radio access network. The tunneling function of LPP/RRLP/RRC in SUPL is rather straightforward. Nonetheless, the standard also needs to ensure privacy and security functions and allow network-initiated location requests to be a full replacement for C-Plane solutions.

So far, there are three major versions of SUPL: 1.x, 2.x, and 3.x. While 2.x introduced several improvements, like LTE support, multi-GNSS, periodic sessions,

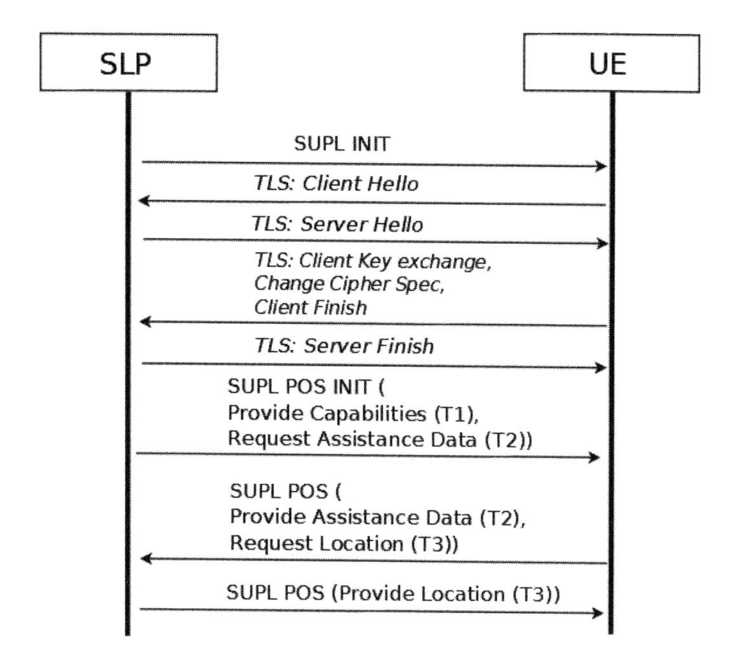

Figure 13.10 SUPL positioning call flow.

SUPL 3.x was intended as a simplification and never found widespread use, as it offers little advantages compared to SUPL 2.x.

13.5.1 SUPL Call Flow

The SUPL call flow presents a couple of differences with respect to the basic call flow seen in Figure 13.1. For a start, the SUPL session needs to be initiated prior to the positioning session. This also includes the required Transport Layer Security (TLS) handshake (as will be explained in the next section).

All these differences can be seen in Figure 13.10. Some of the messages are combined in the same SUPL message: the provide capabilities with the request assistance data, and the provide assistance data with the request location information. The provide capabilities message also comes unsolicited.

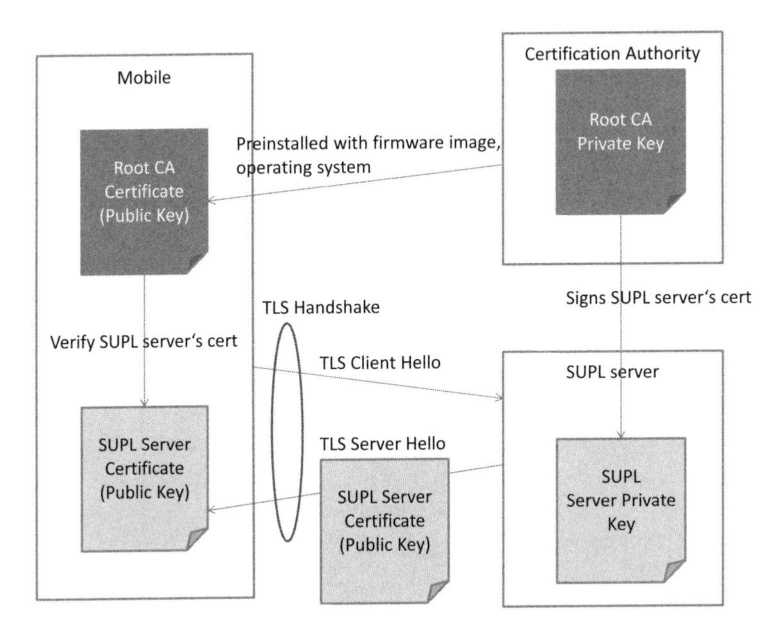

Figure 13.11 Public key infrastructure for SUPL (same as for HTTPS).

13.5.2 Security and Encryption

Security on C-Plane is ensured inherently by the encryption of the EPS/NAS. On the U-Plane, security needs to be applied explicitly by using the TLS protocol. The TLS handshake includes authentication of the location server versus the mobile and is shown in Figure 13.11. The mobile verifies the certificate sent by the SUPL server against a trusted root certificate. This is the same mechanism used for any HTTPS-secured web page. Per specification, SUPL 1.0 mandates TLS 1.0, SUPL 2.0 mandates TLS 1.1, and SUPL 2.0.4 mandates TLS 1.2. In practice, this is interpreted as a "higher than" requirement and a SUPL 1.0 device can also use a TLS 1.2 connection.

13.5.3 Network-Initiated Sessions

As SUPL was designed to be a full replacement for C-Plane positioning, it requires a mechanism to initiate a location session from the network (e.g., for usage in

E911 scenarios). Currently, this can be achieved with relative ease by sending a special UDP message to port 7275 of the mobile, as modern smartphones are always online. 2G and 3G phones might, however, not always have an active IP connection. Therefore, SUPL specifies special binary Push SMS, which can start a packet data network (PDN) session from the mobile. There are two formats for these binary SMSs:

1. MT-SMS (binary SMS with a special header for SUPL)
2. WAP-Push (binary SMS with a special header for WAP, which includes a SUPL content)

These SMSs can in turn be delivered via C-Plane (SMS over SG) or via IMS (SMS over IMS). On reception of such a SMS, the mobile connects to a preconfigured fully qualified domain name (FQDN), which is hardwired in the firmware of the phone. If no such FQDN is configured or is not reachable, the UE can construct a FQDN from the MNC and MCC of the used SIM card in the following pattern:

h-slp.mnc<MNC>.mcc<MCC>.pub.3gppnetwork.org (Regular SUPL Location Platform (SLP))

e-slp.mnc<MNC>.mcc<MCC>.pub.3gppnetwork.org (Emergency SLP)

As the domain 3gppnetwork.org is registered to the GSM Association, it can be ensured that only operators can register a certificate for them. A notable exception to this security mechanism is the SUPL emergency session. If a 911 call is ongoing, UEs accept a FQDN provided by the network.

13.5.4 Set-Initiated Sessions

This variant is the workhorse for providing GNSS assistance data to mobiles. Whenever an app or the mobile's operating system requires a GNSS position fix, the SUPL client connects to the SUPL sever, performs a TLS handshake, and retrieves the required data. While this session type is not used for traditional E911 calls, the advent of ELS and other top-IP emergency applications give this mechanism a new importance.

13.5.5 SUPL Version Compatibility

SUPL 2.x (and 3.x) allows full backward compatibility:

1. The SUPL 2.x UE may fall back to 1.0 if the server is not supporting 2.x
2. The SUPL 2.x server may respond as 1.0 if a legacy device requests a connection

Note that compatibility might be hampered by security requirements, such as if a server does not allow a legacy UE to downgrade to TLS 1.0.

13.5.6 Proprietary Protocols

Almost every GNSS chipset manufacturer operates a proprietary assistance data delivery protocol and its own servers. Wi-Fi based location relied heavily on proprietary protocols until recently. Nevertheless, SUPL is the only industry standard protocol for top-IP Assistance data provision and offers several unique features like E911 support even when roaming. Its backward compatibility allows long-term support in phones independent of the actual chipset manufacturers and allows operators to ensure the quality of A-GNSS in their networks.

13.6 PRIVACY

The privacy concept for 3GPP's and OMA's location-based services foresees that the network operator stores the privacy rules for every user in a database. Depending on regional legislation and the user type (e.g., SIM for smartphone vs. SIM for tracking device), user consent needs to be obtained before a user's location is provided. The user consent may be overridden for emergency calls.

13.6.1 C-Plane Privacy and MO Sessions: LCS Protocol

Besides LPP, RRLP, or RRC messages, the core network's NAS may also carry LCS messages specified in TS 24.080 [20]. These messages can be used to query user consent for a MT-LR.

Since 3GPP decided to introduce location services in 5G Release 15 only for regulatory purposes (NI-LR does not require user consent or mobile originated sessions), the transport of LCS messages over the 5G NAS is not yet specified. The document TS 23.273 [21] summarizes planned LCS services in 5G scheduled for Release 16.

13.6.2 SUPL Privacy

User plane location provides very similar privacy protection measurements as C-Plane. First and foremost, the TLS protocol with its certificate-based handshake ensures that only a preconfigured, trusted SUPL server is accepted by the mobile. The TLS encryption also protects against simple eavesdropping; for example, the UE is connected via an unprotected WLAN hotspot. Only the trusted SUPL server may have access to the phone's location.

In the case of a set-initiated (SI) session, it is under the UE's control when the server is contacted. In case of network-initiated (NI) sessions, the SUPL protocol includes the "notification and verification" feature similar to the LCS protocol on C-Plane. This allows the operator of the SUPL server to ask the user for consent if the network initiates a location request.

13.7 PERIODIC SESSIONS AND GEOFENCING

The most traditional positioning use case is a one-time request for a specific application (e.g., during an emergency call). However, there are also use cases where the position must be calculated periodically (e.g., for navigation) and also geofencing (i.e., a virtual perimeter around a geographic area that must send a certain trigger when it is crossed). Such use cases are also supported in the different positioning protocols.

13.7.1 C-Plane

To set up geofencing or periodic sessions via control plane, the LCS protocol (which also provides the privacy notification features) allows to set up special types of location sessions.

13.7.2 U-Plane

While SUPL 1.0 was limited to GPS and simple NI/SI sessions, SUPL 2.0 brought many extensions like periodic sessions and geofencing.

Most chipsets support these features, but with the advent of smartphones, these features became less relevant since mobile operating systems offer efficient implementations for these use cases on a higher layer.

13.8 CONCLUSION

This chapter has served as an introduction to the different positioning standards defined by 3GPP, 3GPP2, and OMA for cellular network positioning services, briefly describing the legacy standards used for 2G and 3G networks. It has also given an overview of the different encoding mechanisms used and the difference between C-Plane and U-Plane positioning. The following chapters will focus on the most relevant positioning standard, the LPP, which was first introduced for 4G networks and is being reused for 5G.

All these protocols are used for the communication between the base station and the mobile device via the air interface. Chapter 16 will examine the different protocols used for positioning in the core network, such as LPPa.

References

[1] 3GPP TS 24.301, *Non-Access-Stratum (NAS) Protocol for Evolved Packet System (EPS)*, V16.2.0, September, 2019.

[2] 3GPP TS 24.501, *Non-Access-Stratum (NAS) Protocol for 5G System (5GS)*, V16.2.0, September, 2019.

[3] 3GPP TS 32.271, *Telecommunication Management; Charging Management; Location Services (LCS) Charging*, V15.0.0, June, 2018.

[4] 3GPP2 C.S0022-0 *Location Services (Position Determination Service)*, V1.0, March, 2012.

[5] 3GPP TS 24.007, *Mobile Radio Interface Signalling Layer 3; General Aspects*, V16.1.0, June, 2019.

[6] ITU-T X.680 *Information Technology – Abstract Syntax Notation One (ASN.1): Specification of Basic Notation*, August, 2015.

[7] 3GPP TS 36.355, *LTE Positioning Protocol (LPP)*, V15.5.0, September, 2019.

[8] ITU-T X.691 *Information Technology – ASN.1 Encoding Rules: Specification of Packed Encoding Rules (PER)*, August, 2015.

[9] 3GPP TS 23.501, *System Architecture for the 5G System*, V16.2.0, September, 2019.

[10] 3GPP TS 23.502, *Procedures for the 5G System (5GS)*, V16.2.0, September, 2019.

[11] 3GPP TS 29.572, *5G System; Location Management Services; Stage 3*, V16.0.0, September, 2019.

[12] OMA-TS-MLP, *Mobile Location Protocol*, V3.5, December, 2018.

[13] 3GPP TS 44.031, *LCS; MS - SMLC Radio Resource LCS Protocol (RRLP)*, V15.0.0, June, 2018.

[14] 3GPP TS 25.331, *RRC Protocol Specification*, V15.4.0, September, 2018.

[15] 3GPP2 C.S0022-0 *Position Determination Service Standard for Dual Mode Spread Spectrum Systems*, V3.0 , February, 2001.

[16] 3GPP2 C.S0022-A *Position Determination Service for cdma2000 Spread Spectrum Systems*, V1.0, March, 2004.

[17] 3GPP2 C.S0022-B *Position Determination Service for cdma2000 Spread Spectrum Systems*, V1.0, April, 2019.

[18] 3GPP2 C.R0022-B *Position Determination Service for cdma2000 Spread Spectrum Systems Software Distribution*, V1.0, August, 2011.

[19] OMA-TS-ULP, *User Plane Location Protocol*, V2.0.4, December, 2018.

[20] 3GPP TS 24.080, *Mobile Radio Interface Layer 3 Supplementary Services Specification*, V16.0.0, September, 2019.

[21] 3GPP TS 23.273, *5G System Location Services (LCS)*, V16.1.0, September, 2019.

Chapter 14

Positioning Protocol in LTE

14.1 INTRODUCTION

The previous chapter was a first introduction to positioning protocols in cellular networks. This chapter will focus on the LPP, defined by the 3GPP in TS 36.355 [1]. While LTE was introduced in 3GPP Release 8, the location framework followed one cycle later with Release 9. Between Release 9 and Release 12 it was marginally extended (e.g., A-BeiDou and Inter-Frequency OTDOA were added), but there were no fundamentally new features and the working principle remained the same over these releases.

As of Release 13, LPP experienced major updates. With the advent of the IoT, OTDOA and ECID were adapted to support this technology. In addition, to support the stricter FCC indoor positioning regulations, the OMA LPPe, had already added support for noncellular network based positioning technologies (Wi-Fi, Bluetooth) and atmospheric pressure measurements. The 3GPP Release-13 included a similar set of features in the LPP standard. Later, in Release 15, the 3GPP continued to enhance LPP to support high-accuracy GNSS methods (such as RTK, discussed in Chapter 7 of this book) and IMU measurements, discussed in Chapter 12 of this book. In order to accommodate these features, LPP underwent a major upgrade with the definition of a posSIB for broadcasting the assistance data.

This chapter will provide a detailed overview of LPP, from its first version to the latest LTE upgrades in the 3GPP Release 15.

14.2 LPP

The WCDMA RRC positioning protocol was not very successful in penetrating the commercial (IP-based) market. One of the reasons for this is the tight integration of the positioning functionality within the WCDMA control plane protocols, which is not very practical for SUPL. Therefore, the 3GPP decided to define a separate specification for LTE positioning [1] independent of other network functionality.

Nevertheless, the 3GPP followed the same general structure for LPP as all positioning protocols previously seen, with some differences.

14.2.1 Fundamental Transactions and Basic Call Flow

The transactions that form an LPP session are very similar to the ones already defined in the previous chapter of this book. There is a capability exchange to determine the positioning modes and technologies that are supported by the device; then the assistance data (if any data is required or available) is provided to the mobile device, and finally there is an exchange of positioning information, either in the form of measurements or a UE-based calculated position.

However, when used over the C-Plane, the assistance data is typically sent by the network in an unsolicited mode. The UE may still send an assistance data request if the information provided by the network is not sufficient or it requires some additional data. Another particularity is that each LPP message may carry an acknowledgment request, indicating to the receiver that it should acknowledge the correct reception of the message.

These transactions are reflected in the call flow in Figure 14.1. The optional messages (i.e., the LPP request assistance data and the acknowledgments) are marked with dashed lines and square brackets. Each request / response pair is identified by a transaction ID, in the example T1, T2, and T3. The LPP response time, also known as TTFF, is indicated by the network as part of the QoS parameters and represents the maximum time that the mobile device is allowed to take for the measurement.

14.2.2 Description of the LPP Transactions

This section will describe in detail the messages and some of the most relevant information elements in LPP. The reader may refer to TS 36.355 [1] for a description of all other elements not included in this overview.

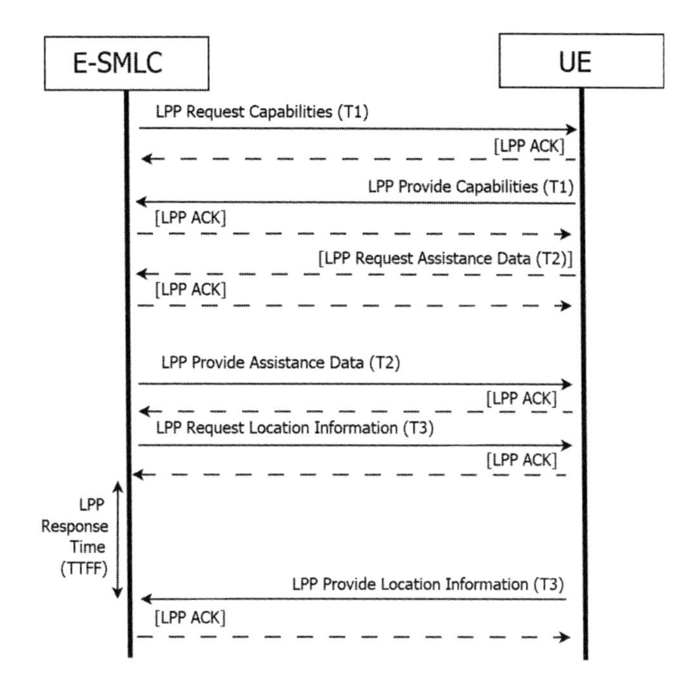

Figure 14.1 Basic call flow for a regular LPP session.

All the information relating to the LPP positioning section is encapsulated into LPP-Message containers, as shown in Figure 14.2. This message is a sequence (i.e., it contains a series of information elements). These include the following parameters:

- Transaction ID: a unique identifier for requests, responses, or error messages that belong to the same transaction and also which entity has initiated the transaction, whether it has been the location server or the target device (the UE).
- End transaction: indicates that this is the last message of a transaction (e.g., the last positioning information report in case of multiple reports).
- Sequence number: indicates the order of the message within the transaction (e.g., the request typically would be the first message of a transaction).

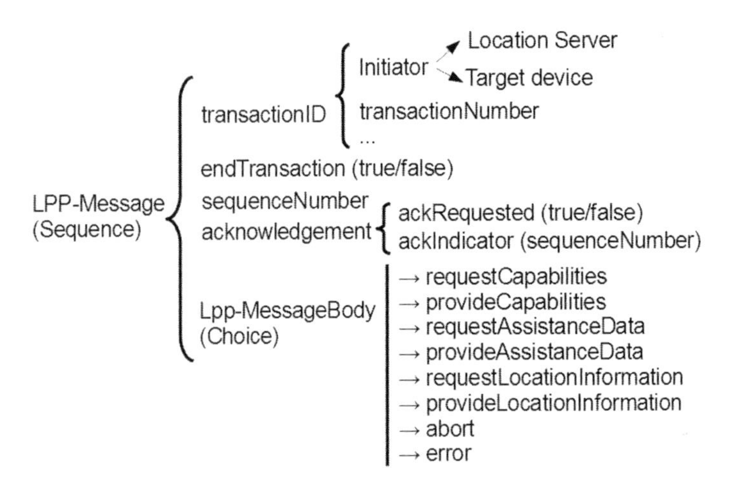

Figure 14.2 Generic LPP-Message container.

- Acknowledgment: this parameter is a sequence number containing a Boolean flag to request an acknowledgment to the message and an indicator containing the sequence number of the message being acknowledged, if any.
- LPP message body: a container with the specific LPP information to be transmitted.

The LPP message body contains a choice between multiple possible messages, including the requests and responses for the capabilities, assistance data and location information, as well as abort and error messages, as will be described in the following sections.

14.2.2.1 Capability Exchange

The first step in the positioning session is the exchange of the UE positioning capabilities. In C-Plane sessions, this transaction typically comprises a request and a response. On the other hand, for LPP over SUPL, the positioning capabilities are sent by the UE unsolicited.

The *LPP request capabilities* was initially defined in Release 9, containing provisions to add future extensions in a backward compatible way. As can be seen

```
RequestCapabilities ::= SEQUENCE {
    criticalExtensions      CHOICE {
        c1                      CHOICE {
            requestCapabilities-r9      RequestCapabilities-r9-IEs,
            spare3 NULL, spare2 NULL, spare1 NULL
        },
        criticalExtensionsFuture    SEQUENCE {}
    }
}

RequestCapabilities-r9-IEs ::= SEQUENCE {
    commonIEsRequestCapabilities        CommonIEsRequestCapabilities        OPTIONAL,   -- Need ON
    a-gnss-RequestCapabilities          A-GNSS-RequestCapabilities          OPTIONAL,   -- Need ON
    otdoa-RequestCapabilities           OTDOA-RequestCapabilities           OPTIONAL,   -- Need ON
    ecid-RequestCapabilities            ECID-RequestCapabilities            OPTIONAL,   -- Need ON
    epdu-RequestCapabilities            EPDU-Sequence                       OPTIONAL,   -- Need ON
    ...
}
```

Figure 14.3 LPP request capabilities message for Release 9 [1].

in Figure 14.3, there are multiple spare elements that can be defined later, and an empty "critical extensions future" sequence. Furthermore, the specific Release 9 message also contains the ... operator, explained in Chapter 13 of this book.

All other messages, such as the *LPP provide capabilities*, the *LPP provide assistance data*, or the *LPP request location information* contain the same extension options. Therefore, these will not be mentioned again explicitly in the other sections.

The initial *request capabilities* message contained only one item, which is the *request capabilities Release 9 information elements (IEs)*. As can be seen in Figure 14.3, this message contains items to query the A-GNSS, OTDOA, and ECID capabilities. As they are labeled "Optional," the network can decide which capabilities need to be requested and turn the specific items on or off. Apart from the three technologies mentioned, there is a container for common IEs, which is so far unused, and something called external protocol data unit (EPDU), which will be described in a later section of this chapter when discussing LPPe.

For OTDOA and ECID, the *request capabilities* is just an empty container without any further information. If present, the UE must include the corresponding information in the *provide capabilities* response. However, for A-GNSS, the *request capabilities* contains further items that are shown in Figure 14.4. The GNSS support list is used to request information about which GNSSs (e.g., GPS, Galileo) are supported by the target device, including also the different GNSS frequency bands, such as L1 and L5, as seen in Chapter 6 of this book. The assistance data support list is used to request information about what kind of assistance data the target device can process. The last item is used to indicate the location coordinates and velocity formats supported by the target device. These formats will be described later in Section 14.2.2.3.

```
A-GNSS-RequestCapabilities ::= SEQUENCE {
    gnss-SupportListReq                  BOOLEAN,
    assistanceDataSupportListReq         BOOLEAN,
    locationVelocityTypesReq             BOOLEAN,
    ...
}
```

Figure 14.4 LPP request capabilities message for A-GNSS [1].

```
OTDOA-ProvideCapabilities ::= SEQUENCE {
    otdoa-Mode        BIT STRING {     ue-assisted (0) } (SIZE (1..8)),
    ...,
    supportedBandListEUTRA        SEQUENCE (SIZE (1..maxBands)) OF SupportedBandEUTRA       OPTIONAL,
    supportedBandListEUTRA-v9a0 SEQUENCE (SIZE (1..maxBands)) OF SupportedBandEUTRA-v9a0
                                                                                             OPTIONAL,
    interFreqRSTDmeasurement-r10          ENUMERATED { supported }                          OPTIONAL,
    additionalNeighbourCellInfoList-r10 ENUMERATED { supported }                            OPTIONAL
}

ECID-ProvideCapabilities ::= SEQUENCE {
    ecid-MeasSupported  BIT STRING {    rsrpSup     (0),
                                        rsrqSup     (1),
                                        ueRxTxSup   (2) } (SIZE(1..8)),
    ...
}
```

Figure 14.5 LPP provide capabilities message for OTDOA and ECID [1].

The *provide capabilities* message structure is identical to the *request capabilities* message, just by replacing all appearances of "Request" with "Provide" in Figure 14.3. What varies is the content of the respective OTDOA, ECID, and A-GNSS messages.

For OTDOA, the *provide capabilities* message indicates the positioning mode supported, although only UE-assisted is possible so far; the list of LTE bands for which OTDOA measurements are supported, and whether interfrequency OTDOA (i.e., performing RSTD measurements between LTE cells in different frequencies) is supported or not. For ECID, the message contains a 8-bit string where the three least significant bits indicate whether RSRP, RSRQ, or UE Rx-Tx measurements are supported, respectively. The other 5 bits remain unused so far. These two messages are shown in Figure 14.5.

For A-GNSS, the *provide capabilities* message contains several items, as shown in Figure 14.6. First of all, it includes a list of all GNSS elements supported. Each of the supported GNSS elements is characterized by an ID, indicating the satellite system (e.g., GPS, Galileo). It also contains, among other parameters, the positioning modes supported for that satellite system, such as UE-based or UE-assisted, and the signals supported, which could be L1, L5, E1, and so forth. The assistance data support list indicates what type of assistance data the target device can process, such as reference time, ionospheric models, and navigation models. The most relevant of these items will be described in Section 14.2.2.2. The location

```
A-GNSS-ProvideCapabilities ::= SEQUENCE {
    gnss-SupportList            GNSS-SupportList              OPTIONAL,
    assistanceDataSupportList   AssistanceDataSupportList     OPTIONAL,
    locationCoordinateTypes     LocationCoordinateTypes       OPTIONAL,
    velocityTypes               VelocityTypes                 OPTIONAL,
    ...
}

GNSS-SupportList ::= SEQUENCE (SIZE(1..16)) OF GNSS-SupportElement

GNSS-SupportElement ::= SEQUENCE {
    gnss-ID                        GNSS-ID,
    sbas-IDs                       SBAS-IDs                  OPTIONAL,    -- Cond GNSS-ID-SBAS
    agnss-Modes                    PositioningModes,
    gnss-Signals                   GNSS-SignalIDs,
    fta-MeasSupport                SEQUENCE {
                                       cellTime    AccessTypes,
                                       mode        PositioningModes,
                                       ...
                                   }                         OPTIONAL,    -- Cond fta
    adr-Support                    BOOLEAN,
    velocityMeasurementSupport     BOOLEAN,
    ...
}

AssistanceDataSupportList ::= SEQUENCE {
    gnss-CommonAssistanceDataSupport    GNSS-CommonAssistanceDataSupport,
    gnss-GenericAssistanceDataSupport   GNSS-GenericAssistanceDataSupport,
    ...
}
```

Figure 14.6 LPP provide capabilities message for A-GNSS [1].

coordinate types indicates the format of location coordinates supported by the target device, such as the ellipsoid point format, with or without the uncertainty. Finally, the velocity types indicates whether vertical velocity in addition to horizontal velocity is supported and whether the velocity uncertainty indication is supported.

14.2.2.2 Assistance Data

The exchange of assistance data is typically the second transaction in a LPP session. Over the control plane, the assistance data is normally sent by the location server of the network without a previous request. However, in SUPL sessions or if the target device requires any further assistance data, it can be requested with the *LPP request assistance data* message. The basic skeleton of the message is shown in Figure 14.7. As can be seen, it is very similar to the request capabilities message, a generic container with multiple possibilities to be enhanced in the future and a specific container for the Release 9 assistance data request. The main difference is that the request is only possible for A-GNSS, OTDOA, common IEs, and EPDU. ECID assistance data cannot be requested. This is because ECID measurements do not require any additional information to be provided by the network.

```
RequestAssistanceData ::= SEQUENCE {
    criticalExtensions       CHOICE {
        c1                       CHOICE {
            requestAssistanceData-r9     RequestAssistanceData-r9-IEs,
            spare3 NULL, spare2 NULL, spare1 NULL
        },
        criticalExtensionsFuture     SEQUENCE {}
    }
}

RequestAssistanceData-r9-IEs ::= SEQUENCE {
    commonIEsRequestAssistanceData       CommonIEsRequestAssistanceData       OPTIONAL,
    a-gnss-RequestAssistanceData         A-GNSS-RequestAssistanceData         OPTIONAL,
    otdoa-RequestAssistanceData          OTDOA-RequestAssistanceData          OPTIONAL,
    epdu-RequestAssistanceData           EPDU-Sequence                        OPTIONAL,
    ...
}
```

Figure 14.7 LPP request assistance data message [1].

```
OTDOA-ProvideAssistanceData ::= SEQUENCE {
    otdoa-ReferenceCellInfo       OTDOA-ReferenceCellInfo          OPTIONAL,    -- Need ON
    otdoa-NeighbourCellInfo       OTDOA-NeighbourCellInfoList      OPTIONAL,    -- Need ON
    otdoa-Error                   OTDOA-Error                      OPTIONAL,    -- Need ON
    ...
}
```

Figure 14.8 LPP provide assistance data message for OTDOA [1].

As was already the case with the capabilities exchange, the *provide assistance data* response is almost identical to the request just by changing all appearances of "Request" with "Provide."

The OTDOA *request assistance data* contains only one parameter: the physical cell ID of the current primary cell of the target device. This cell ID will be used as a reference cell to generate the required assistance data. The corresponding OTDOA assistance data provided by the network is shown in Figure 14.8. It usually contains two elements: the reference cell information and a list with the neighbor cell information. A third possibility is to transmit "OTDOA Error" to indicate that the assistance data is not available or not supported by the location server. The different error messages will be explained later.

The reference cell information contains a set of basic parameters to identify the reference cell, such as the cell ID, the frequency channel (EARFCN) or the length of the cyclic prefix. It also contains information about the PRS in order to help the target device decode it properly. The PRS parameters have been introduced in detail in Chapter 8 of this book, and they include the PRS bandwidth, the configuration index, the number of consecutive downlink frames containing PRS, and the muting information.

The neighbor cell information contains a list of the neighbor cells that the target device is likely to see. For each, similar parameters as for the reference cell are provided (e.g., cell ID, frequency channel, or PRS information). In addition, the location server also transmits the expected RSTD measurement result calculated by the network based on the expected propagation delay. This calculation includes an uncertainty estimation based on the location server's a priori knowledge of the location of the UE. Using these two values, the UE can estimate a search window for the measurement.

For A-GNSS, the number of assistance data elements is very large, and it is out of the scope of this book to analyze each in detail. Luckily, LPP specifies most of the A-GNSS assistance data in a way that closely follows the actual navigation messages specified in the GNSSs' ICDs [4–7] and described in Chapter 6 of this book. For each GNSS, a dedicated information element exists in the protocols, using exactly the same number of bits to represent the data. A summary of the different models used for each particular GNSS will be presented here.

Even though GPS, Galileo, and BeiDou all use a very similar Keplerian representation of the orbits, clock corrections, and UTC time corrections, LPP specifies dedicated information elements as seen in Table 14.2. The reason behind is that the resolution of some parameters differs between the GNSS systems. In some cases (e.g., LPP UTC), multiple systems re-use the same data elements, which can lead to confusion in the model ID numbering. As an example, BeiDou is assigned Model-6 for Clock and Navigation, Model-5 for UTC, and Model-7 for Almanac. Table 14.1 gives and overview of the various model IDs. Table 14.3 summarizes the ionospheric model used for different systems, including the data ID element.

Table 14.1

LPP ASN.1 Model ID Mapping

GNSS	*LPP Nav&Clk*	*UTC*	*LPP Alm.*
Galileo	1	1	1
GPS L1C/A	2	1	2
QZSS L1C/A	2	1	2
GPS/QZSS L2C,L5	3	2	3/4
Glonass	4	3	5
SBAS	5	4	6
BeiDou	6	5	7

There are a few elements that are not part of the GNSSs' ICD. These are the reference location, reference time, and acquisition assistance. While the encoding of reference location and time is straightforward and does not require further explanation, acquisition assistance is more complicated.

Table 14.2

LPP Navigation, Clock, and Almanac Model Names

GNSS	*Nav. Mod. Name*	*Clk Mod. Name*	*Alm. Name*
Galileo	Keplerian	standard	keplerian
GPS L1C/A	nav-Keplerian	nav	keplerianNAV
QZSS L1C/A	nav-Keplerian	nav	keplerianNAV
GPS/QZSS L2C,L5	cnav-Keplerian	cnav	keplerianReduced/Midi
Glonass	glonass-ECEF	glonass	Glonass
SBAS	sbas-ECEF	sbas	ECEF-SBAS
BeiDou	bds-Keplerian	bds	BDS

Table 14.3

LPP Ionospheric Model Mapping

GNSS	*Ionospheric Model*	*Data ID (only for Klobuchar)*
Galileo	neQuick	-
GPS L1C/A	Klobuchar	'00'
QZSS L1C/A	Klobuchar	'11'
GPS/QZSS L2C,L5	Klobuchar	'00'
Glonass	-	-
SBAS	-	-
BeiDou	Klobuchar	'01'

As described in Chapter 6, acquisition assistance specifies a 2D search window with the axes code and Doppler. Both values were generalized in LPP to remove the dependency on the carrier frequency and the code rate. For the Doppler, this was achieved by dividing the value by the signal wavelength to obtain a value in meters per second. Similarly, the code phase is scaled by the signal's code rate. This way, LPP can use a single information element to provide acquisition assistance for all GNSS systems and all signals. For historic reasons, RRLP and RRC initially specified an acquisition element, which was tailored to GPS L1 C/A. When 3GPP extended these legacy protocols, they kept the old format for GPS L1C/A, but decided to use the LPP-like definition for all other signals.

14.2.2.3 Location Information

The last transaction on a regular LPP session is the exchange of location information. The location server of the network will start this transaction by sending a *LPP*

```
CommonIEsRequestLocationInformation ::= SEQUENCE {
    locationInformationType     LocationInformationType,
    triggeredReporting          TriggeredReportingCriteria  OPTIONAL,    -- Cond ECID
    periodicalReporting         PeriodicalReportingCriteria OPTIONAL,    -- Need ON
    additionalInformation       AdditionalInformation       OPTIONAL,    -- Need ON
    qos                         QoS                         OPTIONAL,    -- Need ON
    environment                 Environment                 OPTIONAL,    -- Need ON
    locationCoordinateTypes     LocationCoordinateTypes     OPTIONAL,    -- Need ON
    velocityTypes               VelocityTypes               OPTIONAL,    -- Need ON
    ...
}
```

Figure 14.9 Common IEs for the request location information message [1].

request location information message, following exactly the same format as already shown for the request capabilities and request assistance messages. In contrast to the *request capabilities*, the common IEs here are not empty and contain very relevant information for the positioning session, as can be seen in Figure 14.9.

The most relevant items of this message are:

- Location information type: used to communicate whether a location estimate (i.e., UE-based positioning) is required or just location measurements (i.e., UE-assisted positioning). It also possible to select the preferred method but leave it up to the target device to make the final decision.
- Periodical reporting: used to configure the target device to send periodical measurement reports (e.g., for navigation).
- QoS: this field is used to establish the maximum allowed response time and the required horizontal accuracy. Vertical positioning can be optionally configured, including the desired positioning accuracy. Velocity measurements can also be requested if needed.

The remaining items are of less importance. More information about them can be found directly in the TS 36.355 [1]. One important consideration about the location information type is that if OTDOA and / or ECID positioning is requested, the location information type must be set to the value "locationMeasurementsRequired," as these methods do not support the UE-based mode.

The OTDOA request location information is used to ask the target device for OTDOA positioning measurements. It contains only one additional parameter indicating whether more assistance data can be provided by the network if needed or not. The ECID *request location information* also contains one parameter to specify what type of measurements are required. The network can configure power measurements (both RSRP and RSRQ) and time measurements (UE Rx-Tx). Both OTDOA and ECID request location information messages are shown in Figure 14.10.

```
OTDOA-RequestLocationInformation ::= SEQUENCE {
    assistanceAvailability      BOOLEAN,
    ...
}
ECID-RequestLocationInformation ::= SEQUENCE {
    requestedMeasurements       BIT STRING {    rsrpReq     (0),
                                                rsrqReq     (1),
                                                ueRxTxReq   (2) } (SIZE(1..8)),
    ...
}
```

Figure 14.10 Request location information message for OTDOA and ECID [1].

The A-GNSS *request location information* includes a container for the GNSS positioning instructions to configure the UE. This container has the following configuration possibilities:

- GNSS methods: a bitmap to specify which of the different GNSS are to be used for the position calculation. Each bit position is associated with a particular GNSS as shown in Table 14.4 and can be switched on and off.
- Fine time assistance measurement request: this field is used to request the target device to provide the association between GNSS time and network time.
- Accumulated delta range (ADR) measurement request: this field is used to request ADR measurements (for differential GNSS methods such as the ones seen in Chapter 7 of this book).
- Multi-Freq measurement request: a field to specify whether the target device should include measurements on multiple supported GNSS frequencies.
- Assistance availability: indicates whether the server has additional assistance data that may be requested by the target device if needed.

Table 14.4

GNSS ID Bitmap

GNSS	GPS	SBAS	QZSS	Galileo	Glonass	BDS	Spare
Associated bit	0	1	2	3	4	5	6-15

Upon receiving the location information request, the target device must initiate positioning measurements and provide the necessary location information to

```
OTDOA-ProvideLocationInformation ::= SEQUENCE {
    otdoaSignalMeasurementInformation    OTDOA-SignalMeasurementInformation  OPTIONAL,
    otdoa-Error                          OTDOA-Error                         OPTIONAL,
    ...
}
OTDOA-SignalMeasurementInformation ::= SEQUENCE {
    systemFrameNumber          BIT STRING (SIZE (10)),
    physCellIdRef              INTEGER (0..503),
    cellGlobalIdRef            ECGI                     OPTIONAL,
    earfcnRef                  ARFCN-ValueEUTRA         OPTIONAL,
    referenceQuality           OTDOA-MeasQuality        OPTIONAL,
    neighbourMeasurementList      NeighbourMeasurementList,
    ...,
    [[ earfcnRef-v9a0          ARFCN-ValueEUTRA-v9a0    OPTIONAL
    ]]
}
NeighbourMeasurementList ::= SEQUENCE (SIZE(1..24)) OF NeighbourMeasurementElement

NeighbourMeasurementElement ::= SEQUENCE {
    physCellIdNeighbour        INTEGER (0..503),
    cellGlobalIdNeighbour      ECGI                     OPTIONAL,
    earfcnNeighbour            ARFCN-ValueEUTRA         OPTIONAL,
    rstd                       INTEGER (0..12711),
    rstd-Quality               OTDOA-MeasQuality,
    ...,
    [[ earfcnNeighbour-v9a0 ARFCN-ValueEUTRA-v9a0       OPTIONAL
    ]]
}
```

Figure 14.11 Provide location information message for OTDOA [1].

the network within the specified response time. For E911 calls, this response time is typically set to 20 seconds. Depending on the technologies involved and the information requested, the device may need to calculate a position or just report back the measurements.

For OTDOA, the *provide location information* message includes the OTDOA signal measurement information, as shown in Figure 14.11. This container indicates the cell identification for the reference cell, along with other useful information such as the channel number, a system frame number used for time reference, and a list of neighbor cell measurements. For each of the neighbor cells, the report includes a neighbor measurement element message with the cell identification, the frequency (only if it is different from the reference cell frequency), and the RSTD measurement and its quality. The RSTD measurement is calculated as the difference between the reception time of the PRS from the neighbor cell with respect to the reference cell. The mathematical derivation for the OTDOA measurements was seen in detail in Section 8.6 of this book.

For ECID, the *provide location information* message is shown in Figure 14.12. Similar to OTDOA, it contains the ECID signal measurement information, which in turn contains the primary cell measurement results and a list of neighbor cell

```
ECID-ProvideLocationInformation ::= SEQUENCE {
    ecid-SignalMeasurementInformation   ECID-SignalMeasurementInformation       OPTIONAL,
    ecid-Error                          ECID-Error                              OPTIONAL,
    ...
}
ECID-SignalMeasurementInformation ::= SEQUENCE {
    primaryCellMeasuredResults  MeasuredResultsElement  OPTIONAL,
    measuredResultsList         MeasuredResultsList,
    ...
}

MeasuredResultsList ::= SEQUENCE (SIZE(1..32)) OF MeasuredResultsElement

MeasuredResultsElement ::= SEQUENCE {
    physCellId          INTEGER (0..503),
    cellGlobalId        CellGlobalIdEUTRA-AndUTRA           OPTIONAL,
    arfcnEUTRA          ARFCN-ValueEUTRA,
    systemFrameNumber
                        BIT STRING (SIZE (10))              OPTIONAL,
    rsrp-Result         INTEGER (0..97)                     OPTIONAL,
    rsrq-Result         INTEGER (0..34)                     OPTIONAL,
    ue-RxTxTimeDiff     INTEGER (0..4095)                   OPTIONAL,
    ...,
    [[ arfcnEUTRA-v9a0        ARFCN-ValueEUTRA-v9a0         OPTIONAL
    ]]
}
```

Figure 14.12 Provide location information message for ECID [1].

measurements. In this case, the primary cell and neighbor cell measurement containers follow the same structure, containing the cell identification, the frequency channel, a reference time frame, and the power measurement results for RSRP and RSRQ. Additionally, the primary cell report also contains the UE Rx-Tx time difference measurement, which is only possible in RRC connected state (i.e., the mobile device can only calculate this value with respect to the base station to which it is connected). The mathematical derivation of these measurements was explained in Section 8.5 of this book.

GNSS admits both UE-based and UE-assisted modes. If GNSS measurements are requested in the same transaction as OTDOA or ECID, then only UE-assisted is possible. The network can use UE-assisted GNSS measurements using a similar algorithm to the one defined by 3GPP in Section F.3 of TS 36.171 [2], which has been expanded for the hybrid solution in Section 8.9 of this book. In UE-assisted mode, the A-GNSS *provide location information* message contains the signal measurement information, including the measurement reference time and a GNSS measurement list, as seen in Figure 14.13. The measurement reference time is a container used to calculate the validity time of the GNSS measurements. For more detailed information on the contents of this message, refer to [1].

The GNSS measurement list contains a sequence of GNSS measurements for one GNSS, characterized by its GNSS ID and including a list of GNSS signal measurements. Each signal is also characterized by a specific ID according to

```
GNSS-SignalMeasurementInformation ::= SEQUENCE {
    measurementReferenceTime        MeasurementReferenceTime,
    gnss-MeasurementList            GNSS-MeasurementList,
    ...
}
GNSS-MeasurementList  ::= SEQUENCE (SIZE(1..16)) OF GNSS-MeasurementForOneGNSS

GNSS-MeasurementForOneGNSS ::= SEQUENCE {
    gnss-ID                 GNSS-ID,
    gnss-SgnMeasList        GNSS-SgnMeasList,
    ...
}

GNSS-SgnMeasList ::= SEQUENCE (SIZE(1..8)) OF GNSS-SgnMeasElement

GNSS-SgnMeasElement ::= SEQUENCE {
    gnss-SignalID               GNSS-SignalID,
    gnss-CodePhaseAmbiguity INTEGER (0..127)        OPTIONAL,
    gnss-SatMeasList            GNSS-SatMeasList,
    ...
}
GNSS-SatMeasList ::= SEQUENCE (SIZE(1..64)) OF GNSS-SatMeasElement
GNSS-SatMeasElement ::= SEQUENCE {
    svID                    SV-ID,
    cNo                     INTEGER (0..63),
    mpathDet                ENUMERATED {notMeasured (0), low (1), medium (2), high (3), ...},
    carrierQualityInd       INTEGER (0..3)              OPTIONAL,
    codePhase               INTEGER (0..2097151),
    integerCodePhase        INTEGER (0..127)            OPTIONAL,
    codePhaseRMSError       INTEGER (0..63),
    doppler                 INTEGER (-32768..32767)     OPTIONAL,
    adr                     INTEGER (0..33554431)       OPTIONAL,
    ...
}
```

Figure 14.13 Provide location information message for A-GNSS UE-assisted [1].

Table 14.5. The combination of GNSS ID and GNSS Signal ID results in a unique frequency signal of a specific satellite system. In later releases, with the introduction of differential GNSS methods, these two values have been combined in a GNSS-SignalIDs bitmap.

For each of the measured frequency signals, the report contains a list of satellite measurement elements characterized by the spatial vehicle ID (SVID) and including the carrier-to-noise ratio (C/No) and other parameters such as the code phase, the Doppler measurement, an indicator determining if the target device has detected multipath, and the measurement quality estimations. An explanation of each of these measurements and how to use them to calculate the position of the target device was detailed in Chapter 6 of this book. Further references can be found in specialized GNSS books such as [3].

In the case of UE-based positioning, the A-GNSS provide location information contains a message called GNSS-LocationInformation, as shown in Figure 14.14. This message includes the measurement reference time, the same as for UE-assisted, and the GNSS ID bitmap indicating which GNSSs were used for the positioning measurements. The calculated position is reported in the common IEs

Table 14.5

GNSS Signal ID Mapping

System	*Value*	*Description*
GPS	0	L1 C/A
	1	L1C
	2	L2C
	3	L5
	4-7	spare
SBAS	0	L1
	1-7	spare
QZSS	0	QZS L1 C/A
	1	QZS L1C
	2	QZS L2C
	3	QZS L5
	4-7	spare
Glonass	0	G1 = L1OF
	1	G2 = L2OF
	2	G3 = L3OC
	3-7	spare
Galileo	0	E1
	1	E5A
	2	E5B
	3	E6
	4	E5A + E5B
	5-7	spare
BDS	0	B1I
	2-7	spare

for the provide location information. The common IEs message includes a location estimate, which can take different formats depending on whether the device reports just an ellipsoid point, or whether it can also calculate an uncertainty value:

- Ellipsoid point: the device reports the latitude and longitude of the calculated position.
- Ellipsoid point with uncertainty circle: in addition, the device can estimate the radius of the uncertainty estimation.
- Ellipsoid point with uncertainty ellipse: the device can calculate the latitude and longitude uncertainties separately.
- Polygon: instead of just a point, the device can calculate an area.
- Ellipsoid point with altitude: the device provides a 3D location fix.
- Ellipsoid point with altitude and uncertainty: the device provides a 3D fix plus the respective uncertainties for each coordinate.

Together with the location estimate, the device may include a velocity estimate that can be just the horizontal velocity or include additionally the vertical velocity, with or without uncertainty estimation.

```
GNSS-LocationInformation ::= SEQUENCE {
    measurementReferenceTime        MeasurementReferenceTime,
    agnss-List                      GNSS-ID-Bitmap,
    ...
}

CommonIEsProvideLocationInformation ::= SEQUENCE {
    locationEstimate        LocationCoordinates        OPTIONAL,
    velocityEstimate        Velocity                   OPTIONAL,
    locationError           LocationError              OPTIONAL,
    ...,
    [[  earlyFixReport-r12  EarlyFixReport-r12         OPTIONAL
    ]]
}
```

Figure 14.14 Provide location information message for A-GNSS UE-based [1].

14.2.3 LPP Error

When an *LPP error* message is included in the *LPP message body*, it indicates that a particular LPP message was received with erroneous or unexpected data, or that something is missing. The receiving end must abort any ongoing procedure related to the indicated transaction ID.

The *LPP error* message contains a common IE container that specifies the cause of the error, to be selected between *undefined, message header error, message body error, EPDU error*, or *incorrect data value*.

Apart from the generic *LPP error* message, each of the positioning technologies has its own specific error container that can be sent as part of the provide assistance data or as part of the provide location information LPP messages. The error message can be sent by the location server or by the target device. These specific errors can be used for more detailed information. The list of possible error causes is specified in Table 14.6.

14.2.4 LPP Abort

The *LPP abort* is a special message that allows either the target device or the location server in the network to interrupt the positioning session. Upon reception of the *LPP abort* message, the device or location server must stop any ongoing procedure associated with the indicated transaction ID.

The *LPP abort* message contains a common IEs container to specify the cause for the abort, to be selected between *undefined*, *stop periodic reporting* (if periodic measurements are configured), *target device abort* (the LPP session is aborted by the UE), or *network abort* (the LPP session is aborted by the location server).

Table 14.6
Specific Error Messages

Tech.	*Side*	*Description*
OTDOA	Location server	Assistance data not supported by server
		Assistance data supported but currently not available
	Target device	Undefined
		Assistance data missing
		Unable to measure reference cell
		Unable to measure any neighbor cell
		Attempted but unable to measure some neighbor cells
A-GNSS	Location server	Undefined
		Undelivered assistance data not supported by server
		Undelivered assistance data currently not available
		Assistance data partly not supported and partly not available
	Target device	Undefined
		Assistance data missing
		Not enough satellites received
		Not all requested measurements possible
		Fine time assistance measurements not possible
		ADR measurements not possible
		Multifrequency measurements not possible
ECID	Location server	Undefined
	Target device	Undefined
		Requested measurements not available
		Not all requested measurements possible
		RSRP measurements not possible
		RSRQ measurements not possible
		UE Rx-Tx measurements not possible

14.3 LPP UPDATES UP TO RELEASE 12

During the initial 3GPP releases up to Release 12, the LPP did not experience major modifications. However, a few special LPP features were added to allow more flexible positioning scenarios.

14.3.1 Early Fix

While good GNSS measurements typically can take up to 20s, OTDOA and ECID measurements are available almost instantaneously. The LPP protocol up to Release 11 supported only joint reporting of all measurements, therefore the slowest positioning technology (GNSS) dictated the response time. Thus, it took 20s for a emergency dispatcher to see a location, even if the mobile was located via OTDOA after just one second. In order to resolve this limitation, the 3GPP added a new feature called early fix in Release 12 of LPP. The early fix allows the network to specify two response times, one longer (typically for A-GNSS) and one shorter (typically for ECID/OTDOA). This second one is called *response time early fix*. If the mobile device has location information available before the end of the early-fix response time, it must send a provide location information measurement with an early-fix report. The early-fix report container is also used to indicate whether there are more early-fix messages coming ("moreMessagesOnTheWay") or if this is the last early fix report ("noMoreMessages").

The call flow for a positioning session with early fix is shown in Figure 14.15. As can be seen, the provide location information for early fix and the regular one share the same transaction ID, but with different sequence numbers, represented by S in the figure. The intermediate early-fix response in S2 is optional. Although typically the early-fix report is used for OTDOA or ECID, the target device may also report the GNSS positioning information if it is already available by the expiration of the early-fix timer.

The early fix allows emergency dispatchers to obtain a first fairly accurate location quickly and send the rescue personal toward the person in need, while waiting for the higher-accuracy location received at the expiry of the response timer.

14.3.2 Dual-Technology LPP Flow

Although this is not exactly an LPP feature rather a use case, its wide usage among some operators (especially in the United States) makes it worthwhile to mention. As explained earlier, OTDOA and ECID technologies admit only UE-assisted positioning. However, some operators rely on UE-based GNSS as their main positioning technology. LPP does not allow to mix UE-based and UE-assisted positioning in the same location request. Thus, the dual-technology flow is used to collect all the information possible in one LPP session.

The dual-technology flow is nothing else than an LPP session where the location server of the network sends out two simultaneous location information requests: one for UE-based A-GNSS and another one for OTDOA and/or ECID. Each of the location requests may have a different response time (and it may use additional LPP features like the early fix). The order of the location responses depends on the configured response times. Typically, the OTDOA / ECID positioning information arrives faster.

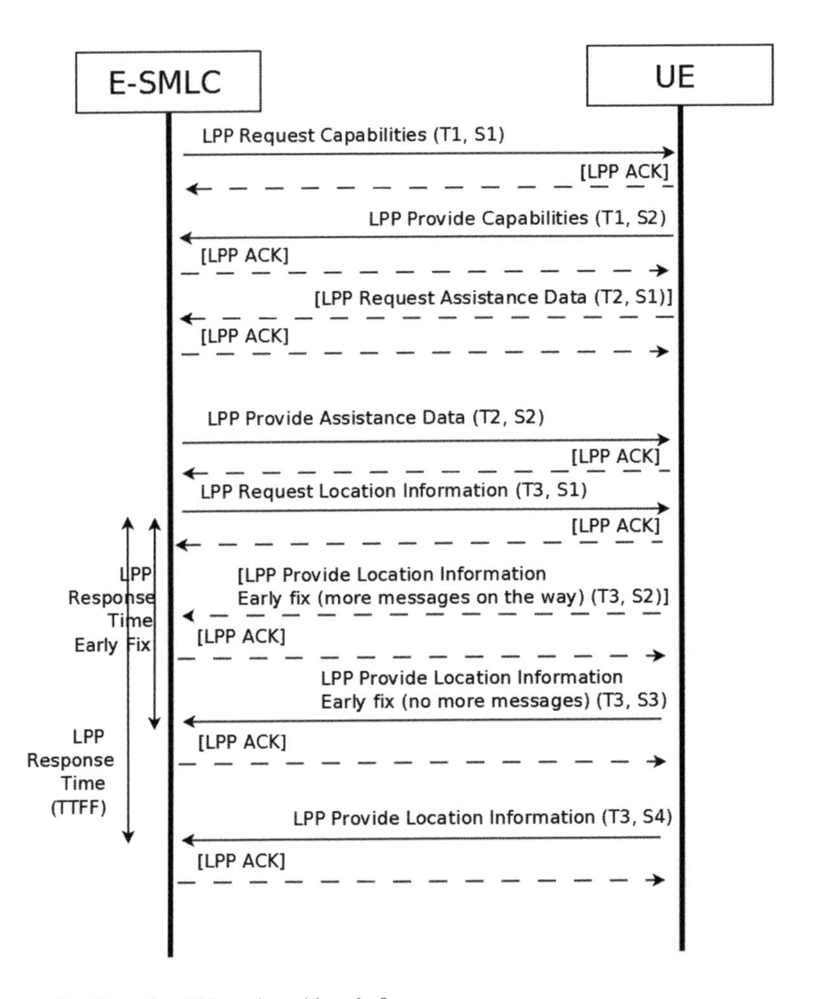

Figure 14.15 Flow of an LPP session with early fix.

This call flow is shown in Figure 14.16, where it can be seen that there are consecutive request location information messages in two different transactions. Each of the location information responses refers to a particular transaction, so it is easy to determine to which technology they correspond. In this case, T3 contains a UE-based A-GNSS positioning request, and T4 contains a UE-assisted OTDOA positioning request. Due to the different response times, the OTDOA

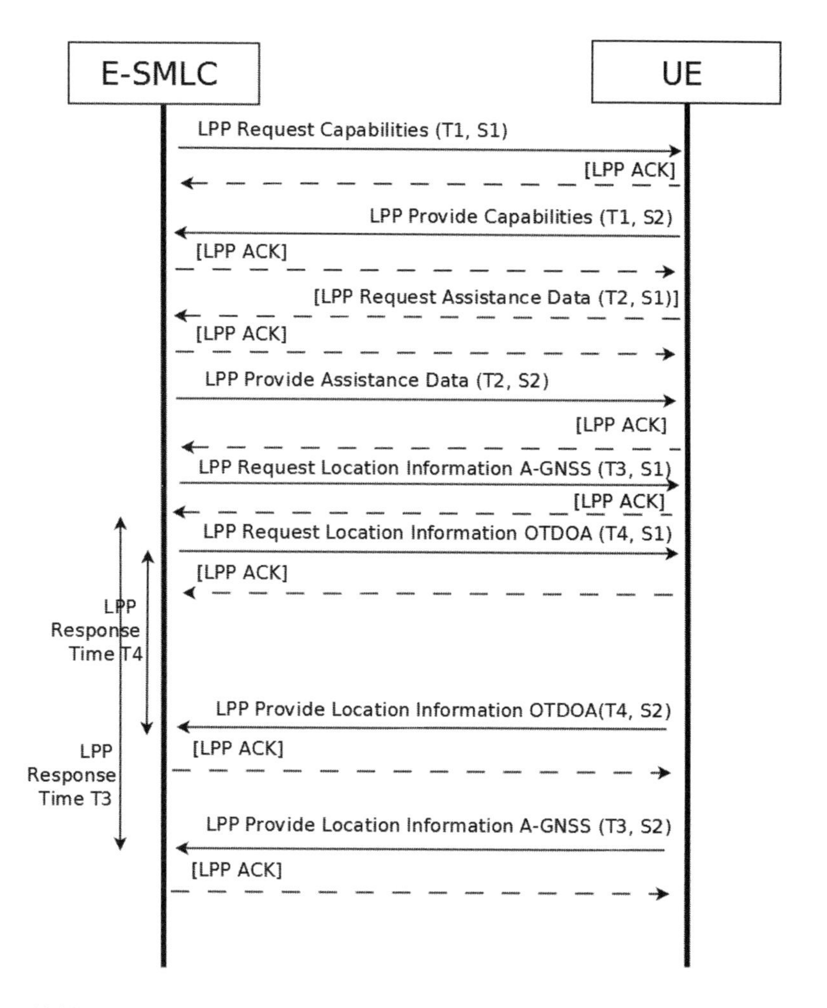

Figure 14.16 Flow of a dual-technology (A-GPS UE-based and OTDOA / ECID) LPP session.

positioning report arrives earlier. A limitation of this type of flow is that as A-GNSS pseudorange measurements are not reported (the mobile device reports a position instead), it cannot be used for a hybrid location fix as described in Section 8.9 of this book.

14.4 OMA LPPE

In the mid 2010s, the OMA, responsible for SUPL, decided to work on a parallel set of features to enhance the 3GPP LPP. The resulting specification was called LPPe [9]. The original intention of OMA LPPe was to augment the LPP features with high-accuracy positioning methods, a tighter integration with SUPL, and to cover the lack of indoor positioning support in LPP. The main differences of LPPe with respect to LPP are the following:

- LPPe offers a tighter integration with the OMA SUPL protocol, since both have been developed by the same organization.
- Therefore, LPPe is also better decoupled from the cellular network technologies. For instance, 3GPP used RRC positioning in WCDMA, and changed to LPP for LTE. By using LPPe over SUPL, the positioning protocol is fully decoupled from the RAT used to transport the positioning data.
- One of the main disadvantages of OTDOA and ECID positioning methods in LTE, is that they do not support the UE-based mode. Thus, the mobile device can only send the positioning measurements to the network, but it cannot calculate its own position. Therefore, OTDOA and ECID are not useful for commercial applications requiring the positioning information to be available in the smartphone. LPPe has added UE-based support for these technologies.
- LPPe added support for high-accuracy position reports (in centimeter-level) and dispatchable locations:
  - OMA LPPe civic location: this format is used to communicate a civic location, as defined in [10]. The civic location is what the FCC also refers to as a dispatchable address, as seen in Chapters 2 and 3 of this book. It contains a country code, formed by two capital letters as defined by ISO 3166 (e.g., US for the United States or ES for Spain); a number from 0 to 255 identifying the civic address type [10], and a string, also known as the civic address value.
  - OMA LPPe high accuracy 3D (HA3D) position: this field augments the LPP position by making the altitude field mandatory and increasing the number of bits used to represent both latitude and longitude to 32 (LPP uses 1 bit for the sign and 23 bits for the value). Thus, it provides an improved resolution of 4.7 mm for the latitude and 9.3

mm for the longitude at the Equator, which is the worst case. Together with the high-accuracy position, LPPe also provides a high-accuracy 3D velocity field that defines the speed in an ENU coordinate system with respect to a reference point.

- LPPe included multiple GNSS assistance data improvements (e.g., more precise ionospheric models) and also support for high-accuracy GNSS methods.
- LPPe addressed the FCC indoor location requirements, by defining Wi-Fi, Bluetooth, and barometric sensor positioning measurements.

Being considered a extension to LPP, LPPe messages need to be embedded into the LPP messages. In order to achieve that, LPPe makes use of the EPDU, already introduced when describing the LPP messages in Section 14.2.1. The EPDU contains two fields, an identifier and a body. The identifier is formed by an ID (an integer representing the type of message contained in the EPDU) and an optional name. For LPPe, the ID must be set to 1. The EPDU-Body contains the LPPe message. An important note is that the LPPe message has its own encoding and is transmitted as a payload within LPP.

As it is always part of an LPP message, LPPe can directly use LPP fields such as the transaction ID without the need of defining its own. Furthermore, LPPe relays on the same fundamental transactions (i.e., the exchange of capabilities, assistance data, and location information plus the abort and error procedures) already defined by LPP. All these transactions have already been explained in Section 14.2.1 of this book, and the basic call flow was introduced in Figure 14.1.

14.4.1 Coexistence with LPP

As mentioned at the beginning of this section, LPPe also provides mechanisms for A-GNSS, OTDOA, ECID, and many other positioning technologies. Nonetheless, these technologies are already supported in the corresponding cellular network protocols (mostly LPP). Thus, instead of reimplementing all the features, most network operators and commercial devices opted to use both LPP and LPPe simultaneously as part of the same or consecutive transactions. This is illustrated in Figure 14.17. The optional LPP ACK messages have been left out for simplicity.

In the left part of Figure 14.17, LPP and LPPe request location information messages are sent in the same transaction. The advantage of this method is that it reduces the amount of transactions needed, simplifying the call flow. On the other hand, both LPP and LPPe will share the same response time. Thus, even though

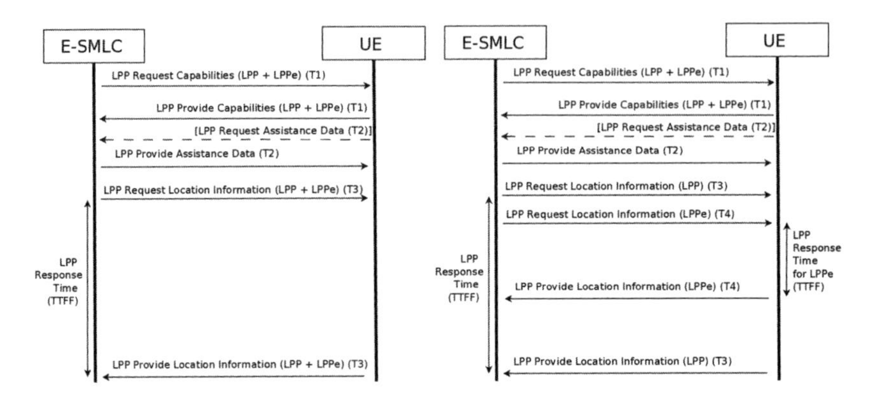

Figure 14.17 Example of call flows with both LPP and LPPe as part of the same transaction (left) or different transactions (right).

LPPe measurements typically are faster than A-GNSS measurements, the target device will not send any location information response until A-GNSS measurements are concluded. This can take up to 20 seconds. On the right part of Figure 14.17, LPP and LPPe are requested in two different transactions, with two different transaction IDs and response times. Hence, even though LPP measurements are requested in an earlier transaction, the LPPe response is received faster. The early-fix feature from LPP could also be used to obtain the same effect (LPPe report coming as early as possible) with a single transaction.

Due to the possibility to combine LPP and LPPe messages in the same positioning session, there was no practical need to duplicate the LPP functionality. Thus, even though LPPe is a large protocol with numerous features, only a handful found their way into commercial implementations.

14.4.2 Closing the Gap Between LPP and LPPe

The 3GPP LPP has evolved with time, and it has by now closed some of the gaps with respect to the OMA LPPe. These LPP updates will be seen in detail in the subsequent sections of this chapter:

- Indoor positioning technologies: LPP has addressed the indoor positioning market in Release 13.

- Higher positioning accuracy: as part of the Release-15 enhancements, the resolution of the reported position has increased, and LPP is also able to report location fixed in the centimeter level.
- High-accuracy GNSS: in the same release, LPP also added support for high-accuracy GNSS methods, such as RTK.
- Decoupling between positioning protocol and RAT: this has been partially improved, since 5G NR reuses LPP.
- UE-based OTDOA: at the time of the writing, LPP does not support UE-based OTDOA. Nonetheless, UE-based support has been added for DL-TDOA and DL-AoD positioning methods in 5G NR, as will be seen in Chapter 15. The 3GPP is discussing whether UE-based support should be extended to LTE OTDOA.

The addition of all these features directly in 3GPP LPP questions the need for implementing a second positioning protocol.

14.5 INDOOR POSITIONING

As seen in Chapter 3 of this book, the FCC was the main driver of indoor positioning accuracy. The requirements in its report [8] made it clear that the existing LPP features would not suffice to meet regulatory expectations. Two different organizations, the 3GPP and the OMA, investigated new positioning methodologies to address indoor positioning efficiently. Wi-Fi, Bluetooth, and barometric sensor positioning support was added to the specifications. In addition, the 3GPP also included TBS.

Since the new features in LPP and LPPe are fairly similar, this book will primarily focus in the LPP Release-13 enhancements.

14.5.1 3GPP Release-13 Indoor Positioning Updates

Release 13 of TS 36.355 [1] saw the definition of four new RAT independent positioning technologies: sensors, Bluetooth, Wi-Fi, and TBS. Table 14.7 captures the fundamental LPP transactions, with the new messages marked in bold. As can be seen, the capability and location information messages were updated in LPP Release 13, while the assistance data extension only came in Release 14.

It is worth noting that the entire of LPP was released with a few years delay with respect to the OMA LPPe. This delay was crucial for the industry, and by

Table 14.7
LPP Release-13 Updates

Request	Request messages	Provide	Provide messages
Request Cap.	commonIEsRequestCapabilities a-gnss-RequestCapabilities otdoa-RequestCapabilities ecid-RequestCapabilities epdu-RequestCapabilities **sensor-RequestCapabilities-r13** **tbs-RequestCapabilities-r13** **wlan-RequestCapabilities-r13** **bt-RequestCapabilities-r13**	Provide Cap.	commonIEsProvideCapabilities a-gnss-ProvideCapabilities otdoa-ProvideCapabilities ecid-ProvideCapabilities epdu-ProvideCapabilities **sensor-ProvideCapabilities-r13** **tbs-ProvideCapabilities-r13** **wlan-ProvideCapabilities-r13** **bt-ProvideCapabilities-r13**
Request Assistance Data	commonIEsRequestAssistanceData a-gnss-RequestAssistanceData otdoa-RequestAssistanceData ecid-RequestAssistanceData epdu-RequestAssistanceData **sensor-RequestAssistanceData-r14** **tbs-RequestAssistanceData-r14** **wlan-RequestAssistanceData-r14**	Provide Assistance Data	commonIEsProvideAssistanceData a-gnss-ProvideAssistanceData otdoa-ProvideAssistanceData ecid-ProvideAssistanceData epdu-ProvideAssistanceData **sensor-ProvideAssistanceData-r14** **tbs-ProvideAssistanceData-r14** **wlan-ProvideAssistanceData-r14**
Request Location Info	commonIEsRequestLocationInfo a-gnss-RequestLocationInfo otdoa-RequestLocationInfo ecid-RequestLocationInfo epdu-RequestLocationInfo **sensor-RequestLocationInfo-r13** **tbs-RequestLocationInfo-r13** **wlan-RequestLocationInfo-r13** **bt-RequestLocationInfo-r13**	Provide Location Info	commonIEsProvideLocationInfo a-gnss-ProvideLocationInfo otdoa-ProvideLocationInfo ecid-ProvideLocationInfo epdu-ProvideLocationInfo **sensor-ProvideLocationInfo-r13** **tbs-ProvideLocationInfor-r13** **wlan-ProvideLocationInfor-r13** **bt-ProvideLocationInfo-r13**

the time of the LPP Release-13 update, most of the U.S. operators had already incorporated LPPe into their network implementation. Nonetheless, LPPe has found little acceptance outside the United States. Even within the United States, the replacement of LPPe through LPP in the midterm or long term cannot be excluded due to the advent of 5G and the continuously increasing number of features in LPP.

14.5.2 Wi-Fi Positioning

Wi-Fi is one of the indoor positioning methods that has been deployed in the United States to achieve the FCC requirements. Wi-Fi positioning methodologies were explained in detail in Chapter 9 of this book. LPP allows primarily two types of Wi-Fi measurements: BSSID + RSSI measurements and RTT measurements. The Wi-Fi *provide location information* message supports a list up to 64 Wi-Fi access points, including the following important parameters:

- The BSSID of the Wi-Fi AP.

- If supported by the mobile device, the RSSI measurement.
- If supported by the mobile device and the AP, the RTT measurement.

Wi-Fi RTT measurements, as seen in Chapter 9, have not been widely used to date, since most commercial Wi-Fi access points do not support this type of measurement.

LPP supports Standalone, UE-based, and UE-assisted positioning modes for Wi-Fi. Since LPP Release 14, the Wi-Fi assistance data information has been defined and can be provided to the mobile phone. However, as seen in Chapter 9, it is unlikely that the cellular network has any information on the Wi-Fi APs in the surroundings of the target device.

14.5.3 Bluetooth Positioning

Bluetooth is another one of the indoor location technologies. The 3GPP has added generic Bluetooth support to LPP. Nonetheless, typically BTLE will be the technology used for positioning purposes.

The Bluetooth positioning messages are very similar to the Wi-Fi ones. The main part of the *provide location information* message is a list of Bluetooth beacons characterized by the MAC address and, optionally, the RSSI measurement. In the case of Bluetooth, the 3GPP directly did not include any assistance data and only Standalone and UE-assisted modes are supported.

14.5.4 Barometric Sensor Positioning

The atmospheric pressure measurements are considered the most promising candidate to achieve the FCC vertical accuracy requirements indoors. The LPP Release-13 update supports the reporting of UBP measurements. Altitude determination based on UBP measurements was described in Chapter 12 of this book.

As of Release 14, LPP also supports the transmission of assistance data information, which mainly consists of a reference atmospheric pressure, measured at a reference location, together with the validity period and area. Optionally, the server can also include the local temperature at the location where the reference pressure was observed.

14.5.5 Terrestrial Beacon System Positioning

As seen in Chapter 9 of this book, TBS is a noncellular network technology that aims to deploy a GNSS-like network of beacons on the surface of the Earth. TBS positioning has been initially added to LPP Release 13. The only TBS implemented so far is the MBS [12] owned by a company called NextNav and deployed in major U.S. cities.

The *provide location information* message is very similar to A-GNSS. It includes a list of TBS beacons, identified by a transmitter ID. Furthermore, the measurements include the code phase in milliseconds, the RMS error associated to the code phase measurement, and, optionally, the RSSI.

LPP supports Standalone, UE-based, and UE-assisted positioning modes for TBS. The assistance data, defined in LPP Release 14, is specific to MBS. It includes an almanac and acquisition assistance elements. These elements are meant to assist the mobile device in detecting the MBS signal.

14.6 OTHER LPP RELEASE-13 AND RELEASE-14 UPDATES

During the course of E-UTRA Release 13 and 14, the LPP enhancements can be split into two well-defined branches. On one side, the regulatory requirements pushed for indoor positioning improvements, which were covered in the previous section. On the other side, E-UTRA Release 13 set up the foundations for the IoT world, with the eMTC and NB-IoT. These new technologies have distinctive features such as the aim for low-cost low-complexity communications with reduced bandwidths and aspects such as battery saving, idle state measurements, and high number of repetitions. Defining eMTC and NB-IoT is not in the scope of this book. Nonetheless, there are a few aspects of these technologies that can have an impact on the positioning methods:

- Bandwidth: both eMTC and NB-IoT work with bandwidths much smaller than conventional LTE. NB-IoT operates on 180-kHz signals that can be allocated within the LTE cell's bandwidth or outside of it. On the other hand, eMTC operates on a 1.4-MHz bandwidth, which can be optionally extended to 5 MHz since Release 14.
- Coverage enhancements: in order to maintain large ranges while keeping low power consumption, both eMTC and NB-IoT devices can operate in CE modes. This basically means increasing the number of repetitions of each message to make sure the other party can receive it. Depending on the CE level, messages can be repeated even 64 or more times.
- Idle mode: NB-IoT devices are meant to be very low cost. Thus, the devices may not have enough processor power to perform measurements while in

connected state. Measurements in NB-IoT are performed in idle state. That has an impact on positioning. For a start, ECID RTT measurements are not supported in NB-IOT, since they require an active connection.

- Frequency hopping: both eMTC and NB-IoT support frequency hopping (i.e., the transmitted signal may change frequency channel over time). This feature can affect OTDOA, since the PRS of an eMTC / NB-IoT base station can also hop in frequency.

14.6.1 ECID Modifications

Apart from the aforementioned limitation that ECID RTT measurements are not supported in NB-IoT, there is only one relevant addition to ECID. The corresponding power measurements for NB-IoT are not called RSRP and RSRQ, but NRSRP and NRSRQ (where the N stands for narrowband). Thus, support for requesting and reporting these two measurements was added to all ECID positioning messages in Release 13.

14.6.2 OTDOA Modifications

14.6.2.1 Enhancements to the PRS

In order to offer a better solution for indoor positioning, the 3GPP proposed a series of improvements to the PRS and the RSTD measurements. The complete list of enhancements, together with simulation results, are collected in TR 37.857 [11]. The main changes are:

- Support for noncollocated RRH: a macrocell (or eNB) may have a number of low power RRHs to serve specific areas, as shown in Figure 14.18. These RRHs share the physical cell ID (PCI) with the eNB. As the PRS is distinguished by the PCI, the mobile device would receive multiple PRS from each RRH and not be able to identify the sender.

- PRS + CRS combined measurements: LPP Release 14 supports the combined use of PRS and CRS for RSTD measurements.

- Denser PRS transmissions: allow to transmit PRS more often in the time domain by increasing the number of consecutive DL subframes or by reducing the periodicity, thus increasing the frequency of the transmissions. As a result

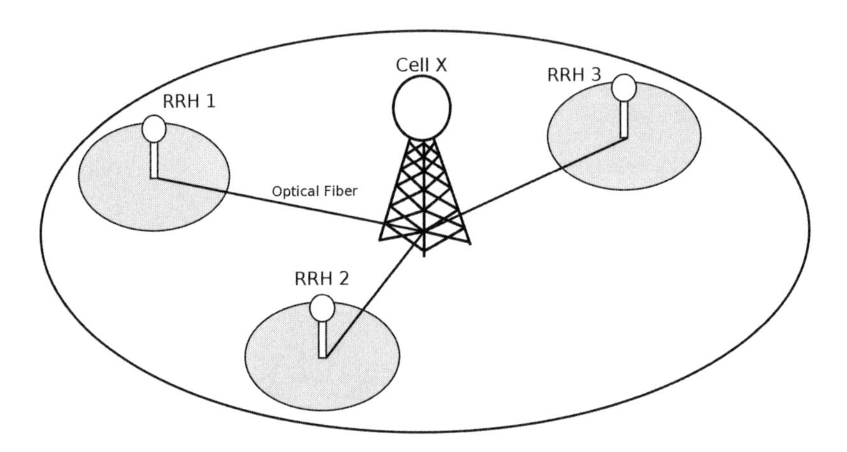

Figure 14.18 Example of a CoMP scenario with an eNB and multiple RRHs.

of these enhancements, it is possible for an eNB or transmission point (TP) to allocate all subframes with PRS transmissions.

- Multipath measurements: allow the target device to report the measured RSTD for additional paths other than the LoS.
- RSTD quantification: LPP Release 14 also allowed higher-resolution RSTD reporting. Instead of the conventional 1-T_s resolution, the device can indicate 0.5 T_s.

Therefore, every OTDOA measurement element (from either serving cell or neighbor cell) and the cell assistance data information have been expanded in LPP Release 14 with multiple parameters, the most relevant listed here:

- tpId: identifier to distinguish which of the TP of a CoMP deployment is associated with a certain PRS transmission or measurement.
- prsId: as an eNB can have multiple transmission points, each with their own PRS configuration (bandwidth, muting pattern, configuration index, etc.), the PRS ID is used to distinguish between the multiple PRSs associated with the same PCI.
- addPRSconfig: additional PRS configurations for a certain cell, since a cell can transmit multiple different PRS configurations (up to a maximum of three).

- prs-only-tp: flag to indicate that a neighbor cell transmits only PRS all the time.
- additionalPaths: relative timing values for one or more additional detected paths with respect to the path used to calculate the RSTD.
- delta-rstd: used to indicate that the reported RSTD is based on the higher-resolution RSTD measurement mapping.

The PRS-Info element, containing the configuration of PRS, also has been enhanced to allow more consecutive downlink subframes for PRS (up to 160) and new PRS muting info patterns of up to 1024 bits. Previously, the number of consecutive DL subframes for PRS was 6, and the muting info pattern could have up to 16 bits.

14.6.2.2 IoT-Related Modifications

For eMTC, the main update affecting OTDOA is that the PRS supports frequency hopping. That means that the mobile device may receive the same PRS in different frequency channels over time. In order to communicate this information, the PRS-Info element has been extended with an additional *prsHoppingInfo* parameter, which can take the value of 2 or 4 indicating that the PRS can hop between 2 or 4 different narrowbands.

Furthermore, NB-IoT features a new positioning reference signal, NPRS, adapted to the 180-kHz bandwidth limitation. Thus, all OTDOA LPP messages have been updated to support the new signal. For the positioning algorithm, it is mostly transparent whether the RSTD measurement comes from PRS or from the new NPRS. Thus, the main update to LPP for NPRS focused on how to communicate to the mobile how to find and decode the NPRS signal. That means an NPRS information assistance data element has been defined. The NPRS information indicates whether the NB-IoT cell is inband (within an LTE cell) or standalone, the carrier frequency, the ID of the NPRS, and the configuration. NB-IoT allows two different NPRS configuration formats, called part A and part B. The combination of part A and part B tells the UE in which subframes to expect NPRS. Part B is the most similar to the legacy PRS configuration, as it indicates the NPRS period, the starting subframe, the number of consecutive subframes, and the muting information. These parameters have been explained for PRS in Section 14.2.2.2. Part A on the other hand provides a bitmap of subframes in which the NPRS is present.

14.7 LPP RELEASE 15

If LPP Release 13 and 14 were mainly focused on indoor positioning for regulatory requirements (with the exception of the updates to support IoT deployments), LPP Release 15 considered the regulatory requirements already well covered and shifted the attention toward commercial applications, with a clear focus on automotive applications. The new technologies researched during LPP Release 15 were motion sensors and RTK, the high-accuracy GNSS enhancement seen in Chapter 7 of this book. However, the most important advance during LPP Release 15 is probably the possibility to broadcast assistance data by using the posSIB. The positioning SIB allows the network to save a lot of resources on unicast assistance data transmissions and enables the efficient use of RTK and other technologies to come, as well as the already existing positioning technologies. Unlike conventional SIBs, the positioning SIB can be encrypted, allowing the network to define subscription models to certain services.

14.7.1 The Positioning SIB

Up to Release 15, LPP assistance data was provided from the network to the mobile device during the LPP positioning session via the LPP provide assistance data message. Thus, the assistance data delivery was a unicast service directed to a specific mobile device. Nonetheless, assistance data, especially for A-GNSS and its enhancements, is not changing from device to device, especially if both devices are within the range of the same serving eNB. Thus, a multicast or broadcast solution is a much more efficient way to distribute assistance data in terms of network resources.

Table 14.8

Mapping of PosSIB Types and LPP Message Types for Common GNSS Assistance Data

posSIBType	LPP message
posSibType1-1	GNSS-ReferenceTime
posSibType1-2	GNSS-ReferenceLocation
posSibType1-3	GNSS-IonosphericModel
posSibType1-4	GNSS-EarthOrientationParameters
posSibType1-5	GNSS-RTK-ReferenceStationInfo
posSibType1-6	GNSS-RTK-CommonObservationInfo
posSibType1-7	GNSS-RTK-AuxiliaryStationData

The broadcast of positioning assistance data was enabled in E-UTRA Release 15 via the positioning SIB defined in TS 36.331 [13] as *SystemInformationBlock-Pos*, shortened as *posSIBs*. The content on the positioning SIB is an assistance data element defined in TS 36.355 [1] and can be divided into three types: GNSS common assistance data, defined in Table 14.8, GNSS generic assistance data, defined in Table 14.9, and OTDOA assistance data, known as *posSibType3-1*. The content of each assistance data container is basically the same as in the equivalent LPP provide assistance data message.

Table 14.9

Mapping of PosSIB Types and LPP Message Types for Heneric GNSS Assistance Data

posSIBType	LPP message
posSibType2-1	GNSS-TimeModelList
posSibType2-2	GNSS-DifferentialCorrections
posSibType2-3	GNSS-NavigationModel
posSibType2-4	GNSS-RealTimeIntegrity
posSibType2-5	GNSS-DataBitAssistance
posSibType2-6	GNSS-AcquisitionAssistance
posSibType2-7	GNSS-Almanac
posSibType2-8	GNSS-UTC-Model
posSibType2-9	GNSS-AuxiliaryInformation
posSibType2-10	BDS-DifferentialCorrections
posSibType2-11	BDS-GridModelParameter
posSibType2-12	GNSS-RTK-Observations
posSibType2-13	GLO-RTK-BiasInformation
posSibType2-14	GNSS-RTK-MAC-CorrectionDifferences
posSibType2-15	GNSS-RTK-Residuals
posSibType2-16	GNSS-RTK-FKP-Gradients
posSibType2-17	GNSS-SSR-OrbitCorrections
posSibType2-18	GNSS-SSR-ClockCorrections
posSibType2-19	GNSS-SSR-CodeBias

Apart from the assistance data element, the positioning SIB also includes an expiration time or validity time for the content of the SIB, segmentation info (in case the content of the SIB is spread over multiple messages) and the ciphering key data to help in the SIB decryption.

14.7.1.1 Positioning SIB Encryption

SIBs are normally not encrypted, meaning that everyone could use this service free of charge. To allow subscription models, 3GPP introduced an optional encryption that allows selective access on a subscription basis. In the case of PPP-RTK, it is even possible to give selective access to particular correction models. This allows various service classes, which differ in terms of accuracy and convergence time.

To allow partial decryption, LPP [1] mandates the use of the Advanced Encryption Standard (AES) with 128-bit key length in counter mode (CM or CTR). Compared to the more widely used cipher-block-chaining mode of AES (CBC), an encrypted block of data does not depend on the previous block's key. Hence it allows random data access (i.e., a receiver may decide to decode only a subset of the data).

The assistance data is delivered from the E-SLMC to the eNB for broadcast as described in [14], Section 9.3a.4, using the LPPa protocol. The E-SMLC is in control of all relevant parameters, such as the number of SIBs used to convey the data, encryption, and periodicity. The encryption is already applied in the E-SMLC, hence the eNB cannot decode the broadcast messages. However, a metadata element indicates to the eNB for which satellite system (e.g., GPS or Galileo) the assistance data is intended.

The key needed for decryption of the assistance data is sent from the E-SMLC to the MME network element. This key is not location-session-specific and it can be the same for large groups of users. The E-SMLC may provide multiple keys, each with an identifier, the partial counter start number c_0, a validity period, the applicable tracking area codes and a set of applicable posSIBs.

The MME may provide the keys to the mobile device during the LTE attach in the ATTACH ACCEPT message. Alternatively, the mobile device may request the keys, for instance when the validity period expires in a TRACKING AREA UPDATE, as defined in [15] Table 8.2.2.1 and Table 8.2.26.1. Depending on the subscription type of a user, which is stored in the HSS, the MME can choose the appropriate key set to differentiate various accuracy levels. The detailed procedure is defined in Section 9.3a.5 of [14] and the NAS element carrying the key data is specified in Section 9.9.3.56 of [15].

To decode a posSIB, the receiver needs to use the first part of the counter from the Attach or TAU (c_0) and combine it with the value d_0, which is broadcast unencrypted in the SIB element *CipheringKeyData*. Since the system information blocks are limited in size, an assistance data container may be segmented and span over multiple posSIBs.

The actual scheduling of the posSIBs is broadcast in SIB1 (which also schedules all other SIBs). The scheduling granularity allows specifying different periodicity per sibPosType and per GNSS. The encryption may also be configured per GNSS and per SIB type. Since the scheduling of the SIBs depends highly on LBS know-how, which is not present in the RAN, these parameters are provided by the E-SLMC via the LPPa protocol. The LPPa protocol will be described in Chapter 16 of this book. If the RAN cannot schedule the SIBs as requested by E-SMLC, it can report an error back to the E-SMLC.

The actual PosSIB provides a binary container for an octet-string, which is defined in the LPP specification [1] and carries the (encrypted) positioning payload along with metainformation. This rather complicated scheme allows fine-grained access rights depending on a user's subscription while still being very efficient in terms of network resource usage.

14.7.1.2 SIB Broadcasting Example

The following example is supposed to draft a possible implementation scheme and is depicted in Figure 14.19:

- PLMN I supports PPP and PPP-AR
- PLMN II supports retwork-RTK with FKP
- User B has a subscription for basic PPP and no roaming
- User C has a subscription for PPP-AR and RTK for roaming

The PLMN I broadcasts PPP data encrypted with one key and the additional PPP-AR (code biases) with another.

When user B(ob) attaches to the network, he receives the key for basic PPP only and can make use of the PPP, although not with the highest accuracy as he cannot decode the Code Biases. When user C(harlie) attaches to the network PLMN 1, she obtains both keys and can make use of all types of data with the highest accuracy and fastest convergence time. The network resources are used efficiently, as Charlie also uses the basic PPP data encrypted with the same key as Bob is using.

This concept also works in roaming scenarios. The E-SMLC in PLMN II broadcasts RTK correction data encrypted with yet another key. When Charlie roams to PLMN II, during attach she is granted access to this key, as her user profile allows roaming and RTK usage. PLMN II does not support PPP-mode, but Charlie's receiver supports RTK as well, hence she can tune to the RTK broadcast and after a short reconvergence time, she can continue to use the precise GNSS.

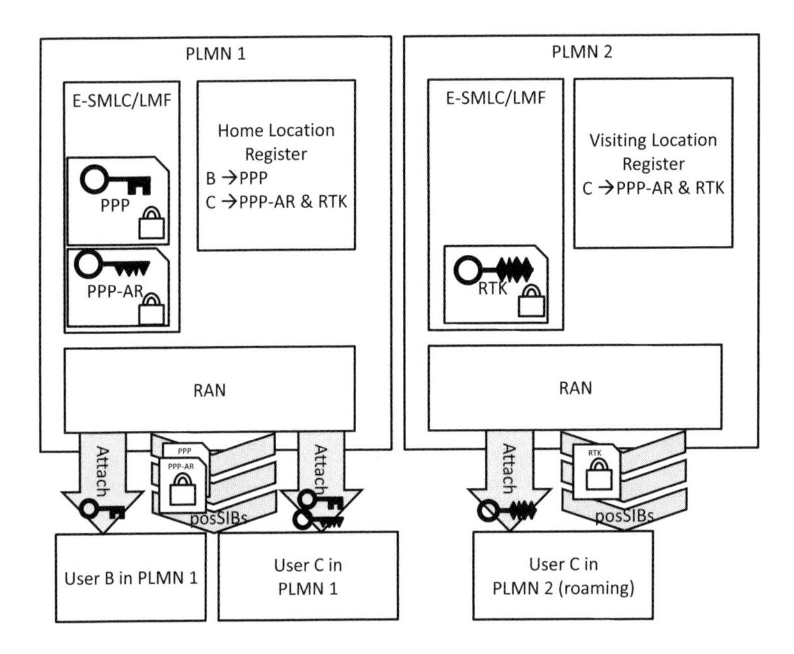

Figure 14.19 Example of SIB broadcasting with different service classes and roaming.

14.7.1.3 Other 4G/5G Broadcast Methods

Alternatively to SIB broadcasting, 4G and 5G both offer a generic broadcast mode, the eMBMS. This mode was intended for video broadcast or firmware upgrades. As eMBMS reserves complete LTE frames for broadcasting, using this method to transmit at the small data rate required by RTK/PPP corrections might not be efficient. Broadcast via eMBMS and NTRIP or LPPe would have been possible with releases prior to Release 15, but never saw practical use. The concept of eMBMS broadcasting might find future application in the V2X context to transport DSRC payloads (see Chapter 5 of this book for more details).

14.7.2 Motion Sensor Positioning

Support for IMU positioning information was added in LPP Release 15 to the sensor positioning messages (defined previously for barometric pressure measurements).

The IMU measurement report provides interfaces to accurately describe the movement of the mobile device. The *provide location information message* contains:

- A reference time.
- A list of displacements, characterized by:
    - Time elapsed with respect to the previous displacement info.
    - Bearing or direction.
    - Horizontal and vertical traveled distance.

14.7.3 RTK

Release 15 of LPP introduced support for high-accuracy GNSS enhancements, in particular RTK and SSR, which were explained in Chapter 7 of this book. The most critical requirement to support these technologies was to define a mechanism to carry all the assistance data needed. Thus, the creation of the posSIBs was a major driver for RTK.

The assistance data enhancements for RTK have already been seen in Tables 14.8 and 14.9 when defining the posSIBs. The GNSS common assistance data has been enhanced to include the RTK reference station, common observation, and auxiliary station information. The reference station information contains the parameters necessary to identify the reference station, including an ID and the precise location of the station's antenna. It indicates as well whether the reference station is physical or virtual (for more details see Section 7.4.3 of this book). The auxiliary station information indicates whether there are other stations, which are part of the same network, and describes them with respect to the reference station. Finally, the common observation info provides parameters that describe how the observations have been performed.

The GNSS generic assistance data (see Table 14.9) contains the RTK observations, the bias information, the residuals, and correction information related to SSR and the network-RTK applicable mode. These parameters are used to provide the mobile device the measurements performed at the reference station and the necessary correction data in order to obtain a high-accuracy GNSS location fix. For a deeper understanding of how to use the reference station measurements, please consult Chapter 7 of this book.

The provide location information messages have also been updated to support RTK and PPP. For UE-based positioning, the request location information can indicate in the positioning instructions whether high-accuracy GNSS methodologies are to be used. If yes, the mobile is expected to utilize the additional assistance data in the position calculation. In UE-assisted, the network calculates the location fix. For that purpose, the network can request ADR measurements from the mobile device. The ADR measurement can be used by the location server of the network to compute the carrier phase measurements, which are needed for RTK. ADR measurement support has been present since the first LPP release, but was extended in LPP Release 15. The set of UE-assisted ADR measurements can report the following parameters:

- The ADR measurement, whose resolution was increased by adding four additional bits in LPP Release 15.
- The estimated RMS of the ADR measurement.
- An indication of the carrier quality, giving information on whether the carrier phase measurement has been continuous or not and whether there is a half-cycle ambiguity in the measurement.
- A higher resolution for the legacy code phase measurement.

14.8 CONCLUSION

This chapter introduced the LPP and explained how a typical LPP positioning session works. It also analyzed in detail each of the messages required for the fundamental transactions, describing the most important parameters. As mentioned in the introduction, the LPP was initially defined in Release 9, one release after the official introduction of LTE. The first part of the chapter covered the LPP features from Release 9 to Release 12. LPP has not changed a lot during that time; the main new features were the addition of new GNSS (e.g., BDS), the early-fix feature in Release 12, and the support for interfrequency RSTD measurements for OTDOA in Release 10. Other LTE features such as carrier aggregation forced minor updates of LPP, especially for OTDOA, but nothing that required a critical LPP update.

The first major LPP revision occurred in Release 13, motivated by the need to meet regulatory indoor positioning requirements. Support for new positioning technologies (WiFi, Bluetooth, etc.) was added. However, the OMA created a new protocol (LPPe), adding similar enhancements as LPP Release 13, triggering a competition between both protocols. Although LPPe was the first one to reach the market and is for the moment the most widely used, the 3GPP solution is likely to turn the tide and become the preferred solution in the long run.

The last part of the chapter focused on the LPP Release-15 updates, the first steps toward supporting high-accuracy (cm-level) positioning technologies for commercial applications. LPP Release 15 laid the foundation for high-accuracy positioning services by introducing RTK and other GNSS enhancements. Furthermore, the definition of the positioning SIB and consequently the improvement of the network efficiency by reducing the amount of assistance data transmissions is likely to play a major role in areas such as autonomous driving for 5G. The possibility to encrypt these SIBs and define a subscriber-based model make it an interesting business case for network operators and other service providers.

3GPP Release 15 was also the release in which 5G NR has been introduced. Instead of creating a new positioning protocol, LPP has been reused for 5G location-based services. The necessary enhancements and updates to LPP to support 5G will be discussed in the next chapter of this book.

References

[1] 3GPP TS 36.355, *LTE Positioning Protocol (LPP)*, V15.5.0, September 2019.

[2] 3GPP TS 36.171, *E-UTRA; Requirements for Support of A-GNSS*, V15.00, July, 2018.

[3] Van Diggelen, F. *A-GPS: Assisted GPS, GNSS, and SBAS*, Norwood, MA: Artech House, 2009.

[4] GPS Directorate, Systems Engineering & Integration, *IS-GPS-200J, NAVSTAR GPS Space Segment/Navigation User Segment Interfaces*, April, 2018.

[5] European GNSS (Galileo) Open Service, *Signal in Space Interface Control Document (OS SIS ICD)*, Issue 1.3, December, 2016.

[6] Global Navigation Satellite System Glonass, "Navigational Radiosignal in Bands L1, L2," *Interface Control Document*, Edition 5.1, Moscow, Russia, 2008.

[7] BeiDou Navigation Satellite System, "Signal in Space Interface Control Document, Open Service Signal," China Satellite Navigation Office, Version 2.0, December, 2013.

[8] Federal Communications Commission, *Wireless E911 Location Accuracy Requirements, Fourth Report and Order*, February 2015.

[9] Open Mobile Alliance, *LPP Extensions Specification*, Candidate Version 1.0, January, 2014.

[10] RFC 4776, *Dynamic Host Configuration Protocol (DHCPv4 and DHCPv6) Option for Civic Addresses Configuration Information*, The IETF Trust, 2006.

[11] 3GPP TR 37.857, *Study on Indoor Positioning Enhancements for UTRA and LTE*, V13.1.0, December, 2015.

[12] Metropolitan Beacon System, Nextnav, http://www.nextnav.com/network#terre, accessed on September, 2019.

[13] 3GPP TS 36.331, *E-UTRA; Radio Resource Control (RRC); Protocol Specification*, V15.7.0, September, 2019.

[14] 3GPP TS 23.271, *Functional Stage 2 Description of Location Services (LCS)*, V15.1.0, September, 2018.

[15] 3GPP TS 24.301, *Non-Access-Stratum (NAS) protocol for Evolved Packet System (EPS); Stage 3*, V16.1.1, June, 2019.

Chapter 15

Positioning Protocol in 5G

15.1 INTRODUCTION

Throughout this book, the importance of LBS in the cellular networks has been extensively discussed. The role of positioning technologies has greatly increased since their addition to the first cellular networks. Their relevance for 5G, not only for regulatory requirements of the smartphone sector, but also for the new industry verticals, is beyond question. Accurate positioning is and will continue to be a crucial element of 5G's success.

Previous chapters of this book have introduced the 5G LBS requirements and its use cases and the whole spectrum of different positioning technologies, both RAT-dependent and RAT-independent. This chapter will explain the positioning protocol and the messages exchanged between the UE and the RAN during a 5G positioning session, building upon the legacy protocols discussed in the previous two chapters.

This chapter will present the Release-15 3GPP updates for NR, and it will describe the new changes being proposed during Release 16, extending LPP to also cover the newly defined 5G RAT-dependent positioning technologies.

15.2 RELEASE-15 5G POSITIONING SUPPORT

The timeline leading up to the first 5G NR official release was quite challenging. The industry needed to showcase the 5G potential and all the nonessential features were postponed to later 3GPP releases, with the aim to focus on the relevant 5G parts. However, the regulatory positioning requirements, mainly driven by the FCC, were one of the fundamental 5G features that could not be postponed.

Therefore, 5G Release 15 supports emergency positioning sessions and location calculation based on A-GNSS, other RAT-independent technologies, and legacy RAT-dependent technologies (e.g., OTDOA). This section will describe the necessary LPP updates to support this basic positioning feature set.

15.2.1 Reusing LPP

For all cellular networks prior to 5G NR, the 3GPP has defined their corresponding positioning protocols between the RAN and the UE. GSM had RRLP; for WCDMA, the positioning messages were part of the RRC; and LPP was defined for LTE. However, for NR, the approach has changed, and LPP has been reused. The reasons for the adoption of LPP as 5G positioning protocol are manifold, some of which include:

- The tight timeline that preceded the first 5G NR release, and the obligation to support basic positioning functions to fulfill the FCC regulatory requirements.
- The 5G deployment options (see Chapter 5 of this book) included multiple combinations between NR and LTE both in the RAN and CN. Some of the deployments have LTE as a primary connection for the UE, while others have NR. Defining a new positioning protocol for 5G NR would result in different positioning flows depending on the deployment option.
- LPP is a mature protocol, including a lot of positioning features, like the posSIB or RAT-independent technologies, which can be beneficial for 5G. Reusing LPP enables these features from the very first 5G release, instead of having to define them all again.
- In LTE, LPP is transported in a generic NAS transport container, which is included in the *DLInformationTransfer* RRC message. Thus, LPP is largely decoupled from the LTE RRC. Following the same approach, LPP can also be injected in the NR RRC messages without further modifications.

Due to all the reasons mentioned above, the 3GPP decided to extend TS 36.355 [1] to include 5G positioning.

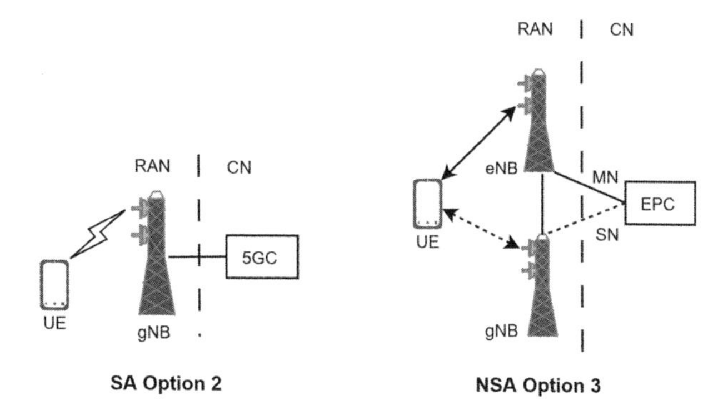

Figure 15.1 SA Option 2 and NSA Option 3 deployments.

15.2.2 Transporting LPP over C-Plane in the Different 5G Deployments

As seen in Chapter 5, there are multiple possible deployment scenarios for the 5G network, based on which core network (EPC or 5GC) and RAN are used, whether to enable dual connectivity or not, and so forth. The deployment selection will impact the way positioning sessions are established and how the LPP messages are transported between UE and location server. Nonetheless, from a positioning point of view, all cases can be explained by looking at the two most popular scenarios, shown in Figure 15.1.

NSA Option 3 is very similar, from the LPP perspective, to the legacy LTE network. The UE has a primary connection (represented by MN) to an eNB, which is in turn connected to the legacy EPC network. The UE also has an additional connection (SN) to an gNB via dual connectivity. However, this additional connection is not used for positioning and does not influence the LPP session.

An LPP session in NSA Option 3 is a communication between the E-SLMC (or the SLP in the case of SUPL) and the UE. The LPP message is transferred via the LTE link. As shown in Figure 15.2, this message is encapsulated in a container part of the EPS *Downlink* (or *Uplink*) *Generic NAS transport*, specified in TS 24.301 [2]. The EPS NAS message is embedded in a E-UTRA RRC *DL* (or *UL*) *information transfer*, which is defined in TS 36.331 [3]. In summary, the whole process works the exact same way as it did for a conventional LTE LPP positioning session.

In SA Option 2, there is no LTE connection. This deployment is a pure 5G network with a gNB connected to 5GC. In this case, to transmit the same

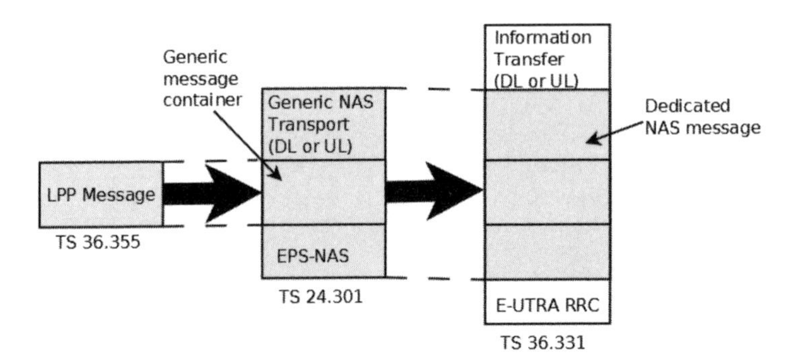

Figure 15.2 Transport of a LPP message via LTE.

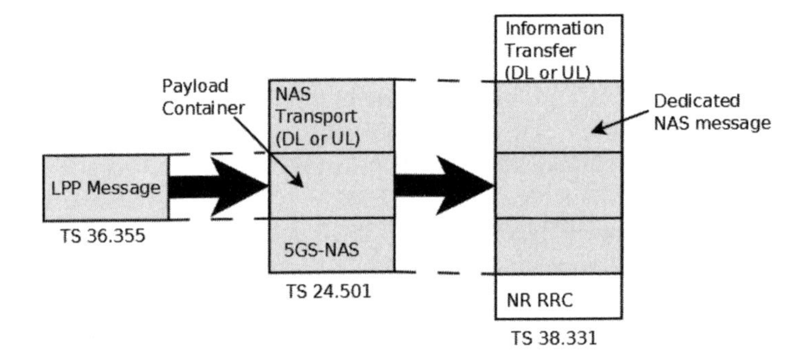

Figure 15.3 Transport of a LPP message via NR.

LPP message as in Figure 15.2, the process is slightly different. This is shown in Figure 15.3.

The first difference is that the LMF replaces the E-SLMC in the location session. Furthermore, the LPP message is added as payload to the 5G System (5GS) *Downlink* (or *Uplink) NAS transport*, as specified in TS 24.501 [4]. The 5GS NAS message (similar to the LTE case) is embedded in a NR RRC *DL* (or *UL*) *information transfer*, which is defined in TS 38.331 [3].

Apart from the above-mentioned process, there are no additional changes required in order to support LPP positioning through 5G NR. As has been seen, none of the changes directly affect the LPP defined in TS 36.355. The only modifications have been performed in the 5GS NAS and NR RRC protocols. Nonetheless, 5G positioning immediately benefited from the other LPP Release-15 modifications seen in Chapter 14 of this book. In particular:

- The definition of the posSIB to broadcast the positioning assistance data, reducing the network load due to dedicated assistance data transmissions and enabling a subscription-based positioning service.
- The LPP enhancements for high-accuracy GNSS methods (e.g., RTK and PPP).

15.2.3 Transporting LPP over SUPL in the Different 5G Deployments

In the case of LPP over SUPL, the SUPL positioning messages are agnostic to the deployment option, since SUPL is an independent protocol. Nonetheless, the establishment of the data bearer can vary for each deployment.

For NSA Option 3, there are three different possibilities to establish the bearer for data transmissions, shown in Figure 15.4. The NR PDCP layer can work independently from the LTE PCDP layer. Thus, the routing of the U-Plane messages can be done through either:

- Master cell group (MCG) bearer: the SUPL messages are transmitted by the LTE lower layers (RLC, MAC and PHY) to the UE. Both NR and LTE PDCP layers are connected to the LTE RLC.
- Secondary cell group (SCG) bearer: the opposite configuration, the SUPL messages are transmitted by the NR lower layers.
- Split bearer: the SUPL messages are split between the LTE and NR lower layers. The data can be sent on either, or even duplicated for robustness.

For SA Option 2, there is no SCG. Thus, the data bearer for SUPL is established through the only connection available.

15.3 RELEASE-16 LPP ENHANCEMENTS FOR 5G

The real enhancements for 5G positioning support are done as part of the 3GPP Release 16, which is scheduled for completion in the second quarter of 2020. Brand-new 5G positioning technologies have been added to TS 38.305 [6] . Thus, LPP needs to be enhanced to support these new positioning technologies and their corresponding measurements.

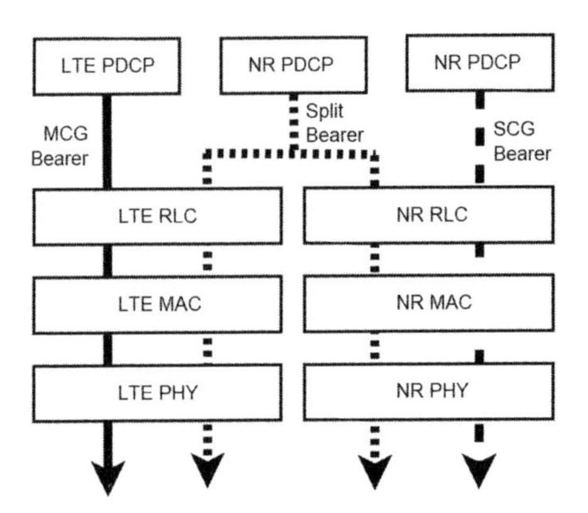

Figure 15.4 Data bearer configurations for NSA Option 3.

The first modification addressed by the 3GPP is not of a technical nature, but is rather formal. TS 36.355 could not be formally extended to include 5G positioning methods. Hence, the 3GPP has terminated TS 36.355 and moved all the contents to a new, RAT-agnostic TS 37.355 [7] .

Formal updates aside, LPP Release 16 will need to address the following positioning methods:

- DL-TDOA
- DL-AoD
- NR ECID
- Multi-RTT

The uplink positioning methods, UL-TDOA and UL-AoA do not need any specific LPP updates, since the position calculation takes place in the network and the positioning session is mainly between the gNB and the LMF. This is covered by NR Positioning Protocol A (NRPPA), which will be explained in the next chapter of this book. Nonetheless, there could be minor LPP updates related to the uplink methods. For instance, to enable a capability request from the network, in order to find out whether the mobile device supports the UL SRS enhancements for positioning.

At the time of the writing, there are two different views on how to extend LPP for 5G positioning methods. One possibility is to add a new item to each of

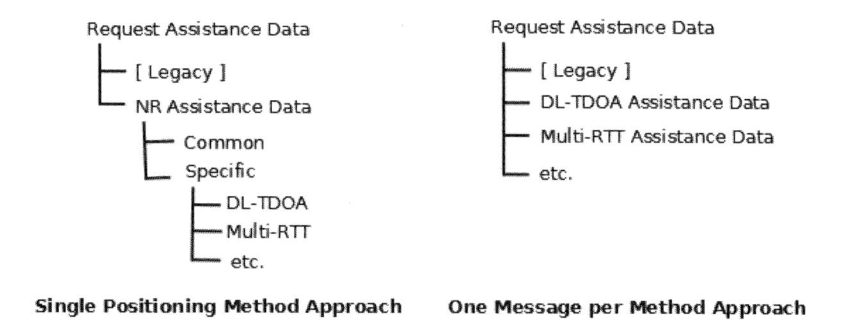

Figure 15.5 Comparison of the two different approaches to add NR LPP messages.

the LPP fundamental transactions for each of the new positioning technologies. This approach follows what has been already done for Wi-Fi, Bluetooth, and sensor positioning. The second possibility is to follow the single-positioning method approach, and add only one new item to each of the fundamental transactions, which covers all 5G-related positioning methods. An example of both approaches is given in Figure 15.5, based on the *RequestAssistanceData* message. For the single-positioning method approach, a generic *NR Assistance Data* message is added, including some common and specific assistance data. On the other hand, on the right side of Figure 15.5, there are multiple new messages added, *DL-TDOA Assistance Data*, *Multi-RTT Assistance Data*, and so forth. Proponents of the single-positioning method approach argue that it is more efficient, since different positioning methods sometimes derive from the same fundamental measurement. Combining the LPP messages, it can be ensured that the measurement is not performed and reported twice.

Independent from the approach that is finally used by the 3GPP, 5G NR RAT-dependent positioning will need a series of parameters to be added to the fundamental transactions. The most important will be analyzed in the subsequent sections.

15.3.1 Enhancements to Common Information Elements

These enhancements should cover basic NR functionality, such as the parameters needed to identify an NR cell (e.g., the physical cell ID) and the synchronization signal block (SSB) configuration. Furthermore, NR has introduced a new PRS, as seen in Chapter 10 of this book, with a structure significantly different from the

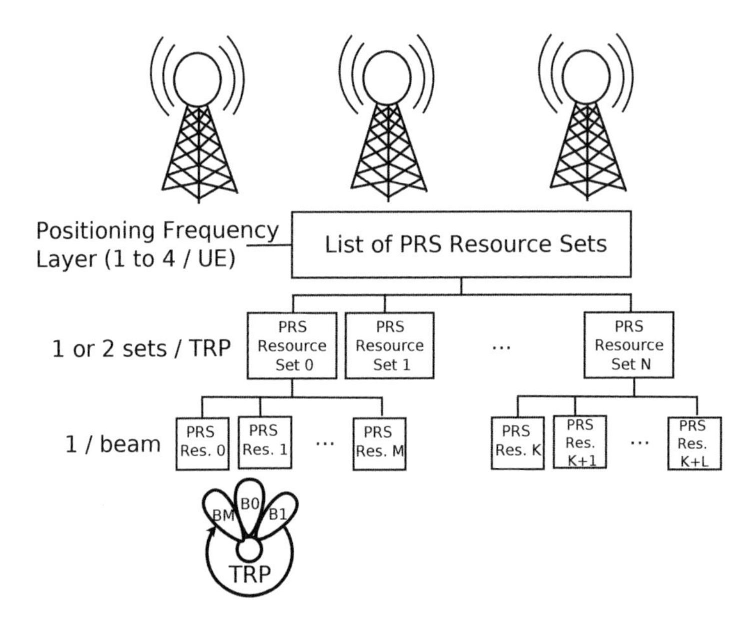

Figure 15.6 Configuration of the PRS resources in 5G.

LTE PRS. Therefore, a *NR-DL-PRS-Config* will be needed in order to provide all the necessary information for identifying and measuring the NR PRS.

It is worth noting at this point the basic PRS structure, shown in Figure 15.6. LPP will have information elements to define the different positioning frequency layers, including the corresponding lists of PRS resource sets and PRS resources. Parameters that need to be considered are, for instance, the PRS bandwidth, the offset, number of symbols, and muting patterns.

15.3.2 Enhancements to the Capability Exchange

The capability exchange is likely the most straightforward transaction to be updated. LPP Release 16 should include messages for the LMF to request information about the positioning methods and positioning measurements supported, and the UE response should be enhanced to include the same corresponding methods and measurements.

15.3.3 Enhancements to the Assistance Data

The enhancements to the assistance data cover two areas: the dedicated assistance data information that can be provided during a positioning session, and the broadcast assistance data information for the posSIBs.

The assistance data is required for DL-TDOA, DL-AoD, and multi-RTT positioning methods. NR ECID does not require any additional data, since it is based only on preexisting measurements. The assistance data for these three methods is fairly similar (hence, one of the reasons for the single-positioning method approach). For UE-assisted, the required assistance data is basically the NR PRS information, in order to help the UE to identify and measure the PRS signal.

For UE-based, supported for DL-TDOA and DL-AoD, the UE will need additional information in order to calculate its own position correctly. The minimum assistance data parameter set required for position calculation will include:

- Information about the TRP for each PRS resource, including its reference location.
- Information about the direction associated with the beam of each PRS resource.

The two parameters mentioned above would be sufficient for the positioning algorithm to compute the UE's location. Nonetheless, given that the network synchronization error is one of the biggest sources for error for DL-TDOA positioning (see Chapter 10), the 3GPP is considering as well the inclusion of a synchronization error correction factor. This correction factor is called RTD, and gives a measurement of the real time difference between the PRS coming from two different TRPs.

The same assistance data messages also need to be included in the posSIBs. It has been agreed that the posSIBs should also support DL-TDOA and DL-AoD positioning methods. The 3GPP is proposing the addition of three new posSIBs, as shown in Table 15.1. The first one, the PRS assistance data, transmits all parameters necessary for UE-assisted positioning. The remaining two are for UE-based, one to indicate the TRP coordinates and related information, and the last one for including the time synchronization correction information.

15.3.4 Enhancements to the Location Information

The last of the fundamental transactions is the location information exchange. The enhancements to the *Request Location Information* are trivial, similar to the updates

Table 15.1

Positioning SIBs for DL-TDOA and DL-AoD Methods

posSIBType	LPP Message
posSibType3-2	NR-DL-PRS-AssistanceData
posSibType3-3	NR-UEB-TRP-LocationData
posSibType3-4	NR-UEB-TRP-RTD-Info

to the capability messages. On the other hand, the *Provide Location Information* message needs to be updated to include each of the measurements performed in each of the new positioning methods. This section will describe the measurement reports to be included for each new technology.

15.3.4.1 NR-ECID

For NR ECID, the following power measurements are required:

- SS-RSRP
- SS-RSRQ
- Channel state information reference signal received power (CSI-RSRP)
- Channel state information reference signal received quality (CSI-RSRQ)

Thus, the positioning measurement report for NR ECID will need to include a list of measurement elements for the four measurements listed above for each visible gNB or TRP.

15.3.4.2 DL-TDOA

The positioning measurements for DL-TDOA should include two different measurement containers. The first, the signal measurement information, will be very similar to the LTE OTDOA measurement report. It will contain the following elements:

- The PRS information for the reference base station.
- A list of measurements for the neighbor base stations or TRPs, including at least:

- The PRS ID for the corresponding PRS resource measured.
- The RSTD, with respect to the reference PRS resource.
- Some indication of the measurement quality.
- Optionally, the positioning reference signal reference signal received power (PRS-RSRP) measurement.

The second measurement container is for UE-based DL-TDOA, and it should include the DL-TDOA location information. The location information comprehends the calculated location of the UE and a reference time at which the measurement was performed. The UE location is reported as part of the common information elements (see Chapter 14) using WGS 84.

15.3.4.3 DL-AoD

From the point of view of the set of measurements expected, DL-AoD is very similar to DL-TDOA. Thus, both positioning methods could be combined nicely into a single report. In the IE signal measurement information, the only difference to DL-TDOA is the absence of the RSTD measurement and the addition of the DL beam index.

- The PRS information for the reference base station.
- A list of measurements for the neighbor base stations or TRPs, including at least:
    - The PRS ID for the corresponding PRS resource measured.
    - The DL beam index, as received by the UE.
    - Some indication of the measurement quality.
    - Optionally, the PRS-RSRP measurement.

For UE-based DL-AoD, a location information container must be provided in addition. Analogous to DL-TDOA, this container comprehends the UE calculated location and the measurement reference time.

15.3.4.4 Multi-RTT

The last method defined for 5G positioning in Release 16 is multi-RTT. The basic structure of the measurement report for multi-RTT is again very similar to the previous two methods. However, in this case, UE-based positioning is not supported. Thus, there is no need to define a multi-RTT location information container.

The multi-RTT signal measurement information container must include a list of measurement elements with the following minimum set of parameters:

- The PRS ID for the corresponding PRS resource measured.
- The UE RxTx time difference measurement.
- Some indication of the measurement quality.
- Optionally, the PRS-RSRP measurement.

15.3.5 Abort and Error LPP Messages

Apart from the fundamental LPP transactions that are part of regular positioning sessions, there are two other LPP messages. These two messages are used in special cases in order to abort an ongoing positioning session or to indicate there has been an error in the process.

The *LPP Abort* is generic for all positioning methods. As such, there are no specific updates required for NR positioning.

The *LPP Error* needs to be extended for each of the new positioning methods. The LPP error should provide a list of possible location service and target device error causes for each technology. Most typical of these errors are, for example, “assistance data not supported” or “measurement not possible.”

15.4 RELEASE-16 COMMON LPP UPDATES

The support of NR RAT-dependent positioning technologies is without doubt the most important feature in LPP Release 16. Nonetheless, there have also been other updates to the LPP protocol during this timeframe. The updates are not necessarily specific to 5G NR, but can also be used in combination with LTE positioning.

15.4.1 Enhancements to SSR Assistance Data

RTK and PPP, two of the high-accuracy GNSS technologies, have been supported since LPP Release 15. However, in Release 16, the assistance data for the SSR model has been improved. The new set of assistance data parameters can be used by the location server to provide further corrections to the UE. Using these corrections, the target device can use PPP-RTK for its location fix with all the benefits of this method mentioned in Chapter 7. Below are some of the new assistance data elements included in Release 16:

- *GNSS-SSR-CorrectionPoints*, which can be used by the location server to provide a list of correction point coordinates.
- User range accuracy (URA) corrections in *GNSS-SSR-URA*, containing information on the quality of the SSR assistance data.
- *GNSS-SSR-PhaseBias*, providing information about the bias affecting the phase of the GNSS signal. The UE can use this value to achieve a PPP-AR fix.
- Parameters to provide ionosphere slant delay corrections, divided in a STEC correction and a grid correction.

These four additional items also require new posSIBs, extending the lists provided in Chapter 14 of this book for the GNSS common assistance data and the GNSS generic assistance data. The list of new posSIB in given in Table 15.2.

Table 15.2

Positioning SIBs for the GNSS SSR Assistance Data Enhancements in Release 16

GNSS Commmon Assistance Data	
posSIBType	LPP message
posSibType1-8	GNSS-SSR-CorrectionPoints
GNSS Generic Assistance Data	
posSIBType	LPP message
posSibType2-20	GNSS-SSR-URA
posSibType2-21	GNSS-SSR-PhaseBias
posSibType2-22	GNSS-SSR-STEC-Correction
posSibType2-23	GNSS-SSR-GriddedCorrection

15.5 CONCLUSION

This chapter explained the extensions of the LTE Positioning Protocol to support NR positioning. The first part of the chapter focused on the Release-15 updates. Release-15 positioning was limited to the minimum set of requirements necessary to meet FCC regulations.

The second part of the chapter introduced the new Release-16 5G RAT-dependent positioning technologies and gave an overview of the LPP enhancements needed to support them. The 3GPP Release 16 is not yet concluded, and the names of certain parameters may still be slightly modified. Nonetheless, this chapter has provided a good overview of the fundamental transactions and the key measurement elements needed to carry on the 5G positioning sessions.

With this chapter, the part of the book dedicated to positioning protocols between the UE and the cellular network is completed. The book has given a general overview of the legacy positioning protocols, such as RRLP or TIA-801, and an extensive explanation of LPP for both LTE and 5G networks. The next and final chapter of the book will take a look at what happens behind the scenes; the link between the RAN and the CN, and the protocols that regulate the positioning exchange within the network components.

References

[1] 3GPP TS 36.355, *LTE Positioning Protocol (LPP)*, V15.6.0, January 2020.

[2] 3GPP TS 24.301, *Non-Access-Stratum (NAS) Protocol for Evolved Packet System (EPS); Stage 3*, V16.3.0, December 2019.

[3] 3GPP TS 36.331, *E-UTRA; Radio Resource Control (RRC); Protocol Specification*, V15.8.0, January 2020.

[4] 3GPP TS 24.501, *Non-Access-Stratum (NAS) Protocol for 5G System (5GS); Stage 3*, V16.3.0, December 2019.

[5] 3GPP TS 38.331, *NR; Radio Resource Control (RRC); Protocol Specification*, V15.8.0, January 2020.

[6] 3GPP TS 38.305, *NG-RAN; Stage 2 Functional Specification of User Equipment (UE) Positioning in NG-RAN*, V15.6.0, April 2020.

[7] 3GPP TS 37.355, *LTE Positioning Protocol (LPP)*, V15.0.0, December 2019.

Chapter 16

Positioning in a Virtualized Network

16.1 INTRODUCTION

Cellular radio networks are specified by 3GPP in two main logical entities: The RAN and the CN. So far, this book has focused on the positioning protocols and functions related to the RAN network (also known as the air interface). However, in a positioning session, a lot of information is also exchanged behind the scenes, in the CN, also referred to as nonair interface, since typically the connections between the different entities of the CN are wired.

Recent trends in CN and RAN implementations use a cloud-based approach, where telecommunications and IT industry concepts are merged. In fact, much of 5G's promise in terms of efficiency and cost savings for network operators stems from its optimization of the CN, introducing new concepts such as network slicing and virtualization. Thus, an understanding of the CN and how it is impacted by these features is paramount. This chapter will clarify the difference between RAN and core and explain the typical functions of each network, especially when it comes to positioning.

The CN also defines communication protocols between the different entities. For instance, the communication protocols between the location server and the base station are the LPPa and its 5G evolution NRPPA. This chapter will give an overview of these protocols, pointing out the main differences between LTE and 5G and the impact of core virtualization in how positioning sessions work in 5G.

Finally, this chapter will also explore virtualization in the RAN and its impact on positioning.

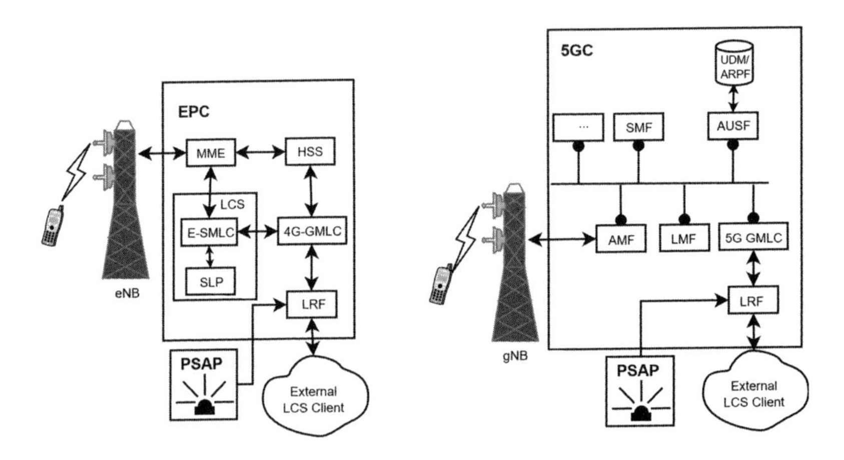

Figure 16.1 EPC and 5GC positioning archictecture.

16.2 THE MOBILE NETWORK CORE

16.2.1 EPC and 5GC Network Architecture

The network architecture for 4G and 5G networks has been explained in Chapters 3 to 5 of this book. This section will just give a summary of the EPC and 5GC architectures, with an emphasis on the network functions that are involved in the positioning sessions. The basic EPC and 5GC positioning architecture in shown in Figure 16.1.

For EPC, the MME takes care of the session management. The HSS, which identifies the GMLC that is associated with a certain user and provides routing information. The 4G-GMLC is the point of access for external location clients (including the PSAPs), through the LRF. The location server is the entity that calculates the target device's location. It contains the E-SLMC and the SLP, for U-Plane positioning.

5GC has NFs equivalent or similar to EPC. The MME is replaced partly by the AMF, which takes care of the initial access and mobility procedures, and partly by the SMF, taking care of the session management. The HSS is replaced by the AUSF and the ARPF. The LMF takes over the role of the E-SLMC. Finally, the 5G-GMLC is the point of access for the external location clients to request positioning information.

16.2.2 Functions of the Core

The CN connects the base stations and all other entities of the network and is responsible for tasks like user authentication, session management, billing, policy control, and location calculation. For these tasks, 3GPP specified several NF for LTE, like the E-SLMC in the case of positioning, or the MME for session management.

Historically, each NF was implemented on specialized hardware and the 3GPP specified the point-to-point protocols between those network elements. These dedicated hardware entities are now referred to as physical network functions. The specialized hardware was necessary in the past to cope with the high computational burden that is required to run networks with up to hundreds of millions of users. However, the approach of physical NFs brings several downsides with it:

- Difficult load balancing.
- High cost for specialized hardware.
- High maintenance effort (e.g., spare parts).
- Potential lock-in to specific vendors.
- Difficult to roll out new features.

However, the appearance of more powerful processors, capable of completing complex tasks in a more efficient and rapid way, reduced the need for specialized hardware. Thus, modern core networks experienced an evolution that moved the complexity from the hardware to the software side. This phenomenon of replacing dedicated hardware units with general-purpose processors (GPPs) is called virtualization, and can be divided into two phases.

16.2.3 First Phase of Core Network Virtualization

With the ever-increasing computational power of server architectures based on GPPs, like x86 and ARM, it became feasible to implement more and more CN elements as virtual network functions (VNFs). The one-to-one replacement of a physical network function by a VNF (e.g., as a virtual machine) is called the first phase of network virtualization and took place in the 4G era. On a protocol layer, the first phase of virtualization has no significant impact, as it replicates the legacy interfaces. While this step already brings benefits by using commercial off-the-shelf (COTS) server and router hardware, other limitations remained, such as the

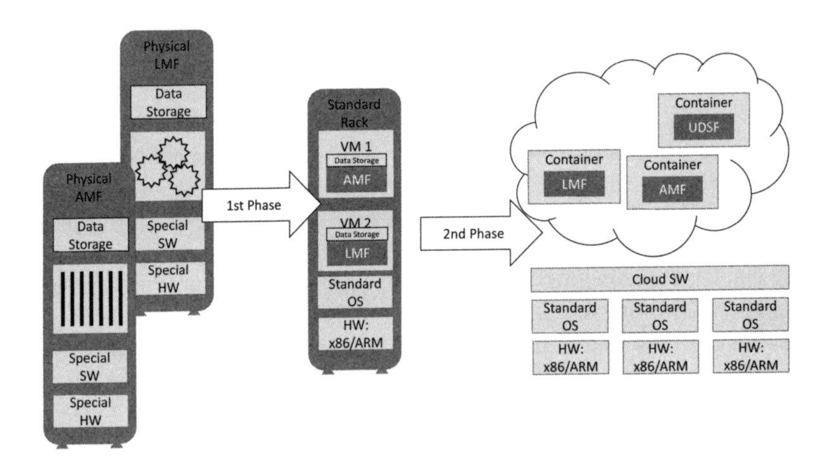

Figure 16.2 The two phases of core virtualization, from dedicated HW to cloud native.

difficulty of load balancing and the general complexity of managing a distributed system. Frameworks like the network function virtualization (NFV) (see use case 5 in [1]) driven by ETSI, are aiming at standardizing the orchestration of VNFs to enable more advanced implementations beyond one-to-one replacements of dedicated hardware by virtual machines. This is a complex task, since it is based on standards that were not initially intended for virtualization.

16.2.4 Second Phase of Core Network Virtualization: Cloud Native

5G revolutionizes the way a cellular network core is organized. By introducing concepts from cloud computing, it starts the second phase of network virtualization. One characteristic feature is the separation of data storage from the network functions. Additionally, the interfaces between VNFs are specified as RESTful APIs based on representational state transfer (REST), which means that every request to a NF contains all information that is required to execute the request, rendering individual NFs stateless. These concepts allow using different instances of the same network function to execute subsequent tasks of a single user. Load balancing is achieved more natively in this concept. Furthermore, the robustness in case of hardware failure increases. If a component like the AMF, the "new" MME in 5G core, fails, all users served by this MME can also be handled by any other MME instance. Network functions may be removed or added in the system on the fly without service interruption, simplifying software upgrades significantly and hence speeding up network innovation. For a comparison of the first and second phase of virtualization, see Figure 16.2.

In order to enable a SBA, the 3GPP defined all 5G interfaces (where both destination and origin are within the 5GC) based on OpenAPI 3.0.0 (see TS 23.501 [2], TS 23.502 [3]). This allows NFs like the AMF or the LMF to be deployed dynamically in virtual machines or containers in a cloud and allows NFs of various vendors to work together. The NFs are connected by a bus rather than point-to-point interfaces. Every node can communicate with every other node (if allowed by policy). The OpenAPI 3.0.0 specification is based on HTTP/2 [4] and the YAML format [5]. YAML is a data serialization language whose purpose is similar to the EPC's ASN.1. YAML can be considered an evolution of JavaScript Open Notation (JSON, [6]) and its syntax is largely compatible. Autogenerated encoders and decoders for OpenAPI and YAML exist for many languages, such as C, C++, JavaScript, ActionScript, Perl, PHP, Python, Java, R, Ruby, and .NET.

16.2.5 Scaling Up and Down

While virtualization brings benefits for large mobile networks, it also scales down to the other end, private mobile networks. A network core for a few hundreds of users may fit on a single physical COTS server today. Along with innovative regulatory frequency allocation schemes like CBRS (see [7]) in the United States or the industry frequency spectrum reserved at 3.7 GHz and 26 GHz in Germany (see [8]), it allows enterprises to operate their own local 5G network (e.g., for industrial automation).

16.2.6 Network Slicing

The flexible, cloud-native 5G core is also central in enabling new features like network slicing, explained in Chapter 5 of this book. This concept foresees several logical networks operated by the same physical network. The intention is to optimize the network for the use case. As an example, a network may provide three slices, one for eMBB (the typical smartphone use case), another one for URLLC (e.g., for autonomous driving support) and a third one for mMTC (also known as the Internet of Things). Each of those slices has very different requirements on the CN and hence may be served by different instances of VNFs. The physical hardware location may differ depending on latency requirements. For eMBB, many VNFs may be running in a central data center per city or region. For URLLC, on the other hand, the latency

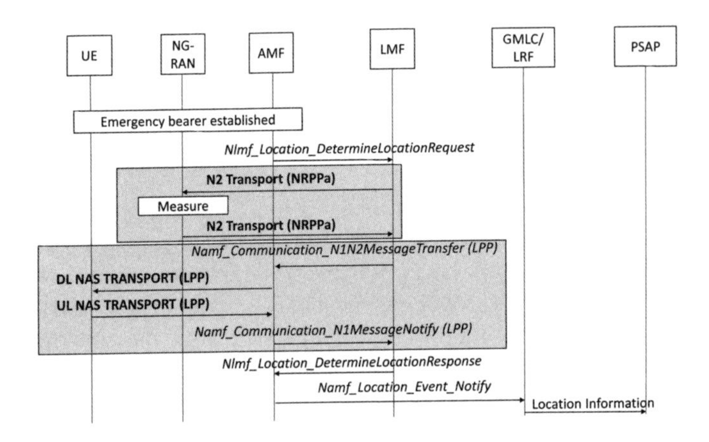

Figure 16.3 NI-LR flow, including UE and network measurements.

critical functions may need to be closer to the individual base stations. Different slices may also utilize different software update cycles, depending on innovation speed and reliability requirements of the slice's use case.

16.3 POSITIONING IN THE CORE NETWORK

16.3.1 Positioning Call Flows

The following section will introduce the CN protocol flows needed to perform an exemplary emergency caller location procedure. This procedure is chosen because it covers all relevant entities for positioning. Both 4G EPC and 5GC flows will be introduced to allow the reader the comparison of the traditional protocol-based core versus the 5G SBA.

There are two fundamental flows for 3GPP emergency caller location: NI-LR and MT-LR. The main difference between them is who initiates the session.

In NI-LR the CN starts the procedure when it notices that the UE initiates an emergency call (e.g., due to the emergency PDN establishment in a E911 VoLTE call). The location is eventually pushed to the GMLC, where it can be requested by the PSAP, as shown in Figure 16.3. In the figure, SBA messages are shown in italic, while NAS messages are printed in bold.

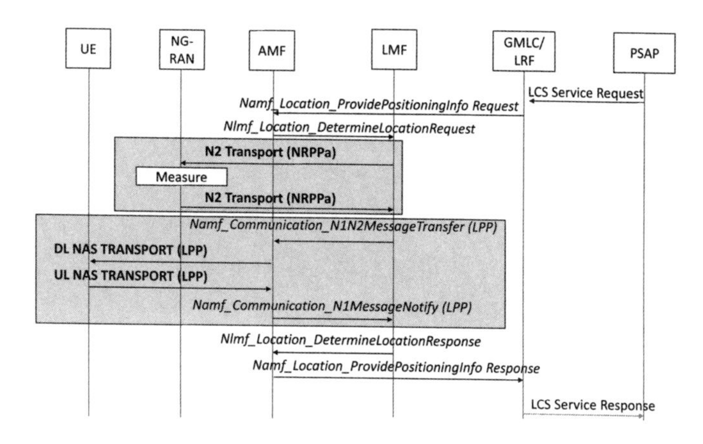

Figure 16.4 MT-LR flow, including UE and network measurements.

In the MT-LR flow, the GMLC initiates the location session (see Figure 16.4), for example, on behalf of a request from the PSAP.

The 3GPP also defines a third location request method, in which the location service client in the mobile device is the one starting the request. This is called MO-LR. However, this type of request is not used during emergency calls.

16.3.2 4G and 5G Core Network Protocols

The protocols introduced in Chapters 13 and 14 are used for communication between a location server and a mobile device. RAT-dependent positioning technologies need to exchange additional information between the base station and the location server (e.g., the location of the base station's antenna). Furthermore, multiple protocols exist that serve functions within the core or toward external applications. One such protocol forwards the LPP messages between the different CN entities, while another allows third parties (e.g., a PSAP) to retrieve the location. Since this communication is internal in a mobile operator's RAN and core, these protocols are out of scope for mobile devices and other user equipment. Figure 16.5 and Figure 16.6 give an overview of the 4G and 5G core interfaces described in the following section. Similar 4G and 5G functionality is grouped together to contrast the traditional telecommunications world 4G CN approach with the 5G cloud-native service-based architecture.

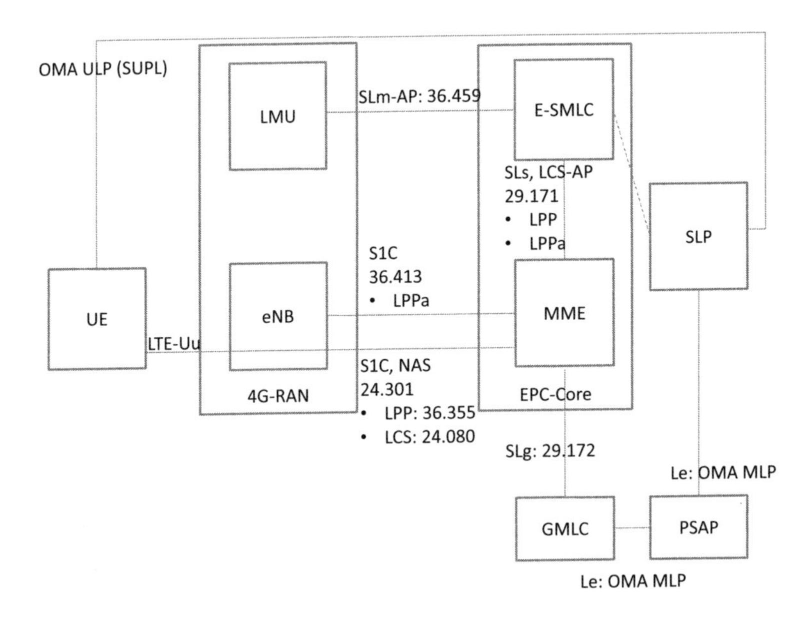

Figure 16.5 4G positioning architecture (see [9]).

For the reader's orientation, it is helpful to know the 3GPP specification nomenclature: 4G RAN specifications are the 36 series (starting with 36.xxx), 5G RAN specifications are the 38 series, and intra-Core specifications are included in the 29 series, with 29.1xx for the 4G intra-Core and 29.5xx for 5G. Protocols that are used to communicate between Core and user equipment start with 24.3xx for 4G and 24.5xx for 5G. However, this nomenclature is not always maintained consistently; for example the LPP protocol is a 37-series protocol,[1] while it might have been specified as a 24 series protocol as well.

16.3.2.1 Location Server : Base Station

4G EPC: LPPa

The location server requires communication with the base stations (e.g., for performing some ECID measurements and retrieving OTDOA configurations). For this

1 TS 36.355 was converted into TS 37.355 in December 2019.

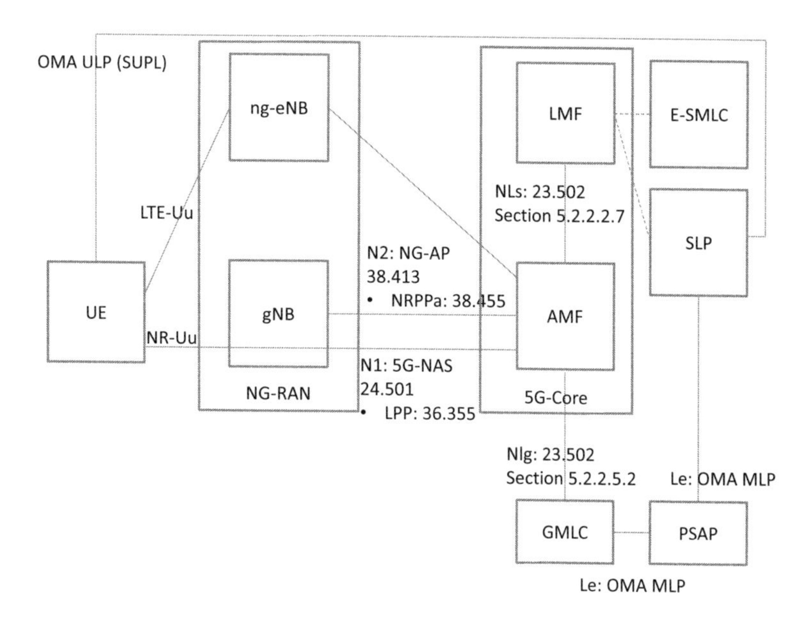

Figure 16.6 5G positioning architecture (See [10]).

purpose, the LPPa protocol (LPP-Annex, TS 36.455, [11]) was defined. It uses the ASN.1 syntax. The data exchanged between E-SLMC and eNodeB includes:

- ECID measurement (UE-associated signaling)
  - Serving cell ID, tracking area code, eNB position, angle-of-arrival, timing advance type 1, timing advance type 2, signal power and quality, and WLAN measurements (see [11])
  - Inter-RAT measurements (GSM RSSI and WCDMA measurements) (see [12])
- OTDOA information (non-UE-associated signaling)
  - Serving cell ID, physical cell ID, tracking area code, OTDOA assistance data, system frame number (SFN) initialization time, eNB position

- UTDOA information (UE-associated signaling)
    - Sounding reference signal transmission characteristics, uplink information, SFN initialization time
- Assistance data broadcast over SIB
    - Periodicity, encryption, ...

Some positioning techniques like UL-OTDOA rely solely on the LPPa protocol and do not involve any LPP transactions with a mobile device.

5G Core: NRPPa

While LPP is reused from LTE for NR Release 15 and 16, the LPPa protocol was not reused. Instead the similar NRPPA protocol (TS 38.455, [13]) was defined for communication between gNodeB and AMF. Like LPPa, it uses ASN.1 syntax to encode the messages. It is not service-based, as one end point (the base station) is outside the core.

16.3.2.2 RAN to Core Interfaces: S1 (4G), N1 and N2 (5G)

Both eNB and gNB reside in the Radio Access Network, while the location server (E-SLMC, LMF) resides in the core (EPC or 5GC). Consequently LPP, LPPa, and NRPPA need to traverse the RAN-Core interface (LTE: S1, 5G: N1 and N2). The entity in the core that communicates with the RAN is the MME for LTE and the AMF for 5G. The LPPa protocol is carried directly in the S1-AP protocol (TS 36.413, [14]) in 4G. NRPPA is analogously carried within the NG-AP protocol (TS 38.413, [15]). These two protocols also carry the NAS protocols (TS 24.301 [16] for 4G, TS 24.501 [17] for 5G) that are used for communication between MME/AMF and UE. The MME/AMF extracts the LPP payload, which is carried in transparent binary NAS containers of 4G and 5G.

To sum up, LPP, LPPa, and NRPPA payloads are extracted in MME/AMF, but not decoded. The MME/AMF just forwards these protocols as described in the following section.

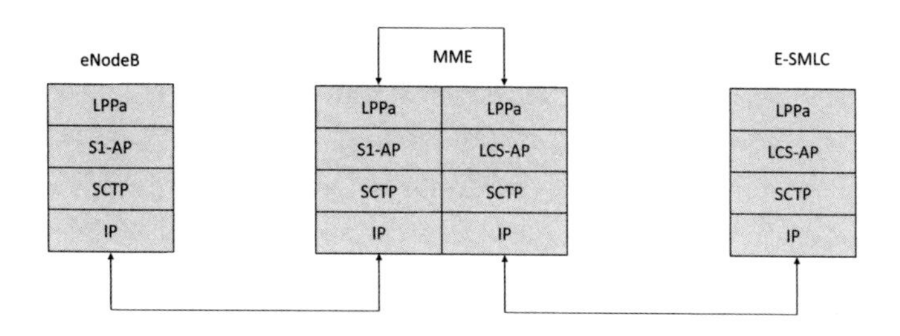

Figure 16.7 LPPa protocol stacks from eNB to E-SLMC.

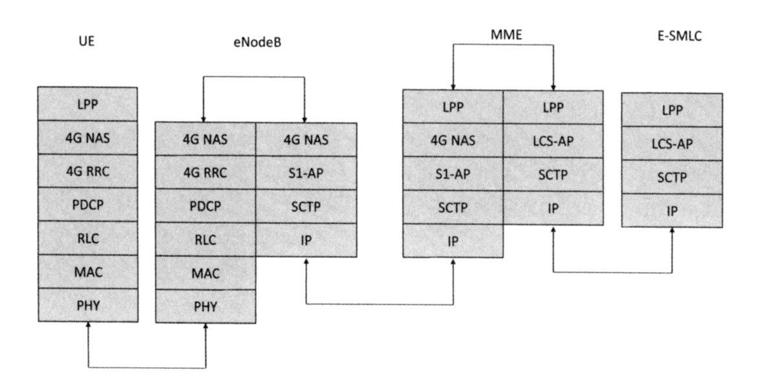

Figure 16.8 LPP protocol stacks from UE to E-SMLC.

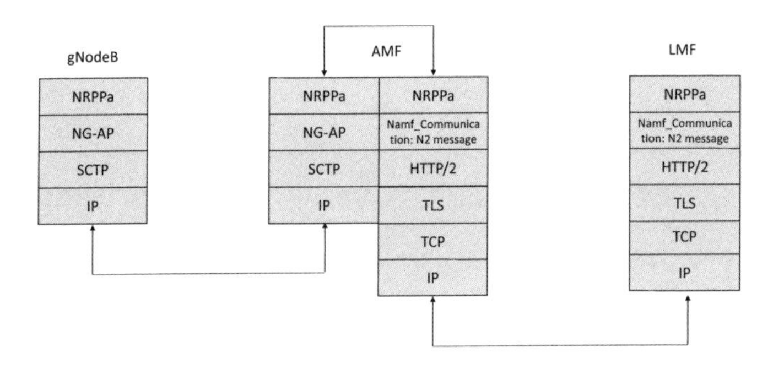

Figure 16.9 NRPPa protocol stacks from gNodeB to LMF.

16.3.2.3 Within the Core: MME/AMF to E-SMLC/LMF

SLs: MME – E-SLMC (EPC)

In the 4G core (EPC), LPP and LPPa messages are packed into the LCS-AP protocol (TS 29.171, [18]) by the MME and forwarded to the E-SLMC, as seen in Figure 16.7 and Figure 16.8.

NLs: AMF-LMF (5GC)

In the 5GC, there is no equivalent to the LCS-AP point-to-point protocol. The forwarding of messages between AMF and LMF is part of the AMF services specified in TS 29.518 ([19]) under 5.2.2.3 (N1N2MessageTransfer). NRPPA messages are carried within a N2 message container as seen in Figure 16.9. The LPP messages are carried in a N1 message container of Type LPP as specified in TS 29.518 [19] section 6.1.6.2.17, which is shown in Figure 16.10. Theoretically, any other core function could send or receive the positioning messages. In practice, the CN policy will ensure that only the LMF will be able to transfer these messages. Nevertheless, in future releases, other core entities could get access to these messages.

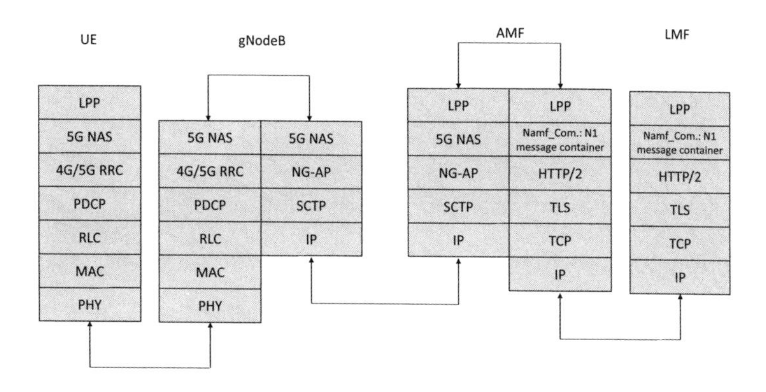

Figure 16.10 LPP protocol stacks from UE to LMF.

16.3.2.4 From the Core to the PSAP: the Role of the GMLC

For E911 applications (as well as commercial applications), it is necessary to offer an external interface for third parties outside the operator's network (e.g., a PSAP requesting a user's location).

Le: GMLC – External location application

It is a priori not known to the PSAP in which core the user is currently registered; the GMLC node abstracts this and offers a standardized and secured interface for external location applications (the Le interface, e.g., OMA's Mobile Location Protocol (MLP) [20]) to 2G, 3G, 4G, and 5G networks. MLP messages are typically transported with HTTP/1.[2] The MLP protocol is defined in XML, but not in YAML or JSON. Although the MLP protocol has some similarities with 5G's service-based architecture (HTTP, XML based), it does not follow OpenAPI 3.0.0.

MLP defines the following fundamental procedures for emergency caller location:

- Emergency Location Immediate Service (equivalent to MT-LR).

2 Earlier versions of the OMA specifications also allowed transport via SOAP or WSP protocols, but these were removed from the later spec revisions.

- Emergency Location Reporting Service (equivalent to NI-LR).

The Emergency Location Immediate Service is the equivalent of 3GPP's MT-LR session. In this case, the PSAP actively queries the GMLC for the location of a user. A PSAP's request consists of the following elements:

- Mobile ID (MSISDN/IMSI/IMEI/IPV4/IPV6/SIPURI,...) - mandatory
- Quality of position (accuracy, response time,...)
- Geographical information, (e.g., WGS84)
- Location type, (e.g., CURRENT or LAST)
- Pushaddr and PSAP URL, potentially with username and password for NI-LR messages
- Transaction ID
- Emergency services routing digits (ESRD), which identify PSAP, cell site, and sector
- Emergency services routing key (ESRK), a temporary ID for the E911 call
- GSM network parameters
- SUPL support parameters (several SUPL-specific extensions)
- MapRequest, a URL or a Base64 encoded map image

Once a position fix is obtained, the GMLC sends a response with the following elements:

- Mobile ID - mandatory
- Positioning data or positioning error, specifying position, method, and result type - mandatory
- ESRD, as seen above
- ESRK, as seen above
- Transaction ID

The emergency location reporting service can be used in the case of 3GPP's NI-LR session. Here, the 3GPP network detects the establishment of an emergency bearer during an emergency call setup, performs the location and forwards the result to the PSAP. The emergency location report contains the following elements:

- EME event (EME trigger, EME ORG, or EME REL) - mandatory
- MSID - mandatory
- Position, method, result type - mandatory

MLP also defines procedures for commercial caller location, but since this feature is no longer in use in the United States, they are not further explained here.

SLg: MME - GMLC (4G EPC)

In the case of LTE/EPC, the GMLC uses the SLg interface (TS 29.172, [18]) to communicate with the MME. There is no direct connection between GMLC and E-SLMC. The protocol transactions are largely similar to the ones defined in Le interface and are hence not detailed here.

NLg: AMF - GMLC (5GC)

In a 5G CN, the NLg protocol is the interface between GMLC and AMF. The communication between GMLC and AMF (NLg and NLs) in the 5GC is again specified as service-based architecture. The GMLC uses the ProvidePositioningInfo service specified in TS 29.518 Section 5.5.2.2. [19]. It is again similar to the Le interface in terms of the content.

16.3.2.5 SLm: Location Measurement Unit - E-SMLC: SLm Application Protocol (4G EPC)

Uplink measurements such as 4G's UTDOA are performed by LMUs. LMUs may be co-located with an eNB. At the time of writing, only the use of LMUs in the EPC is standardized. They are directly connected to the E-SLMC via the SLm Interface. The protocol used is the SLm Application Protocol (SLmAP) specified in TS 36.459 [21]. The use of LMUs in a 5G network is not precluded, but not (yet) standardized.

16.3.3 Virtualized Core: Impact on Positioning

The cloud-native virtual 5GC with network slicing support may be beneficial for positioning applications in several aspects. A 5G network may deploy multiple LMFs with different feature sets in order to support the multiple slices and its specific requirements. For instance, the traditional mobile phone slice (eMBB) the needs to meet the regulatory E911 requirements, the same as in LTE. Thus, this slice may integrate a LMF with similar capabilities as a 4G E-SLMC.

On the other hand, the URLLC slice might require very precise positioning for applications such an autonomous driving, which is SoL-critical. This slide may utilize a LMF with more precise positioning technologies, for instance PPP-RTK.

There could also be slices that require basic positioning, but with the emphasis on low energy and battery consumption. An example of this could be the mMTC slice for applications such as bike sharing or location of wares within a warehouse. In this case, positioning accuracy does not need to be in the centimeter range, but the battery of the user device may need to last several years. Thus, the mMTC slice could run yet another LMF, which is specialized in energy-efficient network-based location techniques. Due to the slicing, new features may be introduced easily in the mMTC slice without endangering the safety-of-life relevant services of E911 and autonomous driving.

A challenge for the core virtualization is the time synchronization of the LMF, as it needs to provide accurate reference time to UEs for A-GNSS. As described in Section 6.7, the required accuracy in the μs range can currently be accomplished even on virtualized platforms, using the Precise Time Protocol (PTP) and operating system extensions.

16.4 VIRTUALIZED RAN

16.4.1 Functions of the RAN

Similar to the pursuit of core virtualization, there is an industry push for RAN virtualization (see use case 6 in [1]). In contrast to the CN functions, RAN functionality, which covers RRC, PDCP, RLC, MAC, and the physical (PHY) layers is significantly more computationally complex and introduces real-time requirements. 3GPP TSG RAN defined several theoretical split options of the RAN, as seen in Figure 16.11.

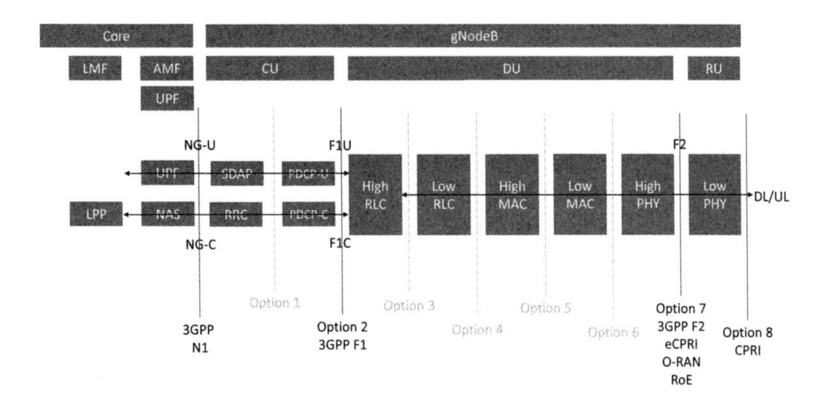

Figure 16.11 3GPP RAN split options.

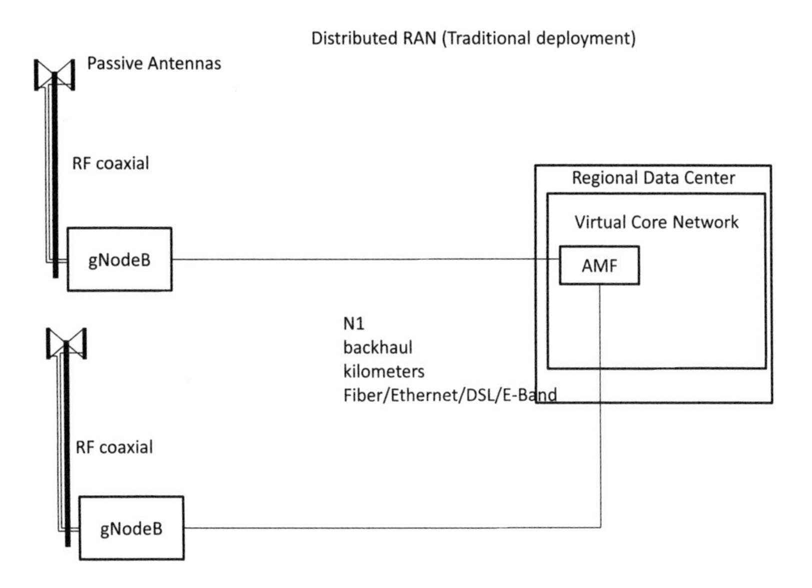

Figure 16.12 Traditional distributed RAN.

The layers below a split remain close to the antenna, while the layers above may be located geographically further away in a more centralized data center. In an early 4G RAN, all the layers were integrated in a single physical entity, which was connected via RF coaxial cables with the antennas and had to be very close to the radiating element, typically at the foot of a cell tower (see Figure 16.12). With the increasing number of bands and MIMO antennas, this concept became more and more challenging. The number of bulky RF cables (one per band and MIMO layer) increased. This approach also limits the phase coherency in the case of MIMO and increases the power consumption, as the RF path loss needs to be compensated for at the baseband unit. As a way out, RAN vendors started to integrate the RF part in the antenna itself (RRH) and connected it via a fiber link with the baseband unit (BBU). The fiber carries the I/Q samples of the RF signal (multiplexing MIMO layers and bands). This equals the split option 8 in Figure 16.13. As the fiber link's latency budget allows distances of up to 10-20 km, some operators started to centralize their BBUs [22], which allowed a smaller footprint at the antenna site. There is also less active equipment at the cell site, which eliminates the need for air conditioning in many cases, saving energy and operational costs.

However, this split option does not scale well for 5G [22]. The advent of massive MIMO brought many more antenna ports. A typical frequency range 1 (FR1, i.e., the frequencies below 7.225 GHz) 5G deployment can have 64 antenna ports. Furthermore, the bandwidths of the 5G signal are also larger than in LTE, with up to 100 MHz in FR1 and up to 400 MHz in frequency range 2 (FR2, also known as millimeter-wave, the frequencies above 24.250 GHz). Consequently, the data-rates for transmitting the raw I/Q samples become prohibitively expensive, if not impossible, for today's fronthaul technology. The mobile industry plans to tackle the problem by introducing two other splits: Option 2 and Option 7.2. They split the gNB's baseband unit further into a central unit (CU), a distributed unit (DU), and a radio unit (RU), as seen in Figure 16.14. In the case of massive MIMO, the RU is typically integrated directly into the antenna module. These splits have the advantage that they do not scale with the number of antennas, but only with the number of MIMO layers (Split 7.2) or user data (Split 2), which is significantly less data.

From an operator's point of view, a standardized RAN split also offers the interesting aspect of being able to use different vendors for CU, DU, and RU.

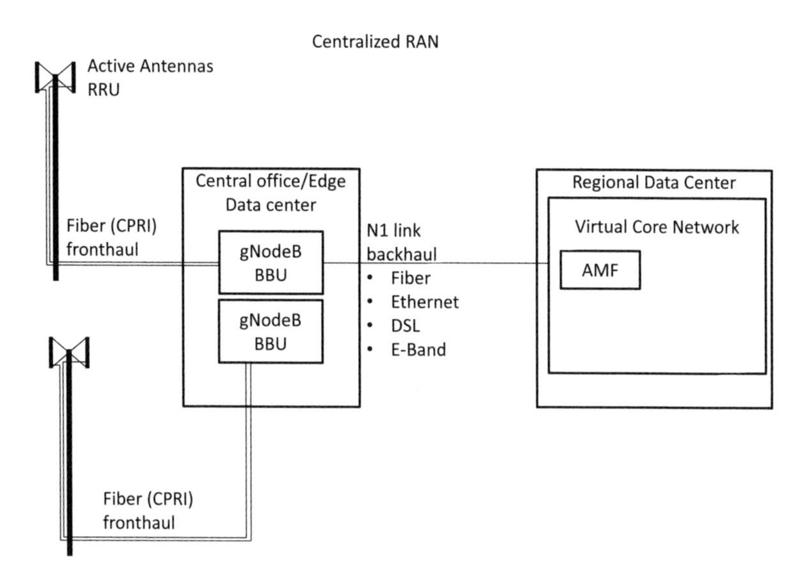

Figure 16.13 Centralized RAN without massive MIMO.

16.4.2 RAN Higher Layer Split (Split Option 2)

The interface between CU and DU is standardized as the F1 interface (TS 38.473, [23]) in Release 15. This split reduces the fronthaul requirements compared to option 8, but still leaves significant parts of the computational complexity (the DU entity) near the antenna. The CU entity does not require real-time capabilities and can be virtualized in a data center.

16.4.3 RAN Lower Layer Split (Split Option 7.2)

As mentioned, a lower layer split offers advantages in terms of antenna site space and energy efficiency. Furthermore, intercell coordination, for example, for CoMP (multiple cell sites jointly transmit a DL signal for one UE or jointly decode the UE's UL signal) is simplified, if multiple DUs are located in the same physical entity. The price to pay is the higher requirement on the fronthaul (i.e., the link between the antenna site and the DU).

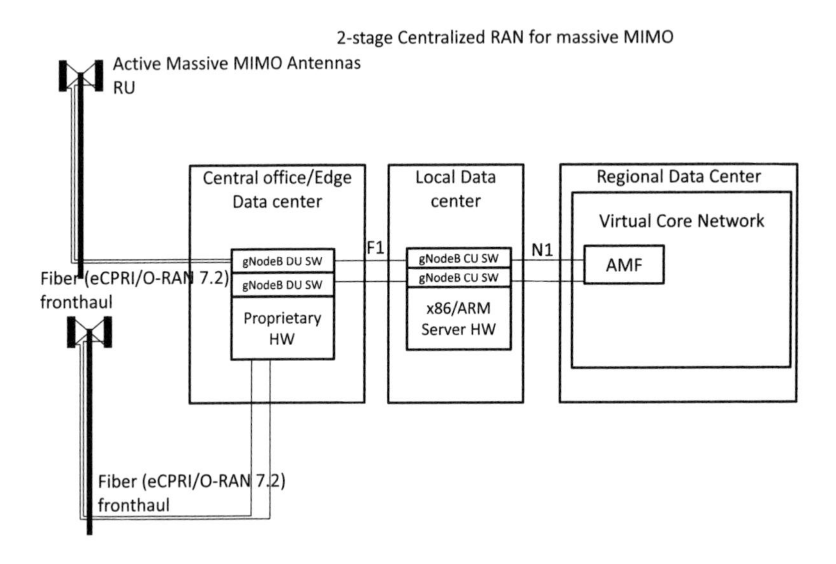

Figure 16.14 Centralized RAN with massive MIMO and further centralization of CU.

The DU is the entity with the highest computational complexity and the most demanding real-time requirements. Consequently, it is still a stronghold for highly optimized application-specific integrated circuits (ASICs), which are designed exactly for this purpose. They are potentially more powerful and energy-efficient than general-purpose implementations, but less flexible.

Nevertheless, several industry players aim at running open-source SW on a DU based on commercial-of-the-shelf hardware. This requires the operating system of the DU to offer real-time features. The O-RAN alliance recently formed a coalition with the Linux foundation. The aim is, among other goals, to enhance the Linux real-time capabilities. Even if the OS supports real-time optimizations (see [24] on current activity), a typical high-performance COTS DU will need to make use of field-programmable gate array (FPGA) accelerator units, which are available as add-on boards for standard data-center servers today.

It remains to be seen if the FPGA or the ASIC approach will find more followers for 5G.

16.4.4 RAN Lower Layer Split Protocols

Option 8 split was traditionally utilizing the common public radio interface (CPRI), which is a de facto industry standard driven by the large RAN vendors. While the CPRI specification is publicly available, each vendor typically added special optimizations, which lead to criticism on multivendor interoperability. Furthermore, CPRI requires a dedicated physical fiber instead of using a standard Ethernet fiber link in fronthaul, and it does not allow any other split than option 8. To address these points, two new specifications were defined as of 2019:

- The CPRI group proposed eCPRI [25], which supports multiple split options and can be carried via Ethernet.
- IEEE standardized the IEEE1914.3 Radio over Ethernet [26], which supports split option 8 and 7.

Notably, 3GPP has not decided to standardize the low-level split yet. To enhance vendor interoperability, the O-RAN alliance specified profiles with focus on Split 7.2. The O-RAN profiles can use either eCPRI or IEEE1914.3.

Table 16.1 gives an overview of the strengths and weaknesses of the protocols.

Table 16.1

Lower Layer RAN Split Protocols

	CPRI	O-RAN/eCPRI	O-RAN/IEEE1914.3
Interoperability	Low	High	High
supported RAN splits	8	7.2	7.2
Fronthaul scales with	Antennas, BW	Data traffic	Data traffic
Ethernet fronthaul	No	Yes	Yes

The use of eCPRI and IEE1914.3 allows operators to use an Ethernet as fronthaul between DU and RU. The fronthaul may have multiple hops/bridges, which has adverse effects on the latency.

16.4.5 Flexible Splits Based on Usage Scenario

Depending on the usage scenario (urban, rural) and the available fronthaul (fiber, Ethernet, DSL, microwave) an operator might decide to use different RAN architectures in the network. A sports stadium might be utilizing many RUs, which are connected to a few centralized DUs, colocated with the CU. This would allow advantages in bandwidth-efficient processing, such as CoMP and small antenna sites. In a rural area, a gNB might still be noncentralized (i.e., including all RAN layers in the same physical entity), allowing it to use a nonideal backhaul link to the core. Suburban areas might use yet another mix, with a centralized CU serving many DU/RU combinations.

16.4.6 RAN Architectures: Impact on Positioning

Virtualized RANs offer new opportunities and new challenges for RAT-dependent positioning. OTDOA is especially impacted by the architectural choices due to the direct impact of the synchronization on its performance.

16.4.6.1 Time/Phase Synchronization

Synchronization has always been a fundamental challenge in digital telecommunication networks (see [27]) and is elementary for the most accurate positioning techniques (i.e., A-GNSS, OTDOA). The need to improve the spectral efficiency from 2G to 5G also increased the synchronization requirements, especially for features like carrier aggregation, TDD, or CoMP. On the other hand, the trend to move from specialized to general-purpose HW with virtualized, centralized network functions adds new challenges for synchronization.

ITU-T published the recommendation G.8271 Amendment 2 [28], which introduces various levels of synchronization accuracy. The eCPRI standard used in the lower-layer split 7.2 reflects these levels and introduces equivalent categories. Table 16.2 gives an overview and shows typical features that mandate the higher classes and levels.

In the eCPRI Requirements for lower layers [29], it is assumed, that Cat A+ is normally only achieved if the RU is connected with an additional synchronization interface. This counteracts the goal of a simple deployment, where only power and fiber are provided to the RU, but would enable OTDOA with an approximate accuracy of 20 m or less.

The less synchronized Cat A may be achieved without this additional sync, but still requires a very precise PTP implementation in the RU. Typically, the RU and DU should be colocated and have a single-hop connection to achieve this synchronization. The accuracy of OTDOA drops to 40 m or less. Using a eCPRI Cat B setup, a simple PTP implementation may be used, but the OTDOA accuracy

Table 16.2

Time Synchronization Levels and Categories

ITU-T G.8271	eCPRI	abs. TE	rel. TE	Typ. applications
Level 1	-	500 ms	-	Billing, alarms
Level 2	-	100 μs	-	IP Delay monitoring
Level 3	-	5 μs	-	LTE-TDD
Level 3A	-	5 μs	5 μs	LTE MBSFN
Level 4	Cat. C	1.5 μs	-	NR-TDD, EN-DC
Level 4A	-	1.5 μs	3 μs	NR non-cont. and Inter-band CA
Level 5	-	1 μs	-	-
Level 6A	Cat. B	-	260 ns	LTE non-cont. CA, Inter-band CA NR (FR1) Intra-band contiguous CA OTDOA: 80 m accuracy
Level 6B	Cat. A	-	130 ns	LTE Intra-band contiguous CA NR (FR2) Intra-band contiguous CA OTDOA: 40 m accuracy
Level 6C	Cat. A+	-	65 ns	LTE/NR MIMO or TX div. transmissions OTDOA: 20 m accuracy

Note 1: TE stands for timing error.

is only around 80 m. If the baseline Cat C is used (which is sufficient for NR-TDD), the OTDOA accuracy is around 450 m, which renders OTDOA virtually useless.

16.4.6.2 OTDOA and Virtualized RAN

As pointed out in the previous section, if the signals visible to a UE are originating from basebands colocated on the same physical entity, which are connected via a single-hop fronthaul to the active antenna, the synchronization might be better than in legacy distributed networks. However, if the signals are generated from physically separate DUs, which are located in separate data centers distant from the cell tower, or if the fronthaul includes several hops, resulting in additional delay and jitter, the synchronization may be worse than in traditional distributed networks.

Operators that take full advantage of centralized RANs might not be able to utilize OTDOA in its original form as both techniques have conflicting objectives. Nonetheless, in private 5G networks for industrial applications, the RAN architecture can be tailored to support very tight local synchronization, for instance within a factory hall. For this purpose, a common DU with multiple synchronized RU could provide meter-level accuracy in a limited space, such as a factory hall.

Compensation schemes are being discussed for RANs that are not tightly synchronized. These schemes rely on measuring timing offsets between neighboring cells and report them via the NRPPA protocol to the LMF. Alternatively, the usage of multi-RTT instead of OTDOA is a solution, as it does not require strict synchronization.

16.4.6.3 Multi-RTT and RAN Architectures

With multi-RTT, 5G introduces a time-based positioning technique, which does not depend on accurate gNB synchronization. Given the complexity and cost of precise synchronization, this may be a very attractive alternative for positioning where GNSS is not available. It comes at the price of additional overhead, as the UE will need to transmit UL-ranging signals, but this may be acceptable, especially if the number of devices that need to be located is limited (e.g., in a factory environment).

16.4.6.4 Signal-Strength-Based Positioning and RAN Architecture

Signal strength-based positioning approaches like ECID are not impacted by RAN choices. They would benefit from a denser network, but it is unlikely that they would reach submeter accuracy, due to the inherent limitations of such methods.

16.5 CONCLUSION

This chapter provided details on cellular network implementation, clarifying the differences between core and RAN and giving an overview of the multiple protocols involved in the communication between the different network entities during a positioning session.

The new, cloud-native 5GC and the advent of RAN virtualization and centralization offer great promise to network operators, beyond plain savings in operational and maintenance expenses. However, from a positioning point of view, the network virtualization and the low-layer split comes with a set of new challenges, especially in the area of synchronization. These challenges must be addressed in order to obtain the high-accuracy requirements that commercial positioning services expect from 5G. This chapter has given a few options that can help overcoming these challenges, from how to improve network synchronization to using synchronization-independent techniques.

This book started with a basic introduction to positioning methodologies and the use cases and requirements for positioning in cellular networks. It explained in detail all the different positioning technologies, both RAT-dependent and RAT-independent, that have been defined for cellular network location-based services. The final part of the book explored the positioning protocols, showing the different call flows and the messages and parameters exchanged during a positioning session. This last chapter brought everything together, analyzing the 5G network and what promises and challenges it brings for positioning.

References

[1] ETSI GS NFV 001, *Network Functions Virtualization (NFV) Use Cases*, V1.1.1, October, 2013.

[2] 3GPP TS 23.501, *System Architecture for the 5G System*, V16.3.0, December, 2019.

[3] 3GPP TS 23.502, *Procedures for the 5G System (5GS)*, V16.3.0, December, 2019.

[4] IETF RFC 7540, *Hypertext Transfer Protocol Version 2 (HTTP/2)*, May, 2015.

[5] Ben-Kiki, O., Evans, C., and döt Net, I., *YAML Ain't Markup Language (YAML)*, Version 1.2, Third Edition, October, 2019.

[6] Standard ECMA-404, *The JSON Data Interchange Syntax*, Second Edition, December, 2017.

[7] FCC 47 CFR Part 96, *Citizens Broadband Radio Service*, 10-1-16 Edition, 2016.

[8] Bundesnetzagentur für Elektrizität, Gas, Telekommunikation, Post und Eisenbahnen, *Verwaltungsvorschrift fuer Frequenzzuteilungen fuer Lokale Frequenznutzungen im Frequenzbereich 3.700-3.800 MHz*, November, 2019.

[9] 3GPP TS 36.305, *Stage 2 Functional Specification of User Equipment (UE) Positioning in E-UTRAN*, V15.4.0, June, 2019.

[10] 3GPP TS 38.305, *Stage 2 Functional Specification of User Equipment (UE) Positioning in NG-RAN*, V15.4.0, June, 2019.

[11] 3GPP TS 36.455, *LTE Positioning Protocol A (LPPa)*, V15.2.1, January, 2019.

[12] 3GPP WI 630132, *LCS_LTE_RFPMT-Core: Positioning Enhancements for RF Pattern Matching in E-UTRA*, June, 2014.

[13] 3GPP TS 38.455, *NR Positioning Protocol A (NRPPa)*, V15.2.1, January, 2019.

[14] 3GPP TS 36.413, *LTE; S1 Application Protocol (S1AP)*, V15.7.1, October, 2019.

[15] 3GPP TS 38.413, *NG Application Protocol (NGAP)*, V15.5.0, October, 2019.

[16] 3GPP TS 24.301, *Non-Access-Stratum (NAS) Protocol for Evolved Packet System (EPS)*, V16.3.0, December, 2019.

[17] 3GPP TS 24.501, *Non-Access-Stratum (NAS) Protocol for 5G System (5GS)*, V16.3.0, December, 2019.

[18] 3GPP TS 29.172, *Evolved Packet Core (EPC) LCS Protocol (ELP) between the Gateway Mobile Location Centre (GMLC) and the Mobile Management Entity (MME)*, V15.2.0, December, 2019.

[19] 3GPP TS 29.518, *Access and Mobility Management Services*, V16.2.0, December, 2019.

[20] OMA-TS-MLP, *Mobile Location Protocol*, V3.5, December, 2018.

[21] 3GPP TS 36.459, *SLm Interface Application Protocol (SLmAP)*, V15.0.0, June, 2018.

[22] Larsen, L.M.P., Checko, A., and Christiansen, H.L., "A Survey of the Functional Splits Proposed for 5G Mobile Crosshaul Networks," in *IEEE Communications Surveys and Tutorials*, vol. 21, no. 1, pp. 146-172, Firstquarter 2019.

[23] 3GPP TS 38.473, *F1 Application Protocol*, V15.7.0, October, 2019.

[24] Stroemblad, P., *High-Performance Real-Time Linux Solution for Xilinx Ultrascale+*, ENEA, https://www.xilinx.com/publications/events/developer-forum/2018-frankfurt/high-performance-real-time-linux-solution-for-xilinx-ultrascale-plus.pdf, accessed on January, 2020.

[25] eCPRI Specification, *Common Radio Public Interface: eCPRI Interface Specification*, V1.2, June, 2018.

[26] IEEE 1914.3-2018, *IEEE Standard for Radio over Ethernet Encapsulations and Mappings*, September, 2018.

[27] Prof. Sutton, A., *The History of Synchronization in Digital Cellular Networks*, https://www.slideshare.net/3G4GLtd/the-history-of-synchronisation-in-digital-cellular-networks, accessed on January, 2020.

[28] ITU-T G.8271/Y.1366 *Time and Phase Synchronization Aspects of Telecommunication Networks*, Amendment 2, November, 2018.

[29] eCPRI Transport Network, *Common Radio Public Interface: Requirements for the eCPRI Transport Network*, V1.2, June, 2018.

List of Acronyms and Abbreviations

3GPP	Third Generation Partnership Project
3GPP2	Third Generation Partnership Project 2
5G NR	5G New Radio
5GC	5G Core
5GS	5G System
A-FLT	advanced-forward link trilateration
A-GNSS	assisted GNSS
A-GPS	assisted GPS
ADAS	advanced driving assistance system
ADR	accumulated delta range
AES	Advanced Encryption Standard
AI	artificial intelligence
ALI	automatic location information
AMF	access and mobility function
AML	Advanced Mobile Location
AMPS	Advanced Mobile Phone Service
AN	access network
ANI	automatic number identifier
AoA	angle of arrival
AoD	angle of departure
AP	access point
AR	augmented reality
ARIB	Association of Radio Industries and Businesses

ARNS Aeronautical Navigation Service Bands
ARPF authentication credential repository and processing function
ASIC application-specific integrated circuit
ASN.1 Abstract Syntax Notation One
ASP aggregated service provider
ATIS Alliance for Telecommunications Industry Solutions
AUSF authentication server function

BDS BeiDou Navigation Satellite System
BER basic encoding rules
BSC base station controller
BSM basic safety message
BSSID basic service set identifier
BTLE Bluetooth Low Energy
BTS base transceiver station
BW bandwidth

C-ITS cooperative Intelligent Transport Systems
C-SGN C-IOT serving gateway node
C/A coarse/acquisition
CAM cooperative awareness message
CBRS Citizens Broadband Radio Service
CDF cumulative distribution function
CDMA code division multiple access
CDMA2000 Code Division Multiple Access 2000
CE coverage enhancement
CEP circular error probability
CID cell ID
CLAS centimeter-level augmentation service
CMA cellular market area
CN core network
CoMP coordinated multipoint
COTS commercial off-the-shelf
CPRI common public radio interface
CRC cyclic redundancy check
CRS cell reference signal
CS circuit switched
CSFB circuit switched fallback
CSI-RS channel state information reference signal

CSI-RSRP channel state information reference signal received power
CSI-RSRQ channel state information reference signal received quality
CTIA Cellular Telecommunications and Internet Association
CU central unit
CUPS control and user plane separation
CW continuous wave

D-AMPS Digital Advanced Mobile Phone Service
D-GNSS differential GNSS
D-GPS differential GPS
D2D device-to-device
DBN dynamic Bayesian network
DC dual connectivity
DENM decentralized environmental notification message
DL downlink
DL-AoD downlink angle of departure
DL-TDOA downlink time difference of arrival
DME distance measuring equipment
DOP dilution of precision
DRB dedicated radio bearer
DSRC Dedicated Short Range Communication
DU distributed unit

E-CSCF emergency call session control function
E-FLT enhanced forward link trilateration
E-OTD enhanced observed time difference
E-SLMC enhanced serving mobile location center
E-UTRA Evolved UMTS Terrestrial Radio Access
EC-GSM-IoT Extended Coverage GSM IoT
ECEF Earth-centered Earth-fixed
ECI Earth-centered inertial
ECID enhanced cell ID
EENA European Emergency Number Association
EGNOS European Geostationary Navigation Overlay Service
EIRP effective isotropic radiated power
EKF extended Kalman filter
ELS Emergency Location Service
eMBB enhanced mobile broadband
eMBMS evolved multimedia broadcast/multicast services

eMTC enhanced machine type communications
eNB enhanced Node-B
ENU east north up
EPA extended pedestrian model A
EPC evolved packet core
EPDU external protocol data unit
EPS evolved packet system
ESC environmental sensing capability
ESN emergency service number
ETSI European Telecommunication Standard Institute
ETU extended typical urban
EVDO Evolution-Data Optimized

FCC Federal Communication Commission
FDD frequency division duplex
FDE fault detection and exclusion
FDMA frequency division multiple access
feMTC further enhanced machine type communications
FMCW frequency-modulated continuous-wave
FOV field of view
FPGA field-programmable gate array
FQDN fully qualified domain name
FR1 frequency range 1
FR2 frequency range 2
FSPL free space path loss
FTM fine time measurement
FWA fixed wireless access

GAA general authorized access
GAGAN GPS-aided GEO augmented navigation
GDOP geometrical dilution of precision
GEO geostationary orbit
GLONASS Globalnaya Navigatsionnaya Sputnikovaya Sistema
GMLC gateway mobile location center
gNB next generation Node-B
GNSS global navigation satellite system
GPP general-purpose processor
GPRS General Packet Radio Service
GPS Global Positioning System

GSM	Global System for Mobile
GSMA	Global System for Mobile Communications Association
GSO	geosynchronous orbit
GTD	geometrical time difference
GTP	GPRS Tunnelling Protocol
HA3D	high accuracy 3D
HSPA	high-speed packet access
HSS	home subscriber server
IAB	integrated access and backhaul
ICD	interface control document
ICIC	intercell interference coordination
IE	information element
IEEE	Institute of Electrical and Electronics Engineers
IERS	International Earth Rotation Service
IF	intermediate frequency
IFFT	inverse fast Fourier transform
IGSO	inclined geosynchronous orbit
IIoT	Industrial Internet of Things
ILS	Instrument Landing System
IMEI	International Mobile Equipment Identity
IMS	IP Multimedia Subsystem
IMSI	International Mobile Subscriber Identity
IMT-2000	International Mobile Telecommunication 2000
IMT-2020	International Mobile Telecommunication 2020
IMU	inertial measurement unit
IoT	Internet of Things
IP	Internet Protocol
IRNSS	Indian Regional Navigation Satellite System
IS-95	Interim Standard 95
ITM	interference tolerance mask
ITS	Intelligent Transportation Systems
ITU	International Telecommunication Union
IVS	in-vehicle system
JSON	JavaScript Object Notation
KPI	key performance indicator

LBS location-based services
LCS location service
LEO low Earth orbit
LIDAR light detection and ranging
LMF location management function
LMU location measurement unit
LoS line of sight
LPP LTE Positioning Protocol
LPPa LTE Positioning Protocol A
LPPe LPP enhancements
LPWA low power wide area
LRF location retrieval function
LRR long range radar
LS location server
LTE Long Term Evolution
LTE-A LTE Advanced

M2M machine-to-machine
MAC medium access control
MAC master auxiliary concept
MBS Metropolitan Beacon System
MCC mobile country code
MCG master cell group
MEMS micro electro mechanical systems
MEO medium Earth orbit
MHSS multiple hypothesis solution separation
MIMO multiple input, multiple output
MLP Mobile Location Protocol
mm-wave millimeter-wave
MME mobility management entity
mMTC massive machine type communications
MN master node
MNC mobile network code
MO-LR mobile originating location request
MOTD multilateration observed time difference
MRA measurement rejection approach
MRR medium range radar
MSA metropolitan statistical area
MSAG master street address guide

MSAS multi-functional satellite augmentation system
MSC mobile switching center
MSISDN mobile station international subscriber directory number
MSS maximum solution separation
MT-LR mobile terminated location request
MTA multilateration timing advance
MTC machine type communications
MTU maximum transmission unit
multi-RTT multi-round-trip time

N-RTK network-RTK
NAS non-access stratum
NB-IoT narrowband IoT
NEAD National Emergency Address Database
NF network function
NFC near-field communications
NFV network function virtualization
NG-eCall Next Generation eCall
ng-eNB next generation enhanced Node-B
NG-IVS next generation in-vehicle system
NG-RAN next generation radio access network
NGC next generation core
NI-LR network induced location request
NLoS non-line of sight
NPRS narrowband positioning reference signal
NR New Radio
NRPPA NR Positioning Protocol A
NSA non-standalone
NSI ID network slice instance identifier
NSSF network slice selection function
NTRIP network transport of RTCM via Internet Protocol

OFDM orthogonal frequency-division multiplexing
OFDMA orthogonal frequency-division multiple access
OMA Open Mobile Alliance
OSR observation state representation
OTDOA observed time difference of arrival

P-CSCF primary call session control function

P-CSCF proxy-call session control function
PAL priority access license
PCI physical cell ID
PDC Personal Digital Cellular
PDCP Packet Data Convergence Protocol
PDN packet data network
PDoA phase difference of arrival
PDU protocol data unit
PEMEA Pan-European Mobile Emergency APPs
PER packed encoding rules
PF particle filter
PGW PDN gateway
PHY physical layer
PLMN public land mobile network
PoA phase of arrival
posSIB positioning system information block
PPP precise point positioning
PPP-AR precise point positioning ambiguity resolution
PPP-RTK precise point positioning real-time kinematics
PRB physical resource block
PRN pseudo-random noise
ProSe proximity services
PRS positioning reference signal
PRS-RSRP positioning reference signal reference signal received power
PS packet switched
PSAP public safety answering point
PSP PSAP service provider
PSS primary synchronization signal
PSTN public switched telephone network

QFI QoS flow identifier
QoS quality of service
QZSS Quasi-Zenith Satellite System

RADAR radio detection and ranging
RAIM receiver autonomous integrity monitoring
RAN radio access network
RAT radio access technology
REST representational state transfer

RF radio frequency
RFID radio frequency identification
RLC radio link control
RMS root mean square
RNC radio network controller
RNSS regional navigation satellite system
RRC radio resource control
RRH remote radio head
RRLP Radio Resource Location Protocol
RRM radio resource management
RSA rural service area
RSRP reference signal received power
RSRQ reference signal received quality
RSS received signal strength
RSSI reference signal strength indicator
RSTD reference signal time difference
RSU roadside unit
RTCM Radio Technical Commission for Maritime Services
RTD real-time difference
RTK real-time kinematics
RTT round-trip time
RU radio unit

S-GW serving gateway
SA standalone
SAE Society of Automotive Engineers
SBA service-based architecture
SBAS satellite-based augmentation system
SBI service-based interface
SC-FDMA single carrier frequency-division multiple access
SC-RTT serving cell round-trip time
SCG secondary cell group
SCS subcarrier spacing
SDAP Service Data Adaptation Protocol
SFN system frame number
SIB system information block
SIG Special Interest Group
SIP Session Initiation Protocol
SL sidelink

SLAM simultaneous localization and mapping
SLAS sub-meter level augmentation service
SLP SUPL location platform
SMF session management function
SMLC serving mobile location center
SMS short message service
SMSC short message service center
SN secondary node
SNR signal-to-noise ratio
SoC system-on-chip
SoL safety-of-life
SPS semi-persistent scheduling
SRB signaling radio bearer
SRR short-range radar
SRS sounding reference signal
SS synchronization signal
SS-RSRP synchronization signal reference signal received power
SS-RSRQ synchronization signal reference signal received quality
SSB synchronization signal block
SSID service set identifier
SSR state space representation
SSS secondary synchronization signal
STANAG Standardization Agreement
STEC slant total electron content
SUPL Secure User Plane Location
SVID spatial vehicle ID
SVN space vehicle number

TBS terrestrial beacon system
TD-SCDMA time division synchronous code division multiple access
TDD time division duplex
TDMA time division multiple access
TDoA time difference of arrival
TIA Telecom Industry Association
TLS Transport Layer Security
TLV type length value
ToA time of arrival
ToF time-of-flight
TP transmission point

TRP transmission and reception point
TTFF time-to-first-fix

U-TDOA uplink time difference of arrival
U-TOA uplink time of arrival
UAV unmanned aerial vehicle
UBP uncompensated barometric pressure
UDM unified data management
UE user equipment
UKF unscented Kalman filter
UL uplink
UL-AoA uplink angle of arrival
UL-RTOA uplink relative time of arrival
UL-TDOA uplink time difference of arrival
UMTS Universal Mobile Telecommunication System
UPF user plane function
URA user range accuracy
URLLC ultra-reliable low latency communications
UTC Coordinated Universal Time
UUID universally unique identifier

V2V vehicle-to-vehicle
V2X vehicle-to-everything
VNF virtual network function
VoIP voice over IP
VoLTE voice over LTE
VOR VHF omnidirectional radio range
VRS virtual reference station
VTEC vertical total electron content

WAAS Wide Area Augmentation System
WADGNSS Wide Area Differential GNSS
WADGPS Wide Area Differential GPS
WAVE Wireless Access for Vehicular Environment
WCDMA Wideband Code Division Multiple Access
WGS World Geodetic System
WiMAX Worldwide Interoperability for Microwave Access
WLAN wireless local area network
WLS weighted least squares

WPAN wireless personal area network

YAML YAML Ain't Markup Language

About the Authors

Adrián Cardalda García is a predevelopment engineer and 3GPP RAN5 Representative for Location Based Services and Radio Resource Management at Rohde & Schwarz GmbH & Co. KG. He holds two master's degrees in telecommunication engineering and mobile network communication from the University of Oviedo, Spain. His master thesis "RailSLAM: Simultaneous Location and Mapping for Railways" was awarded four international prizes, including the Scientific Award of the 2012 ITS World Congress in Vienna. He holds multiple patents in the field of LBS. In his role as 3GPP Representative, he is involved in the definition of 5G NR specifications and he is the Rapporteur of the 5G NR Positioning and RRM subwork items in RAN5.

Stefan Maier is project lead and senior development engineer in the Location Based Services Lab at Rohde & Schwarz GmbH & Co. KG. He studied electrical and computer engineering at Technische Universität München and at the University of Arizona. He holds a bachelor of science and a Dipl.Ing. (M.Sc. equivalent) degree with honors from TUM and has 10 years of professional experience in the fields of A-GNSS and VoLTE/E911. He holds multiple patents in the field of LBS, is a member of the Institute of Navigation, and is currently working on test systems for 5G emergency caller location.

Abhay Phillips is a Director for Location Based Services Test Systems at Rohde & Schwarz GmbH & Co. KG and has over 15 years of experience in the field of cellular communication systems and location-based services systems. He holds a bachelor's degree in electronics and telecommunication and master's degree in telecommunication and software engineering. He has worked as a software developer and project manager for mobile devices and test systems in 3G, 4G, and 5G cellular networks. He has industry experience in the field of cellular positioning systems and is actively engaged in the roll-out of 5G positioning technology updates.

Index

The GNSS Technology and Applications Series

Elliott Kaplan and Christopher Hegarty, Series Editors

A-GPS: Assisted GPS, GNSS, and SBAS, Frank van Diggelen

Applied Satellite Navigation Using GPS, GALILEO, and Augmentation Systems, Ramjee Prasad and Marina Ruggieri

Digital Terrain Modeling: Acquisition, Manipulation, and Applications, Naser El-Sheimy, Caterina Valeo, and Ayman Habib

Geographical Information Systems Demystified, Stephen R. Galati

GNSS Applications and Methods, Scott Gleason and Demoz Gebre-Egziabher

GNSS Interference Threats and Countermeasures, Fabio Dovis, editor

GNSS Markets and Applications, Len Jacobson

GNSS Receivers for Weak Signals, Nesreen I. Ziedan

GNSS for Vehicle Control, David M. Bevly and Stewart Cobb

GPS/GNSS Antennas, B. Rama Rao, W. Kunysz, R. Fante, and K. McDonald

Implementing e-Navigation, John Erik Hagen

Introduction to GPS: The Global Positioning System, Second Edition, Ahmed El-Rabbany

Location-Based Services in Cellular Networks from GSM to 5G NR, Adrián Cardaldo García, Stefan Maier, Abhay Phillips

MEMS-Based Integrated Navigation, Priyanka Aggarwal, Zainab Syed, Aboelmagd Noureldin, and Naser El-Sheimy

Navigation Signal Processing for GNSS Software Receivers, Thomas Pany

Principles of GNSS, Inertial, and Multisensor Integrated Navigation Systems, Second Edition, Paul D. Groves

RF Positioning: Fundamentals, Applications, and Tools, Rafael Saraiva Campos, and Lisandro Lovisolo

Server-Side GPS and Assisted-GPS in Java™, Neil Harper

Spread Spectrum Systems for GNSS and Wireless Communications, Jack K. Holmes

Understanding GPS/GNSS: Principles and Applications, Third Edition, Elliott Kaplan and Christopher Hegarty, editors

Ubiquitous Positioning, Robin Mannings

Wireless Positioning Technologies and Applications, Second Edition, Alan Bensky